Gastro-Esophageal Cytology

Monographs in Clinical Cytology

Vol. 10

KARGER

S. Karger · Basel · München · Paris · London · New York · Tokyo · Sydney

Gastro-Esophageal Cytology

Michael Drake

Director of Anatomical Pathology, Prince Henry's Hospital Director,
Victorian Cytology (Gynaecological) Service, Melbourne, Australia

466 figures, 2 color plates, 1 table, 1985

KARGER

S. Karger · Basel · München · Paris · London · New York · Tokyo · Sydney

Monographs in Clinical Cytology

National Library of Medicine, Cataloging in Publication
Drake, Michael
Gastro-esophageal cytology
Michael Drake. – Basel; New York: Karger, 1985.
(Monographs in clinical cytology, v. 10)
1. Cytodiagnosis 2. Esophageal diseases – diagnosis 3. Esophageal neoplasms – diagnosis 4. Stomach – diseases – diagnosis 5. Stomach neoplasms – diagnosis I. Title
W1 MO567KF v. 10 [WI 300 D762g]
ISBN 3–8055–3931–2

Drug Dosage

The author and the publisher have exerted every effort to ensure that drug selection and dosage set forth in this text are in accord with current recommendations and practice at the time of publication. However, in view of ongoing research, changes in government regulations, and the constant flow of information relating to drug therapy and drug reactions, the reader is urged to check the package insert for each drug for any change in indications and dosage and for added warnings and precautions. This is particularly important when the recommended agent is a new and/or infrequently employed drug.

Printed in Switzerland by Thür AG Offsetdruck, Pratteln
ISBN 3–8055–3931–2

Contents

Acknowledgements

When a book is written in the setting of a routine diagnostic department of a major teaching hospital it owes its content inevitably to a vast number of people who cannot possibly be identified, let alone named, individually. Accordingly I must simply acknowledge my debt to my many clinical colleagues who, over the past twenty years or more, have referred their patients for investigation or sent specimens for interpretation. I am particularly grateful for the opportunities I have had for clinico-pathological discussions and for the readiness with which follow-up information has been provided.

To my fellow pathologists I also express my appreciation. Many have contributed to the assessment of the diagnostic material and again it is difficult to itemise specifically the many occasions on which discussion with colleagues has yielded information and ideas that are incorporated in this book. I am particularly indebted to Dr. *Gabriele Medley* for her many comments on various aspects and, specifically, for her assistance with the section on malignant lymphomas and I am also indebted to Dr. *Peter Wallis* for details of some of the unusual esophageal neoplasms. The assistance received from Professor *Bhathal,* Dr. *Essex* and Dr. *Tang* of Victoria, Dr. *Waters* and Dr. *Armstrong* of Western Australia and Dr. *Leiman* and Dr. *Berry* of South Africa is acknowledged in the appropriate areas of the text and I am most appreciative of this assistance. Dr. *Elaine Waters* of Western Australia was, as always, particularly generous in making available her case material.

I am most grateful to Dr. *Shirley Roberts,* Director of Radiology, Prince Henry's Hospital, who provided me with the X-rays that are reproduced throughout this work. Dr. *Roberts* and her colleague, Dr. *Margaret Kinnaird,* gave generously of their time in the explanation of this radiological material. Dr. *Wayne Hancock,* Research Scientist, Department of Nephrology, Prince Henry's Hospital, provided material and advice on the immuno-histological techniques discussed in Chapter 5.

In every cytology department the contribution of the cytotechnologists is a major one. Again it would be difficult to name individually the many people who, over the years, have contributed to the evaluation of our material. However, Miss *Megan Doran* has made a major contribution to this evaluation by her skill and dedication to the techniques of gastro-esophageal cytology.

The secretarial work associated with a production of this sort is tedious and time-consuming. I thank my former Secretary Miss *Betty Blomfield* who initiated this work and I am deeply grateful to my present Secretary, Mrs. *Carole Korsten,* who so ably completed the various drafts and the final manuscript and assisted so competently with the organization of the material.

Finally it is my pleasure to acknowledge the enormous contribution made by my colleague Miss *Patricia Archer.* Many of the concepts embodied in this text have resulted from our discussions and our evaluation, in collaboration, of cytological material over the past twenty years. Miss *Archer* is personally responsible for many of the illustrations and made a major contribution to the organization of the materials used in the compilation of this book. I am most grateful to her.

Michael Drake

This book is dedicated to

John F. Funder
with respect and affection to my teacher, friend and former colleague.

1. Gastro-Esophageal Cytology – Its Place in Laboratory Medicine

Historical Notes – The Evolution of a Technique

It is of interest to note that the development of appropriate surgical techniques for the treatment of gastric cancer preceded by many years the establishment of those diagnostic procedures that would now be regarded as mandatory before surgery was undertaken. Thus *Billroth* (fig. 1), in 1881, performed the first successful partial gastrectomy for gastric cancer [12] some fourteen years before *Roentgen* (fig. 2) announced his discovery of a 'hitherto unknown manifestation of force or energy' [9] – the now widely used X rays so fundamental to the investigation of the upper gastro-intestinal tract. A further fifteen years were to pass before barium was first used as a contrast medium for outlining the stomach and facilitating the radiological demonstration of lesions within that organ. Subsequently the rapid development of radiological techniques for the diagnosis of gastric cancer tended to overshadow the development of other procedures.

Nevertheless within one year of *Billroth's* first partial gastrectomy, *Rosenbach* [13] made the following rather prophetic statement: 'Without respite, methods have to be developed which allow an exact diagnosis of gastric carcinoma at its earliest stage'. He advocated techniques of gastric lavage with examination of the lavage fluid and was successful in demonstrating particles of tumour tissue in several cases. This, and other similar efforts at about the same time, represented a considerable improvement on previous rather sporadic attempts at gastric cytodiagnosis mostly based upon the examination of vomited material.

The development of effective microscopes was, of course, fundamental to the emergence of techniques such as diagnostic cytology. The origin of the microscope is somewhat controversial although it is generally accepted that the Dutch spectacle maker, *Zacharias Janssen* was the first to use a combination of lenses to produce a compound microscope. However, it was workers such as *Marcello Malpighi* (fig. 3) and *Antony van Leeuwenhoek* (fig. 4) who were responsible for developing the microscope and applying it to the study of biological science. *Leeuwenhoek,* who was a cloth merchant in Delft, Holland, was able to grind microscope lenses that were not bettered for over a century. He observed a wide variety of biological objects being the first, for example, to describe such fundamental structures as red blood cells, the capillary circulation, and spermatozoa. *Malpighi,* a contemporary of *Leeuwenhoek,* also made a major contribution to biological science by way of his microscopic studies.

The word cell is generally attributed to *Robert Hooke* who designed and built a microscope with which he studied very thin slices of cork. He named the individual structural units within this material 'cells or little boxes'. *Hooke* published his findings in his book 'Micrographia' in 1665 [6] but his findings were largely overlooked for more than a century. In the early 19th century several anatomists and biologists made references to cell structures but it was not until 1839 that *Schleiden* and *Schwann* proposed what is now known as the cell theory stating 'all living things are single cells, cell aggregations, or products of cells'. This concept was rapidly accepted and is today the fundamental proposition on which all biology rests.

The beginnings of clinical cytology are usually attributed to *Johannes Müller* (fig. 5), a German physiologist, who in 1840 published a monograph on the nature and structural characteristics of cancer [10]. *Müller* had made a detailed study of the cellular structure of malignant neoplasms in preparing the material for his monograph. Another major influence in the development of clinical cytology was the English physician *Lionel Smith Beale* (fig. 6) whose report of the diagnosis of a pharyngeal carcinoma by sputum cytology represented one of the earliest uses of the technique [2]. In 1867 *Beale* described also the cytological diagnosis of gastric cancer by the examination of vomitus [3].

It is generally accepted that the work of *Giovanni Marini* [7, 8] represents a milestone in the development of gastric cytology. He was the first to attempt a comprehensive survey of benign and malignant cells obtained from gastric washings, his observations being based on a microscopic study of the centrifuged deposit from these washings. His two papers, published in 1909 and 1910, contain excellent drawings of both normal and cancerous cells including probably the first depiction of a malignant 'tadpole' cell. These papers also included the following statement: 'When doctors can be persuaded of the advantage of the cytological examination of the lavage water ... they will not wait to give a diagnosis of gastric cancer until it has become palpable and surgery is, if not actually harmful, at least to no purpose'. *Gibbs* [4], in his excellent historical review of the development of

1

2

3

4

Fig. 1. Theodor Billroth (1829–1894) who is generally regarded as the pioneer of visceral surgery. Courtesy of W.B. Saunders Company.

Fig. 2. Wilhelm Konrad Roentgen (1845–1923), the discoverer of X rays. Courtesy of The New York Academy of Medicine Library.

Fig. 3. Marcello Malpighi (1628–1694), regarded as the founder of biological microscopy. Courtesy of World Health Organization, Geneva.

Fig. 4. Antony van Leeuwenhoek (1632–1723), who developed microscope lenses that were unsurpassed until the 19th century. Copyright Museum Boerhaave, Leiden, The Netherlands.

gastric cytology, quotes this statement in full but tends to downgrade its prophetic significance. Thus he states that 'it is now appreciated that such a view overstates the practical importance of gastric cytodiagnosis, but it should be remembered that *Marini* made his observations before reliable radiological investigation had become available for the investigation of patients in whom gastric cancer was suspected'. *Gibbs,* in turn, could not have foreseen the enormous advances that would be made in the development of endoscopic methods, with the consequent increased use of the cytological examination of specimens collected by endoscopically directed brushing and washing techniques, and the realization of some of *Marini's* hopes regarding the detection of early gastric cancer.

Undoubtedly the major stimulus to the development of gastric and esophageal cytology was the work of *George N. Papanicolaou* (fig. 7). His pioneering work in the application of cytological techniques to the detection of uterine cancer is too well known to need repetition but this work led to renewed interest in the use of similar methods for the investigation of other body systems. In 1947 *Papanicolaou* himself, in association with *Cooper* [11], reported on the use of lavage techniques for the cytological diagnosis of gastric cancer emphasizing the importance of rapid collection and preparation of material to ensure optimal diagnostic results. A large number of papers followed this initial publication and much attention was directed to improving results by various methods of specimen collection and preparation. Simultaneously the methods were applied to the investigation of esophageal lesions. In 1949 *Anderson* and his colleagues [1] described such studies and this paper was followed by a large number of similar publications. Again attention was directed in these initial papers to improving methods of specimen collection and interpretation.

Undoubtedly some of the most significant of these publications were those of *Schade* (fig. 8) who, in several papers and a monograph [14] described his success in the diagnosis of 'early' or superficial gastric cancer. Thus he was able to diagnose cytologically 29 cases of gastric cancer that were not diagnosable by radiology, this number including 16 cases which,

---

Fig. 5. Johannes Peter Müller (1801–1858), a physiologist who is regarded as the initiator of clinical cytology. Photograph courtesy of G-I-T Verlag Ernst Giebeler.

Fig. 6. Lionel Smith Beale, one of the earliest exponents of clinical cytology who, in 1858, claimed to have diagnosed gastric cancer in vomitus. Photograph courtesy of G-I-T Verlag Ernst Giebeler.

5

6

7

8

Fig. 7. George N. Papanicolaou (1883–1962), generally regarded as the father of modern diagnostic cytology. Courtesy of The Editor-in-Chief, Acta Cytologica.

Fig. 8. Rudolf Otto Karl Schade, whose pioneering work demonstrated that gastric carcinoma could be diagnosed in its earliest phases by cytological means. Photograph courtesy of Dr. R.O.K. Schade.

on both clinical and radiological grounds, were not even suspected of malignancy, surgery being carried out on the basis of positive cytology alone. In his introduction to the monograph *Schade* expresses the view that, by cytological means, 'it is possible to detect early carcinoma when the tumour is still confined to the mucosa and therefore not detectable by radiology or gastroscopy'. Supported by his diagnostic achievements this statement by *Schade* was a most significant one in the evolution of the techniques of gastric cytodiagnosis.

The final major advance came with the development of the flexible fiberoptic gastroscopc. Gastric intubation had been practised during the late 18th and early 19th centuries but these initial applications were restricted largely to therapeutic purposes. As early as 1868 a rigid gastroscope had been designed by *Kussmaul* and subsequently this instrument, with various technical improvements, was used increasingly for diagnostic purposes. The scope of the gastroscope was restricted, however, by its rigidity, large areas of the gastric mucosa being inaccessible to direct vision or to direct sampling. In 1958 this investigational technique was revolutionized by the introduction, by *Hirschowitz* and his colleagues [5] of a new gastroscope – the 'fiberscope'. Flexible fiberoptic endoscopes were developed rapidly with numerous improvements and modifications and are now fundamental to the practice of both gastric and esophageal cytology.

It is difficult to foresee whether any further technical achievements in gastric cytodiagnosis are possible. However, as will be made abundantly clear throughout the remainder of this text, the ultimate results depend not so much on the technology employed as on the skill and experience of the people carrying out the examination.

Applications, Value and Limitations of Gastro-Esophageal Cytology

In general there are two broad areas of application of the techniques of diagnostic cytology.

(a) As a diagnostic procedure in those patients who have symptoms and/or signs referable to a particular body system.

(b) As a screening procedure to detect disease, specifically cancer, in asymptomatic people or in patients with non-specific symptoms and/or signs not indicative of cancer.

The techniques of esophageal and gastric cytology have been, and still are, used predominantly for the investigation of patients with symptoms or signs referable to the upper gastro-intestinal tract. They are thus primarily diagnostic procedures and form a valuable adjunct to other procedures such as endoscopy and radiology. As will be seen a combination of all three methods of investigation results in a very high degree of diagnostic accuracy.

That gastric cytology may be of value in the detection of clinically unsuspected cancer was demonstrated by some of the earlier workers in this field, notably *Schade* whose results have already been cited. Nevertheless the techniques are relatively complex, time consuming, and costly and hence their application to population screening must be subject to considerable reservations. In general there are three requirements for a successful cytological screening programme:

It must be possible to define a high risk group.
The area under study must be accessible to cytological investigation.
The area must also be amenable to further investigation so that the presence of a carcinoma predicted cytologically can be confirmed and that the lesion, if present, can be localized and treated.

It is because the uterine cervix satisfies these criteria that cancer of that organ is so amenable to screening programmes. Nevertheless any such screening programme requires careful planning and meticulous operation if it is to achieve worthwhile results. The requirements of a mass survey for the detection of cervical cancer were set out by *Wied* [16] and all of these requirements can be readily applied to screening for both esophageal and gastric cancer. Undoubtedly the major problem that limits screening for cancer in these sites is the identification of the high risk population. Thus to date screening programmes have been virtually limited to those groups with a known ethnic or racial predilection to the disease. As far as esophageal cancer is concerned cytological surveys have been carried out in South Africa, China and Iran, all countries with a high incidence of this disease. These surveys will be discussed in a later chapter. Similarly mass screening for gastric cancer has been carried out quite extensively in Japan where this form of cancer is a major cause of death. Admittedly the primary screening is usually based on radiological techniques often in association with examination with the gastrocamera. Cytology is reserved for the further investigation of those patients in whom an abnormality has been detected by radiology and/or gastrophotography. The principles of such screening programmes and the results achieved is set out in some detail by *Takahashi* [15].

In addition to Japan, countries such as Iceland, Finland and Czechoslovakia also have a significantly elevated incidence of gastric cancer compared to most other countries. In those countries where gastric cancer is not so elevated wide scale population screening would neither be practical nor economical. However, in these countries it may be eminently reasonable to concentrate on those segments of the population known to be at risk. This could include people with pernicious anemia, known chronic gastritis, a history of partial gastrectomy or indeed anyone with symptoms referable to the upper gastro-intestinal tract no matter how vague or non-specific these may be. This was, of course, the general approach adopted by *Schade* in achieving the success already discussed.

In summary, therefore, the techniques of esophageal and gastric cytology form an extremely sensitive diagnostic test for the investigation of patients with symptoms and/or signs referable to the upper gastro-intestinal tract. Their application to population screening for the detection of unsuspected gastric or esophageal cancer should be restricted to populations with a known predilection for these diseases or, alternatively, to selected groups of high risk people in an otherwise low risk population.

References

1 Anderson, H.A.; Macdonald, J.R.; Olsen, A.M.: Cytologic diagnosis of carcinoma of the oesophagus and cardia of the stomach. Proc. Staff Meet. Mayo Clin. *24:* 245 (1949).

2 Beale, L.S.: Examination of sputum from a case of cancer of the pharynx. Archs med. Athenes *2:* 44 (1861).

3 Beale, L.S.: The microscope in clinical medicine; 3rd ed. (Churchill, London 1867).

4 Gibbs, D.D.: Exfoliative cytology of the stomach (Butterworth, London 1968).

5 Hirschowitz, B.I.; Curtiss, L.E.; Peters, C.W.; Pollard, H.M.: Demonstration of a new gastroscope, 'the fiberscope'. Gastroenterology *35:* 50 (1958).

6 Hooke, R.: Micrographia. Jo. Martyn and Ja. Allestry (1665).

7 Marini, G.: Über die Diagnose des Magenkarzinomas auf Grund der cytologischen Untersuchung des Spülwassers. Eigene Beobachtungen über den normalen und pathologischen Zelleninhalt des Magens. Arch. VerdauKrankh. *15:* 251 (1909).

8 Marini, G.: Ancora sulla diagnosi del carcinoma dello stomaco in base all'esame citologico dell'acqua di lavatura. Clinica med. ital. *49:* 65 (1910).

9 Metcalf, W.B.: Original X-ray work and its value to stomach diagnosis. Philad. med. J. *4:* 401 (1899).

10 Müller, J.: On the nature and structural characteristics of cancer and those morbid growths which may be confounded with it (translated by C. West) (Sherwood, Gilbert & Paper, London 1840).

11 Papanicolaou, G.N.; Cooper, W.A.: The cytology of the gastric fluid in the diagnosis of cancer of the stomach. J. natn. Cancer Inst. *7:* 357 (1947).
12 Priestley, M.D.: The early development of surgical treatment for carcinoma of the stomach. Proc. Staff Meet. Mayo Clin. *15:* 641 (1940).
13 Rosenbach, O.: Über die Anwesenheit von Geschwulstpartikeln in dem durch die Magenpumpe entleerten Mageninhalte bei Carcinoma ventriculi. Dt. med. Wschr. *8:* 452 (1882).
14 Schade, R.O.K.: Gastric cytology (Arnold, London 1960).
15 Takahashi, K.: Outline of gastric mass survey by X-ray. Gann Monograph on Cancer Research, vol. II: Early Gastric Cancer, pp. 207–222 (University of Tokyo Press, Tokyo 1971).
16 Wied, G.L.: Quality control mechanism for cytology programs (editorial). Acta cytol. *9:* 407–412 (1965).

2. Anatomy, Histology and Physiology

The following comments are intended to provide a general outline of the anatomy, histology and physiology of the stomach and esophagus with particular emphasis on those features which are relevant to the cytological investigation of these organs and the pathological lesions that are most commonly revealed by this investigation. There are a number of excellent standard texts available if more detail is required [1–5].

Esophagus

Anatomy

The esophagus is a hollow muscular tube which measures approximately 25 cm or 10 inches in length. It extends from the pharynx, beginning at the level of the sixth cervical vertebra, to the cardia of the stomach joining that structure 2.5 cm to the left of the midline opposite the eleventh thoracic vertebra. The esophagus pierces the diaphragm at the level of the tenth thoracic vertebra and has an intra-abdominal portion which measures 1.25 cm in length and is conical in shape, the base of the cone being continuous with the cardiac orifice of the stomach. In its thoracic extent the esophagus passes through the superior and posterior portions of the mediastinum lying largely anterior to the vertebral column. Although its general direction is vertical it has two slight curves. Thus, although it is situated in the midline at its commencement it curves slightly towards the left side as it passes to the root of the neck then curves back to the median plane at the level of the fifth thoracic vertebra, then deviates slightly to the left again before finally inclining anteriorly to pass through the esophageal opening of the diaphragm.

The esophagus is constricted at four sites, namely,

At its commencement 15 cm (6 inches) from the incisor teeth.

Where it is crossed by the aortic arch 22.5 cm (9 inches) from the incisor teeth.

Where it is crossed by the left main bronchus 27.5 cm (11 inches) from the incisor teeth.
Where it pierces the diaphragm 40 cm (16 inches) from the incisor teeth.

These constrictions, which are depicted in figure 9 and demonstrated radiologically in figure 10, are of considerable importance when instruments are being passed along the esophagus and they also appear to be of importance in the localization of the lesions to be discussed.

The relationships of the esophagus during its course through the thoracic cavity are also of considerable importance, particularly when considering the consequences of the local spread of esophageal cancer. As seen in figure 11, the esophagus is closely related to a number of vital structures such as the aortic arch and some of its major branches and also the trachea and main bronchi. In addition, the recurrent laryngeal nerves lie in the groove between the trachea and esophagus.

Lymphatic drainage is also of importance since, by the time the diagnosis of esophageal carcinoma is confirmed, metastases have occurred in 50–80% of cases and the most common sites of metastasis are the regional lymph nodes. The lymphatic channels from the upper third of the esophagus and the pharyngeal region drain to the deep cervical nodes either directly or by way of the paratracheal nodes. From the lower two thirds of the esophagus the lymphatic vessels drain to the para-esophageal nodes in the posterior mediastinum and thence to the thoracic duct. Finally the terminal or intra-abdominal portion of the esophagus has its lymphatic drainage to a ring of nodes around the gastric cardia and to the left gastric nodes.

Histology

The histological structure of the esophagus follows the same general pattern as the rest of the alimentary tract having four layers or coats (fig. 12). However, it has no serosal coat the outermost tissues or adventitia blending with the mediastinal connective tissue. The muscle coat, which is particularly thick, is composed of an outer longitudinal layer and an inner circular one. In man striated muscle is generally limited to the upper two thirds of the esophagus, the lower one third containing smooth or non-striated muscle only. The submucosa is broad and is composed of coarse

9
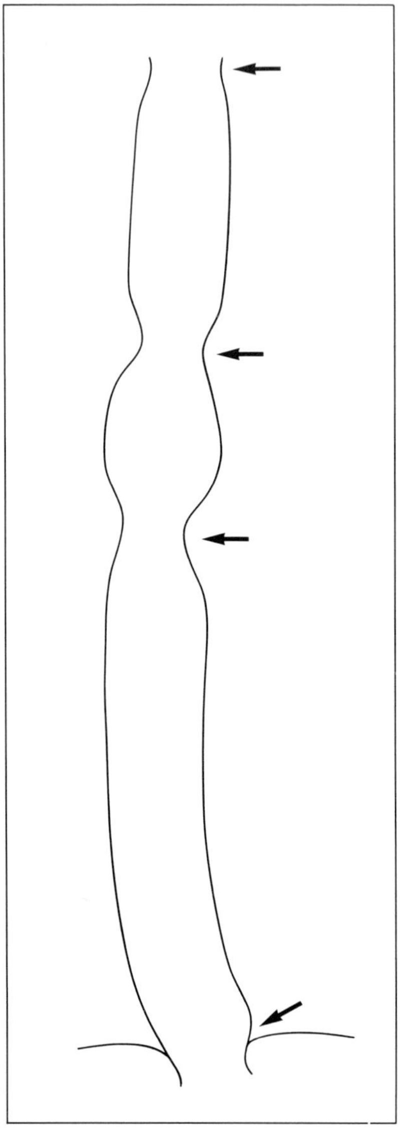

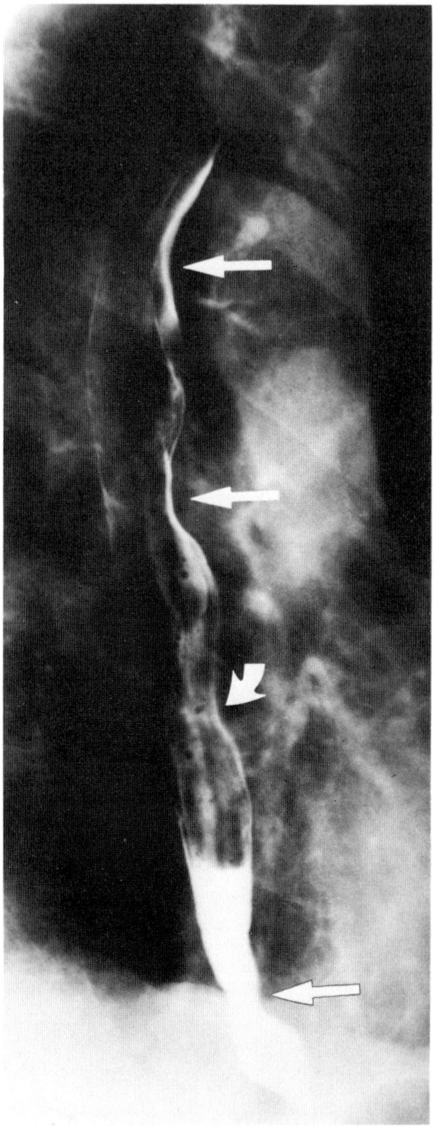
10

Fig. 9. Diagrammatic representation of esophagus showing four constrictions.

Fig. 10. X ray of esophagus showing three of the normal constrictions as indicated by the straight arrows. The most proximal constriction is not visible. The rather poorly defined area of narrowing, indicated by the small curved arrow, is a 'transient' constriction due to peristalsis.

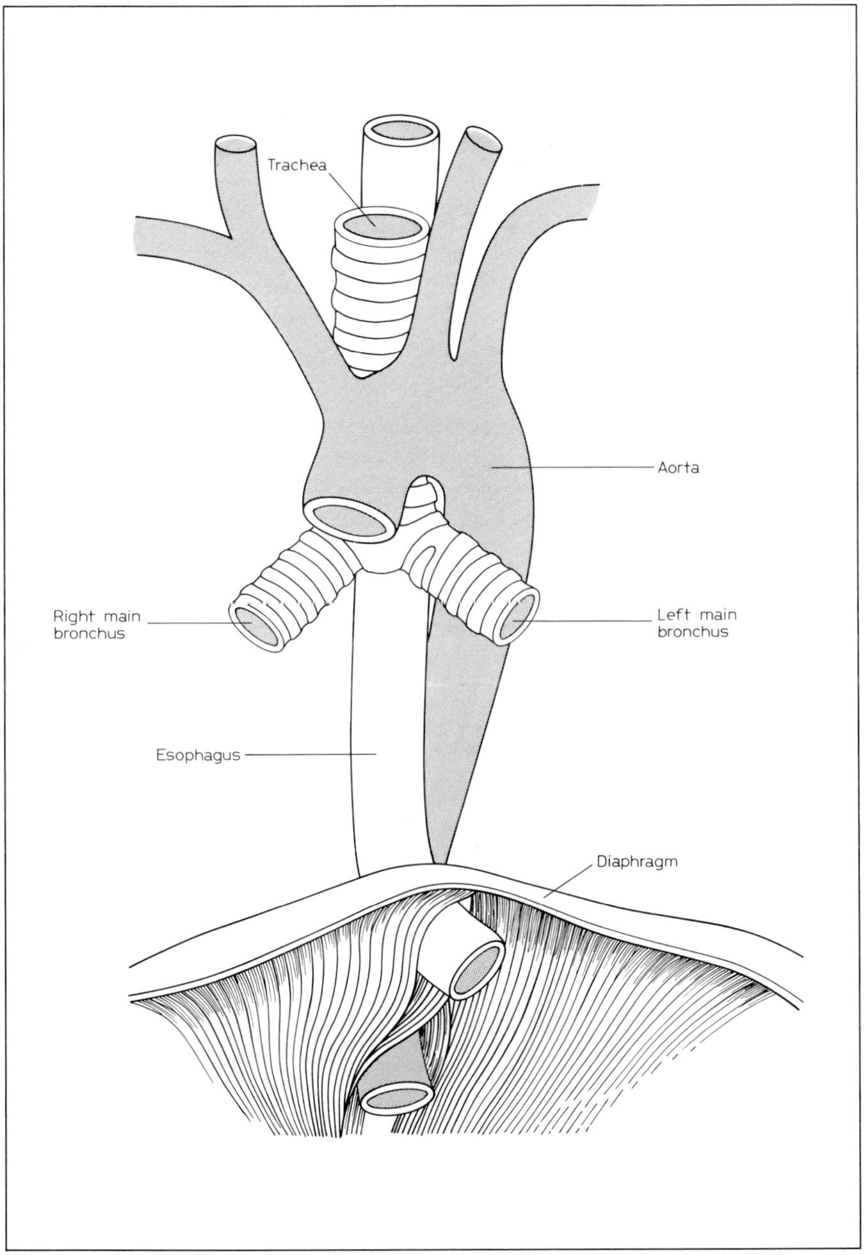

Fig. 11. Diagrammatic representation of relationships of esophagus within thoracic cavity. Note the close association with the trachea, major bronchi and the aorta.

collagenous and elastic fibres. Within this broad submucosa are blood vessels, nerves and a ramifying plexus of lymphatic channels. Although these are relatively few in comparison to the rest of the digestive tract they account for the early and extensive submucosal spread of esophageal carcinoma (fig. 13). The submucosa also contains the esophageal glands (fig. 14). These are small compound racemose glands of mucous secreting type which drain into the lumen of the esophagus by way of a long duct. The cells comprising the glandular acini have basally placed nuclei and abundant vacuolated cytoplasm (fig. 15).

In addition to these esophageal glands proper, which lie within the submucosa, esophageal cardiac glands are found in the lamina propria of the mucosa. These glands, which closely resemble the cardiac glands of the stomach, are found in the upper part of the esophagus, and in the lower part near the gastric cardia. In the areas of the esophageal mucosa that contain the upper and lower groups of cardiac glands the stratified squamous epithelium may be replaced by small foci of simple columnar epithelium of gastric type. Occasionally these foci are of considerable size, may show pit-like invaginations, or even tubular glands like those of the gastric fundus. It has been suggested by some workers that this 'ectopic' gastric epithelium may be of importance in relation to a number of diseases including esophageal cancer.

Of greatest importance to the diagnostic cytologist is the mucosa which is composed of stratified squamous non-keratinizing, or non-hornifying, epithelium (fig. 16). The epithelium rests on a thin basal lamina or membrane. Beneath this basal membrane is a layer of fibrous connective tissue which forms the lamina propria of the mucosa. This connective tissue projects into the surface epithelium in the form of narrow papillae which are, however, always separated from the epithelial cells by the basement membrane (fig. 17). Within the squamous epithelium which forms the mucosal surface are scattered melanoblasts. The stratified squamous epithelium changes abruptly to glandular epithelium of gastric type, this change usually occurring at the gastric orifice (fig. 18). However, it may take place at the level of the diaphragm, the distal or intra-abdominal portion of the esophagus being lined by columnar epithelium. In addition, the stratified squamous epithelium of the lower esophagus may be replaced by metaplastic glandular epithelium of gastric type (fig. 19). If this process is extensive it is referred to as Barrett's syndrome, a condition that will be discussed in a subsequent chapter. The lamina propria of the mucosa is separated from the submucosa by the muscularis mucosae, a layer of non-striated muscle.

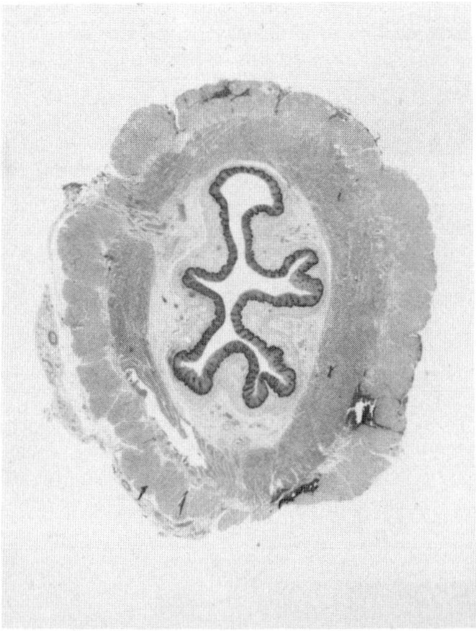

12

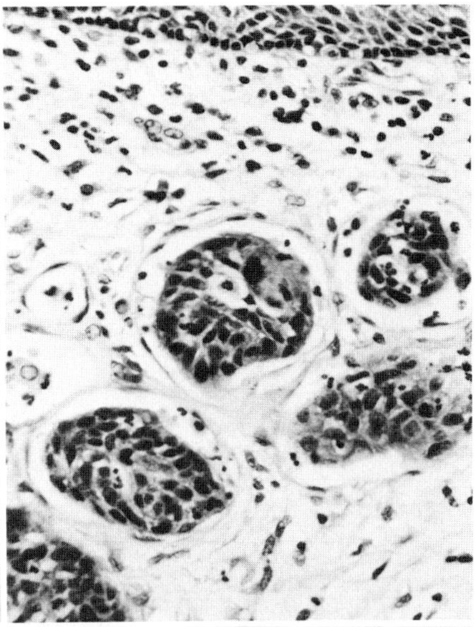

13

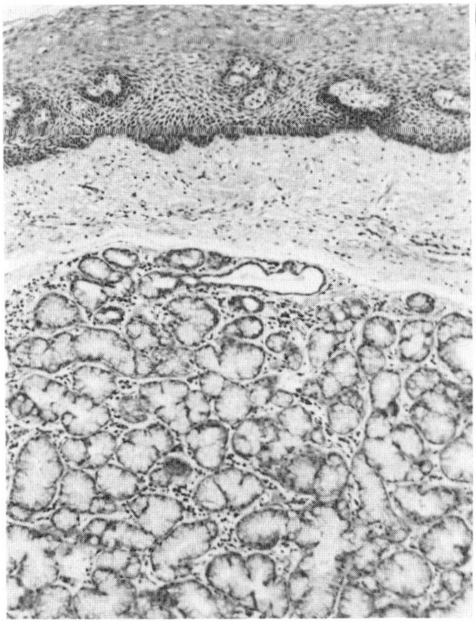

14

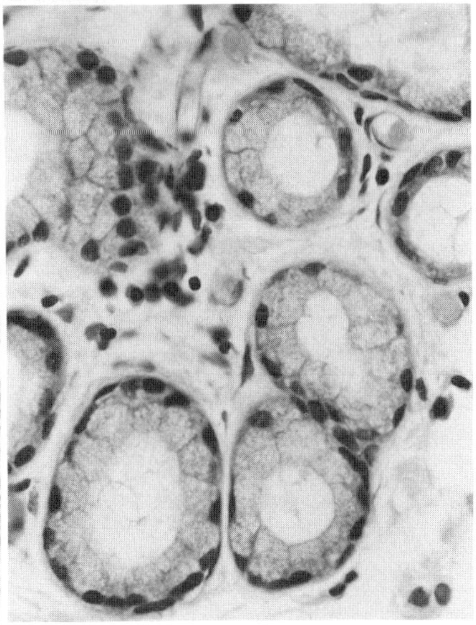

15

Fig. 12. Cross-section of esophagus. Note the four coats, the outer or adventitia being poorly defined. × 2.

Fig. 13. Esophageal submucosal lymphatics permeated by carcinoma tissue. × 220.

Fig. 14. Esophageal submucosal glands. × 55.

Fig. 15. Esophageal submucosal glands. × 345.

Physiology

The esophagus is essentially a muscular tube that conveys food rapidly from the pharynx to the stomach. This is achieved by an orderly peristaltic contraction that passes through the length of the esophagus and is associated with a co-ordinated opening and closure of the sphincters situated at its upper and lower extremities. These motor phenomena have been described in considerable detail using radiological, manometric and electrical recording techniques but, in man, detailed understanding of the control mechanisms responsible for the co-ordinated esophageal motor activity is limited. In particular, the mechanisms of swallowing are not fully understood nor are those mechanisms that control the passage of esophageal content into the stomach. The latter is particularly important as some of the conditions to be discussed subsequently are related to pathological disturbances of this sphincteric mechanism. Radiological studies show that swallowed food is momentarily held up at the esophago-gastric junction and that this is associated with a transient elevation of esophageal pressure. Relaxation then allows the food to enter the stomach. It would appear, therefore, that some form of sphincteric mechanism, capable of contraction and relaxation must be present at the esophago-gastric junction. However, morphologists have been unable to identify any particular aggregation of muscle tissue to account, with certainty, for the functional communication between esophagus and stomach. Many explanations have been offered but it is unnecessary to consider these in detail in this discussion. What is necessary, and indeed very relevant to future discussions, is to recognize that a 'physiological cardiac sphincter' does exist and that disturbances of its mechanism may lead to a hold up of food or fluid in the lower esophagus or, perhaps more importantly, the regurgitation of gastric content into the lower esophagus.

Stomach

Anatomy

The stomach is the most dilated part of the digestive tract, and is situated between the end of the esophagus and the beginning of the small intestine. Its mean capacity varies with age, being about 30 ml at birth, increasing gradually to about 1,000 ml at puberty, and usually reaching approximately 1,500 ml in the adult. The shape and position of the stomach are

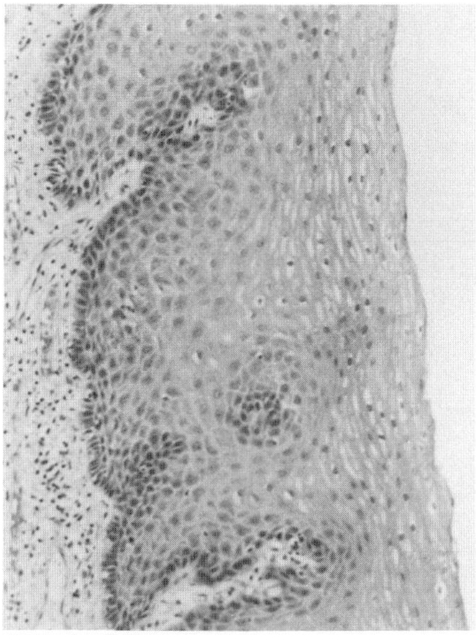

16

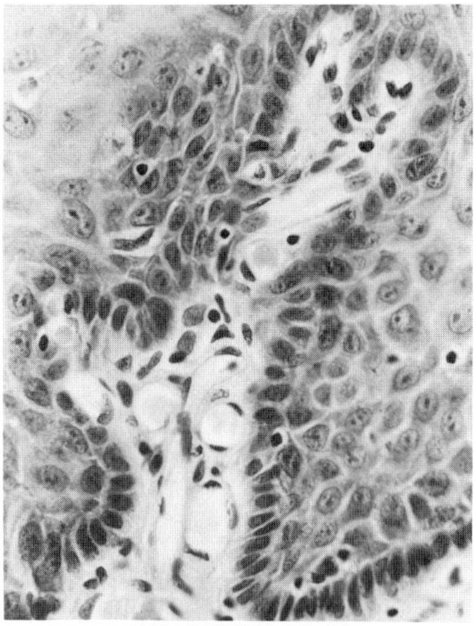

17

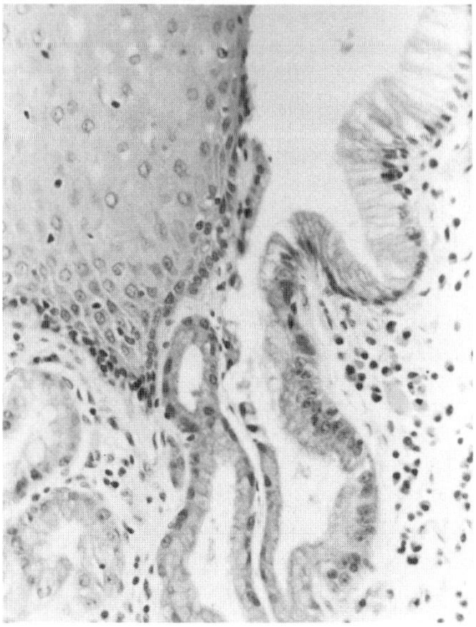

18

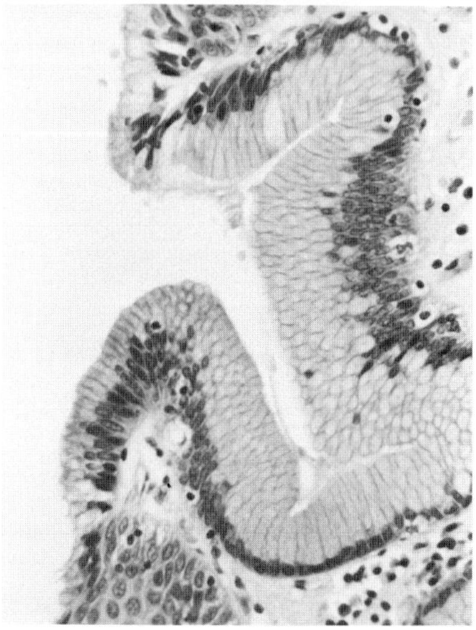

19

Fig. 16. Stratified squamous epithelium forming esophageal mucosa. × 175.

Fig. 17. Deeper portion of esophageal mucosa showing papilla. × 275.

Fig. 18. Abrupt transition from squamous to glandular epithelium at gastro-esophageal junction. × 175.

Fig. 19. Metaplastic glandular epithelium of gastric type in lower esophagus. × 275.

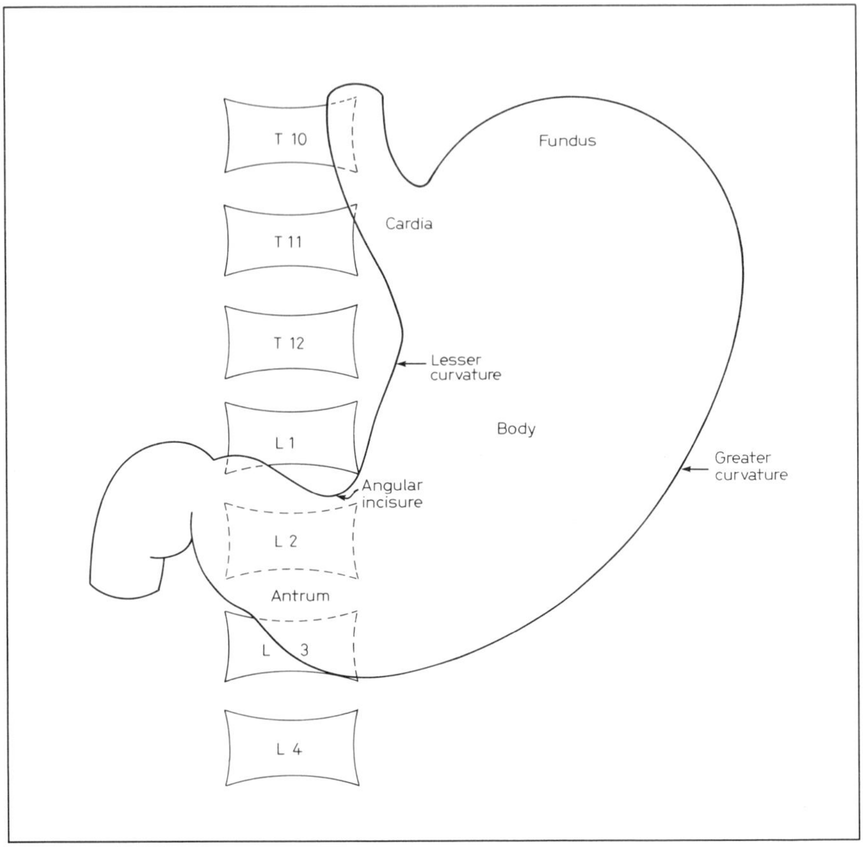

Fig. 20. Relationships of stomach to vertebral column and subdivisions of stomach.

modified by changes within itself and the surrounding viscera and no one form or position is typical. However, most commonly it is seen as a J-shaped organ which extends from the lower end of the esophagus at the level of the eleventh thoracic vertebra, about 2.5 cm to the left of the mid-line, to end in the duodenum just to the right of the lower border of the first lumbar vertebra (fig. 20).

Although the external surface of the stomach is, in fact, a continuum it is convenient to describe the stomach as having two borders or curvatures and two surfaces. The lesser curvature, which extends between the cardiac

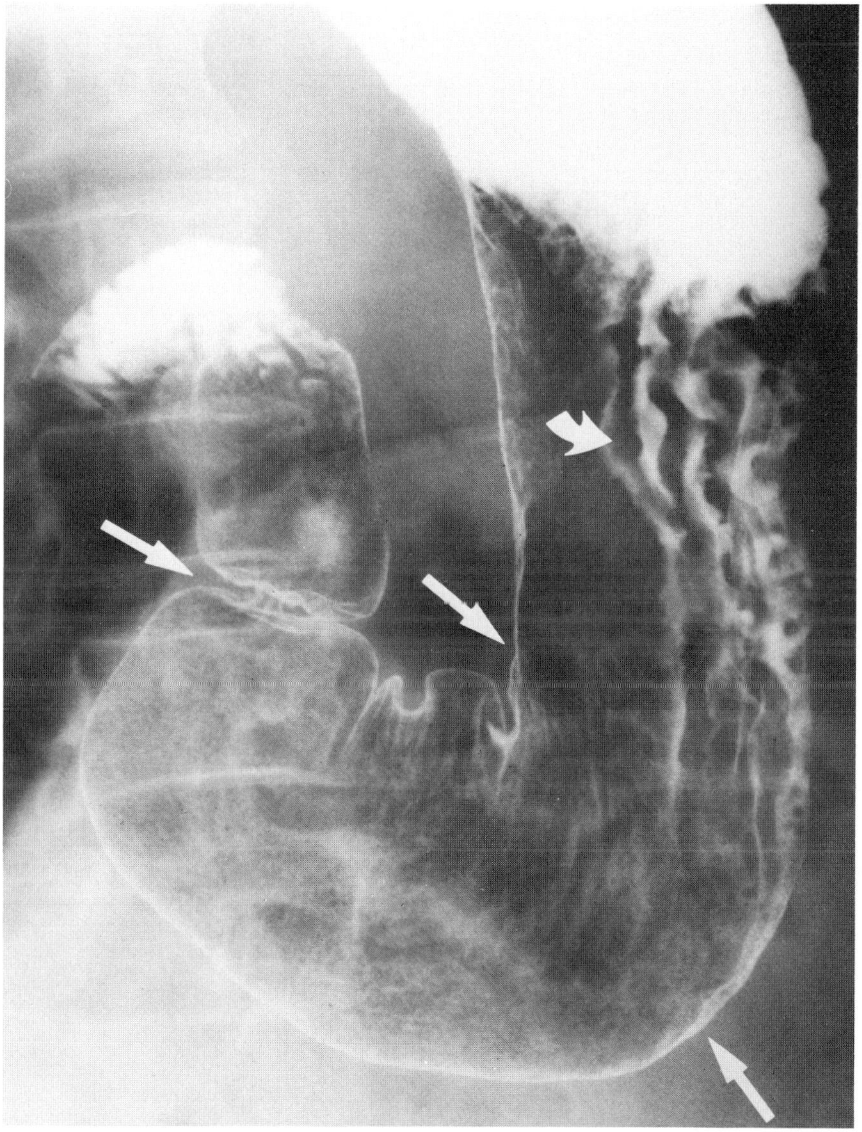

Fig. 21. X ray of normal stomach showing features illustrated in previous diagram. The greater and lesser curvatures and the pylorus are indicated by the straight arrows. Note also the prominent rugal pattern as indicated by the curved arrow.

and pyloric orifices, forms the right or postero-superior border of the stomach. It is a direct continuation of the right margin of the esophagus passing internally from that structure then curving to the right to end at the pylorus. The most dependent part of the lesser curvature may form a notch which is called the angular incisure. The greater curvature is four to five times as long as the lesser curvature and commences where the left margin of the esophagus joins the stomach. This junction forms an acute angle which is known as the cardiac notch. From this cardiac notch the greater curvature arches upwards, posteriorly and to the left, then runs downwards and anteriorly to finally curve to the right and end in the pylorus. This characteristic structure may be demonstrated radiologically (fig. 21).

The stomach is divided rather arbitrarily into four parts. The cardia is the region about 2–3 cm in width immediately distal to the esophagus. The fundus is that part of the stomach which lies above a line drawn horizontally through the gastro-esophageal junction. The body comprises roughly the proximal two thirds of the remainder, and the pyloric antrum, the distal one third which leads, by way of the pyloric sphincter, into the duodenum. Demarcation of the body of the stomach from the antrum can be based on the notch in the lesser curvature, the angular incisure, already described. The separation is achieved by drawing a line from the notch downwards and to the left to the greater curvature. The portion proximal to this line is the body whilst the antrum is distal to it.

A more rational division may be made on the basis of the mucosal glands, three regions being distinguished in this way. Thus the narrow ring-shaped area around the cardia is called the cardiac area and contains the cardiac glands. The fundus and proximal two thirds of the stomach, which contain the gastric glands proper, form the second zone. The third zone is called the pyloric region being characterized by the presence of the pyloric glands. It occupies the most distal portion of the stomach and extends further along the lesser curvature than along the greater. The actual structure of the glands defining these three zones will be discussed below but it should be noted at this point that the zones are not sharply demarcated, the gland types intermingling to some extent at the sites of transition.

The stomach has a very rich lymphatic system. Numerous lymphatic channels form a plexus in the submucosa from which small vessels penetrate the muscularis mucosae and ramify in the lamina propria of the mucosa. Lymphatic drainage occurs to the left gastric, right gastric and subpyloric lymph nodes, and also to the paracardial, pancreaticosplenic and right gastro-epiploic nodes (fig. 22).

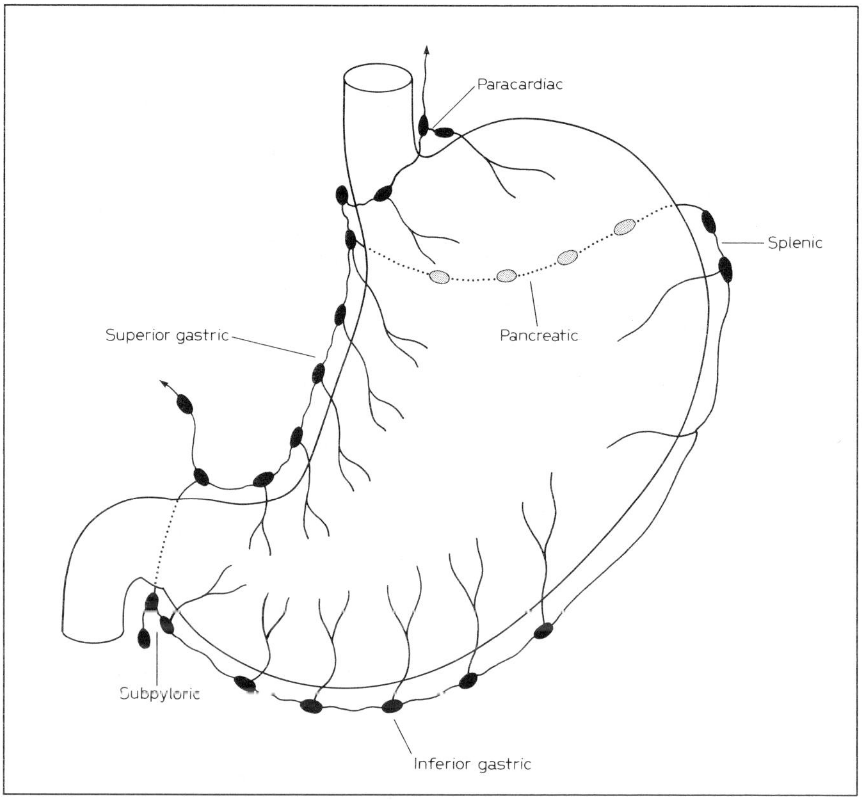

Fig. 22. Diagrammatic representation of lymphatic drainage of stomach.

Histology

As in the remainder of the alimentary tract the wall of the stomach consists of four layers – the serosa, the muscle coat or muscularis externa, the submucosa and mucosa (fig. 23).

The serosal coat is formed by the visceral layer of the peritoneum which covers virtually the whole of the stomach except for the areas along the greater and lesser curvatures which represent the lines of attachment of the greater and lesser omenta. The muscle coat is composed of three layers of smooth muscle which from the outside inwards are arranged in a longitudinal, circular and oblique fashion. The submucosa is a well defined zone comprising loose connective tissue.

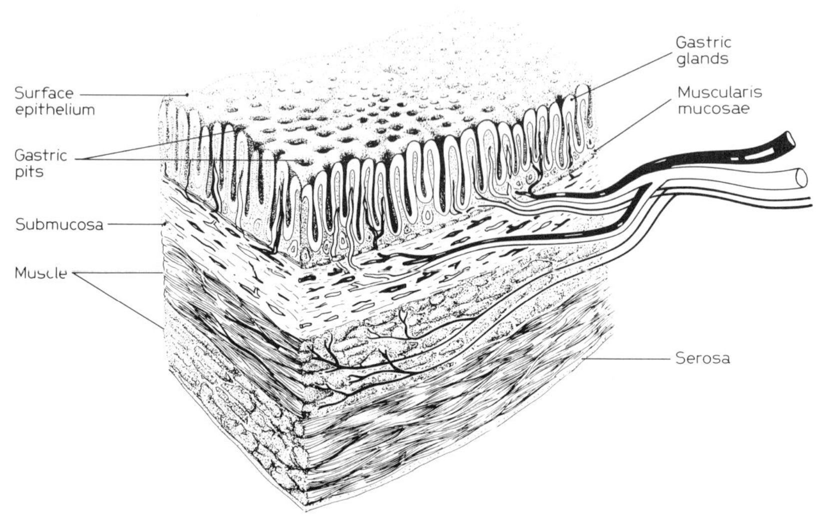

Fig. 23. Diagrammatic representation of full thickness of stomach wall. Note the four coats and the rather complex mucosal structure.

Again it is the mucosa that is of primary interest to the diagnostic cytologist. When the stomach is full the mucosa is evenly stretched and appears smooth. However, more commonly it is examined whilst empty or nearly so and the surface is then seen to be thrown into numerous folds, most of which are longitudinally directed, known as the gastric rugae. When examined with a hand lens the mucosal surface has a honeycomb appearance being covered with small depressions. These are the gastric pits and at the bottom of each are the openings of the gastric glands.

The gastric glands are of three types – cardiac, main gastric glands, and pyloric glands. The cardiac glands are few in number and confined to a small area near the cardiac orifice. Most are simple tubular glands being made up predominantly of mucus secreting cells.

The main gastric glands are of greatest significance to the diagnostic cytologist (fig. 24). They occupy the mucosa of the body and fundus of the stomach with three to seven glands opening into each gastric pit. It is cus-

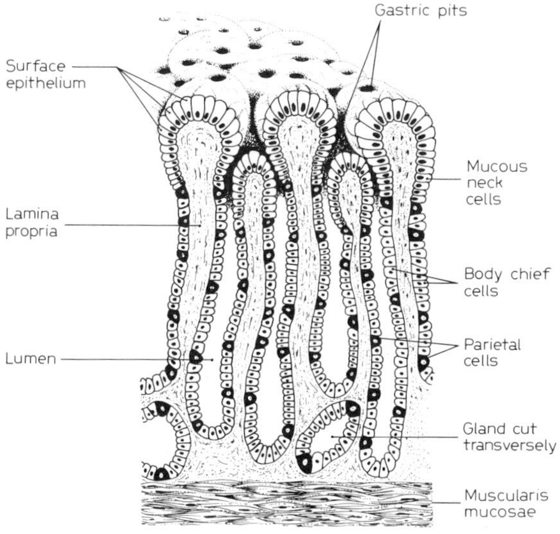

Fig. 24. Diagrammatic representation of main gastric glands.

tomary to divide the normal fundus and body mucosa into two zones – superficial and deep (fig. 25). The superficial zone, which comprises 25% of the mucosal thickness, consist of the surface epithelium which dips inwardly to form the gastric pits or foveolae (fig. 26). Evidence of regeneration is seen only in the deeper parts of the gastric pits where mitoses may be seen in less well differentiated cells which contain only occasional mucigen granules in their cytoplasm (fig. 27). Completely undifferentiated cells are present also in this region. It would appear that these cells are the source of renewal of all other epithelial cells of the gastric mucosa. The deep zone, which comprises the other 75% of the mucosal thickness, consists of straight perpendicular tubules extending downwards from the pits to the muscularis mucosae where they undergo some coiling and appear as acini in cross section (fig. 28). Lining the neck of the glands, but also occurring singly in the body of the gland, are the so-called mucous neck cells. These cells are columnar or flask-shaped with a narrow apex and a broad base.

Their cytoplasm contains droplets of mucus which is strongly PAS positive.

In addition to the scattered mucous neck cells, the gastric glands of the fundus and body contain two other main cell types – the parietal and chief cells (fig. 29). The parietal or oxyntic cells are concentrated mainly in the central portion of the gland. They are large spheroidal or pyramidal cells which have abundant, deeply eosinophilic cytoplasm. Each cell usually contains a single large nucleus (fig. 30) but sometimes two, or even more than two, nuclei are seen within the one cell. The chief or zymogenic cells predominate in the body of the glands but are also seen towards the neck of the gland where they intermingle with parietal and mucous neck cells. They are pyramidal in shape, have a single nucleus, and abundant basophilic cytoplasm (fig. 31). Scattered singly between the basement membrane and the zymogenic cells are the argentaffin cells. They are rounded or sometimes flattened in appearance and their cytoplasm contains granules which stain selectively with silver salts.

The epithelium that lines the gastric pits or foveolae and covers the free surface between the pits is uniformly of the same structure. The cells are tall columnar, mucus secreting, cells with their nuclei towards the base of the cell. The cytoplasm may appear eosinophilic and finely granular (fig. 32). More frequently the cytoplasm is vacuolated (fig. 33). The vacuoles are small, cells with large secretory vacuoles – the so-called goblet cells – not occurring within the normal gastric mucosa.

The mucosa of the antrum or pyloric region is also divisible into superficial and deep zones but here the superficial zone forms 50% or more of the total mucosal thickness (fig. 34). The epithelial cells are predominantly mucus secreting (fig. 35) with occasional parietal or oxyntic cells but no chief cells.

The lamina propria of the mucosa consists of delicate connective tissue which occupies the narrow spaces between the glands and the muscularis mucosae and forms larger accumulations between the necks of the glands and between the pits or foveolae. Scattered throughout this connective tissue are eosinophils, plasma cells and numerous small lymphocytes. Small accumulations of lymphoid tissue occur normally particularly in the pyloric region. These are sometimes referred to as the gastric lymphoid follicles. The mucosa is separated from the submucosa by the muscularis mucosae, a thin layer of smooth or non-striated muscle which is usually divided into inner circular and outer longitudinal layers. The inner layer sends strands between the mucosal glands.

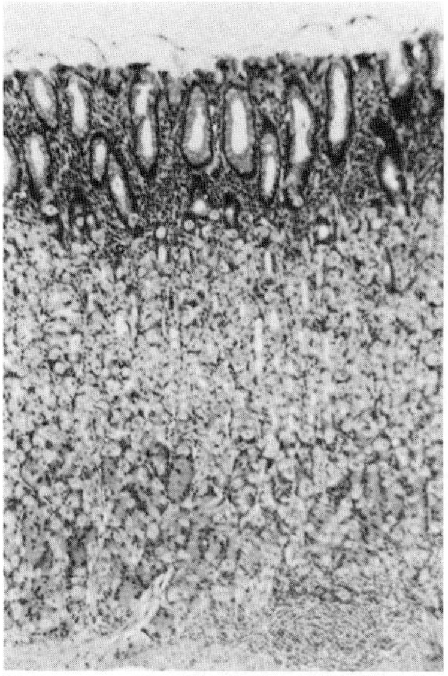

25

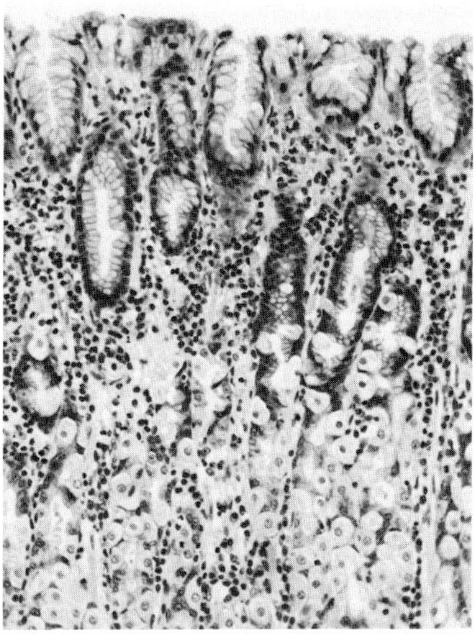

26

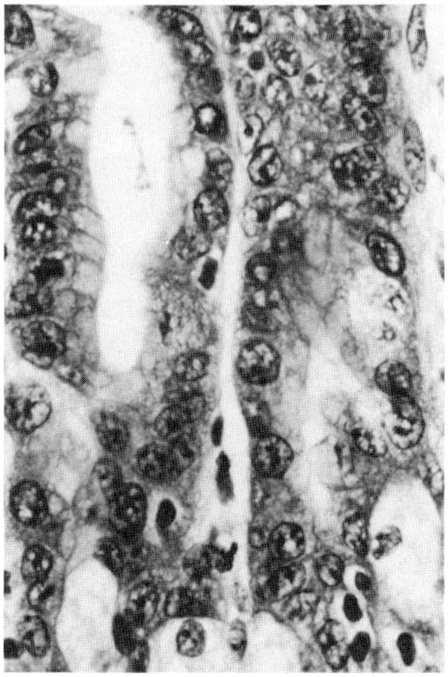

27

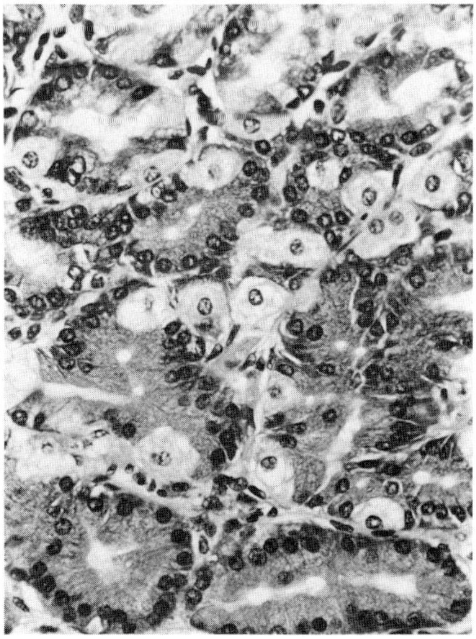

28

Fig. 25. Mucosa of body of stomach showing superficial and deep zones. × 55.
Fig. 26. Gastric pits or foveolae of mucosa of body of stomach. × 110.
Fig. 27. Deeper part of gastric pits and neck of glands showing regenerative zone. × 435.
Fig. 28. Deepest portion of gastric glands showing coiled, acinar, appearance. × 220.

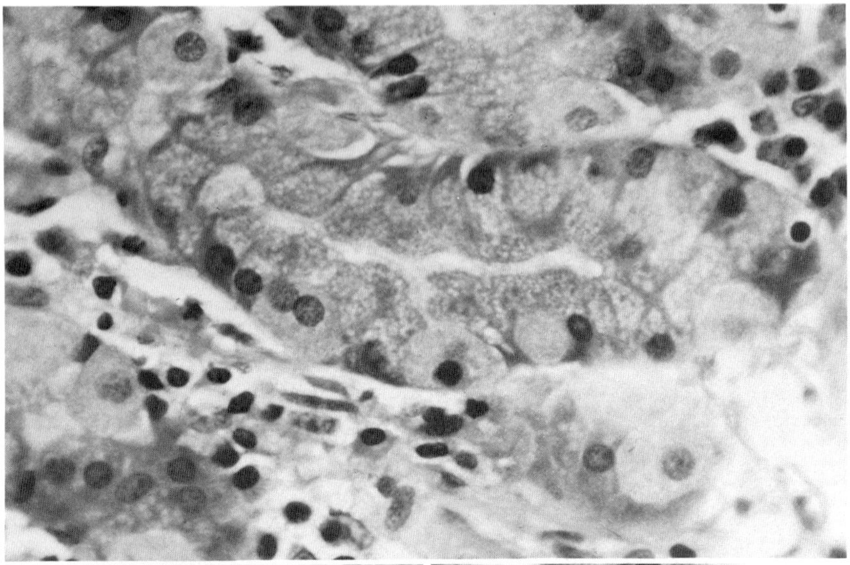

29

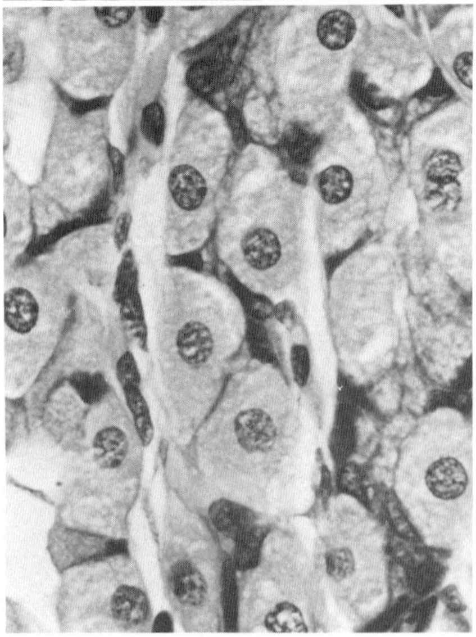

30

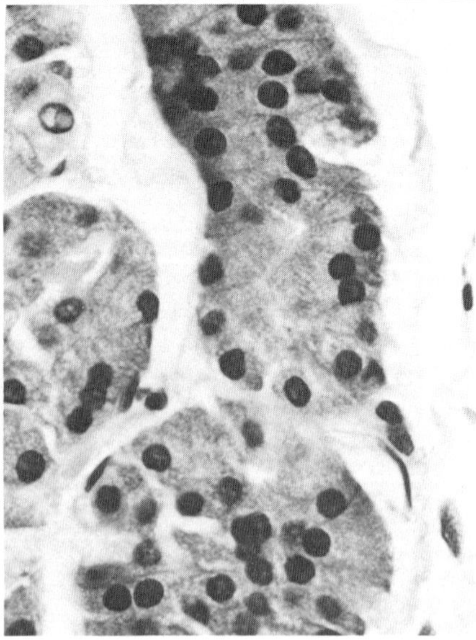

31

Fig. 29. Gastric gland showing mixture of parietal and chief cells. × 370.

Fig. 30. Parietal or oxyntic cells with round nuclei and abundant granular cytoplasm. × 545.

Fig. 31. Chief or zymogenic cells with smaller round nuclei and granular cytoplasm. × 545.

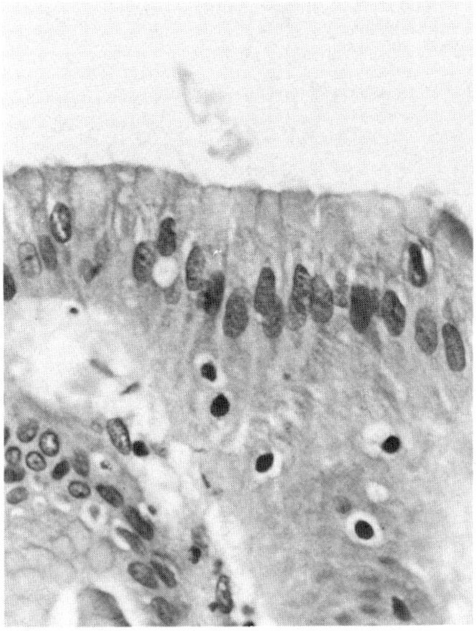

32

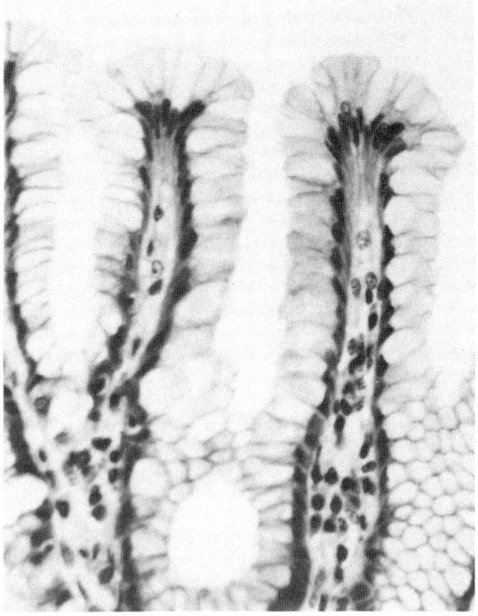

33

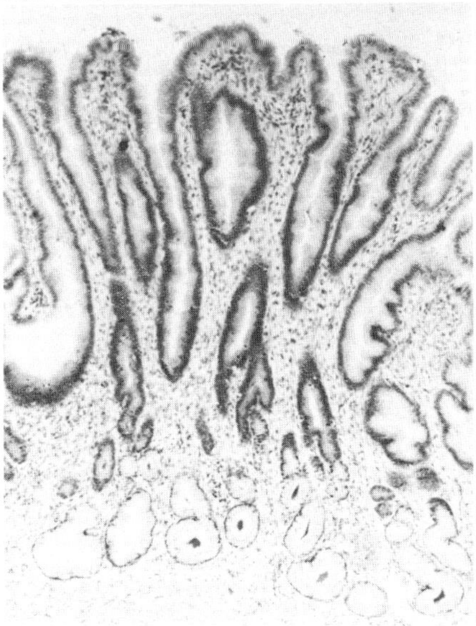

34

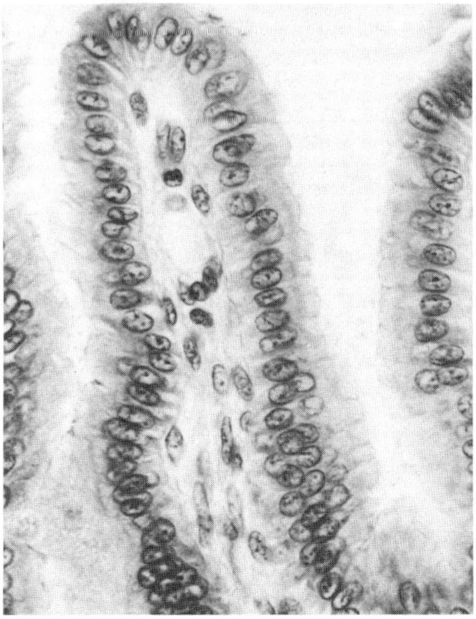

35

Fig. 32. Tall columnar surface mucosal cells with finely granular cytoplasm. × 370.
Fig. 33. Surface epithelial cells with vacuolated cytoplasm. × 275.
Fig. 34. Antral mucosa showing superficial and deep zones. × 55.
Fig. 35. Mucus secreting surface cells of antral mucosa. × 345.

Physiology

Within the stomach the ingested food is stored temporarily and during this period of storage it is churned up and mixed with the gastric juices to form a semifluid mass known as chyme. The gastric juice contains mostly protein splitting enzymes and hydrochloric acid. The gastric mucosa also secretes considerable quantities of mucus the prime purpose of which is to protect the mucosa from peptic and acid digestion. Mucus secretion is the function of the surface and mucous neck cells. Some absorption of water takes place in addition to other fluids such as alcohol and some drugs.

The parietal or oxyntic cells are the source of the hydrochloric acid although the exact mechanism of secretion is uncertain. The proteolytic enzyme pepsin is secreted by the chief or zymogenic cells the enzyme being secreted in an inactive form, pepsinogen, activation to pepsin being due to the hydrochloric acid. Hydrochloric acid is also important for the digestive function of pepsin which requires a very low pH for optimal activity. The so-called gastric intrinsic factor, which is essential for the absorption of vitamin B12, is also secreted by the chief cells.

Gastric secretion and motility are controlled by a variety of neural and hormonal influences. The main hormone involved is the substance gastrin which is released from the G cells in the antrum. This release occurs in response to distension of the stomach, protein digestion products, and an alkaline pH of the stomach contents.

The argentaffin cells secrete serotonin which is known to relax gastric muscle but its role in normal digestion and gastric motility is doubtful.

References

1 Bloom, W.; Fawcett, D.W.: A textbook of histology; 10th ed. (Saunders, Philadelphia 1975).
2 Warwick, R.; Williams, P.L.: Gray's Anatomy; 36th ed. (Longman, Edinburgh 1980).
3 Guyton, A.C.: Textbook of medical physiology; 6th ed. (Saunders, Philadelphia 1981).
4 Ham, A.W.; Leeson, T.S.: Histology; 8th ed. (Lippincott, Philadelphia 1979).
5 Rhodin, J.A.G.: Histology – a text and atlas; 1st ed. (Oxford University Press, New York 1974).

3. Specimen Collection, Preparation, Evaluation

Specimen Collection

As indicated in the previous chapter the gastro-esophageal cytologist is concerned essentially with two types of epithelium – a relatively uncomplicated squamous epithelium and a more complex glandular epithelium, the latter covered by tall columnar epithelium. The two organs are dealt with together in this chapter as the sampling techniques for each have much in common. These techniques may be summarized as follows:

Blind Lavage
- Using physiological fluid, such as saline, only.
- Using physiological fluid with an added mucolytic agent.

Endoscopic Techniques
- Washing under direct vision.
- Brushing under direct vision.
- Biopsy under direct vision with subsequent preparation of smears, imprints, or squash preparations.

It can be seen that those who practise gastro-esophageal cytology tend to fall into two groups – the blind and the sighted. The former relies on a blind lavage of the esophagus or stomach with subsequent aspiration of the washings. The sighted either wash or brush suspicious areas of the mucosa under direct vision, a modified fiberoptic endoscope being used. There has been a tendency to regard the direct washing and brushing techniques as a replacement for blind lavage. It is claimed that well preserved material is obtained, that cytological evaluation is relatively easy, and, perhaps most importantly of all, only one or two slides need to be examined instead of the twelve to twenty yielded by the average wash. Whilst much of this is true one technique does not replace the other – rather they should be regarded as complementary to each other. The direct technique requires a visible lesion to brush or wash and hence the ability of cytology to detect unsuspected cancer may be lost. Indeed, as emphasized by *Takeda* [4] the lavage methods have a special place in finding minute or small cancers at an early stage, particularly the flat lesions, that are difficult to visualize endoscopically.

In summary, it is recommended that brushing under direct vision should be the routine method of collection but that a combination of brushing and blind lavage should be considered whenever a diagnostic problem is encountered or when some form of screening for early gastric cancer is contemplated. It is further recommended that endoscopically derived biopsies be sectioned not smeared, imprinted, or squashed. Many of these biopsies are extremely small and fragile yet, when embedded in paraffin and stained, are readily evaluated (fig. 36, 37). Indeed such evaluation depends as much upon cytological skills as upon histological technique.

Blind Lavage Techniques

The blind lavage techniques are described and illustrated in considerable detail as they are frequently carried out very poorly indeed. Inevitably the results are unsatisfactory and the techniques are unfairly discredited. It cannot be stressed too strongly that the success of the various lavage techniques is largely dependent on meticulous attention to detail. The patient must be carefully prepared, the intubation and washing must be carried out with skill by a person who is dedicated to obtaining the optimum specimen, and the specimen thus obtained must be prepared with great care. Rapidity of preparation is particularly important as is the preservation of the specimen from the time of collection to such preparation. If the patient is not well enough to come to the laboratory, the laboratory must go to the patient (fig. 38).

Gastric Washing

Blind gastric lavage can usually be carried out using isotonic saline only. The technique is a relatively simple one (fig. 39) and should be performed in the following way:

The patient fasts for 12 h prior to the lavage but is encouraged to drink water on the morning of the test. At the commencement of the procedure the patient is given a sputum mug or kidney dish and instructed to clear his mouth and throat of all nasal and bronchial mucus. Throughout the test he retains the receptacle and is asked to expectorate, rather than swallow, similar material (fig. 40). All dentures are removed. With the patient in a seated or semi-reclining position a number 16 French gauge plastic Levin tube is passed, preferably via the mouth (fig. 41). With adequate explanation and reassurance most patients can swallow the tube without undue difficulty and there is seldom need to anaesthetize the throat. It may be necessary to

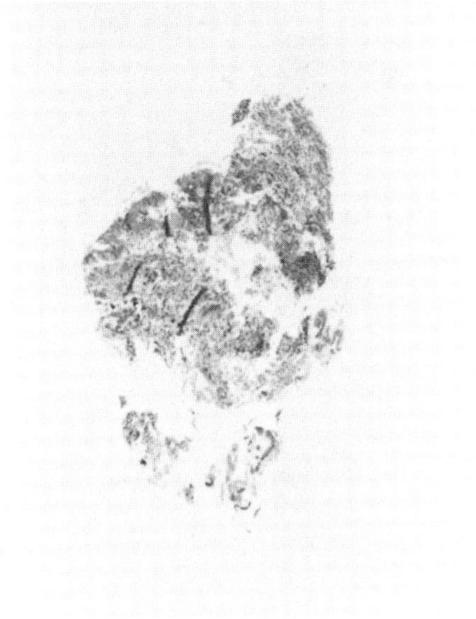

36

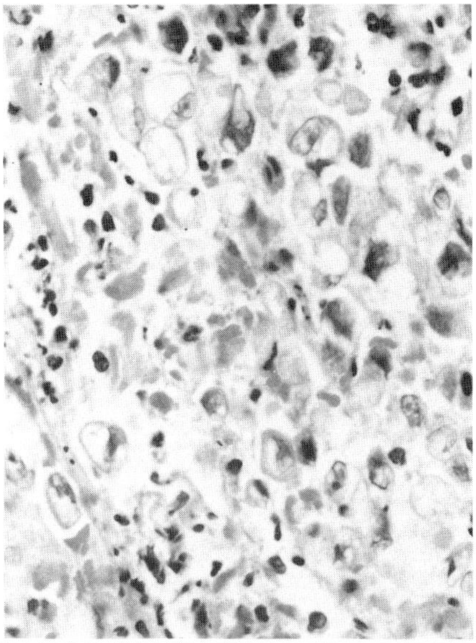

37

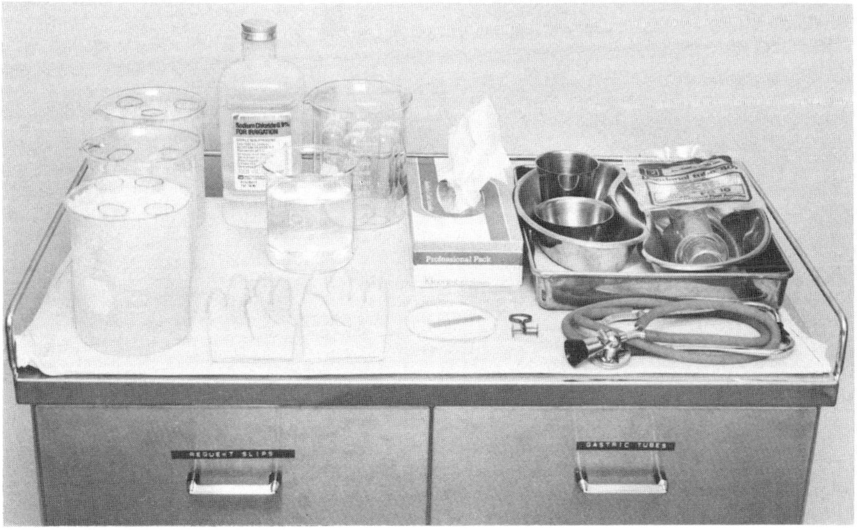

38

Fig. 36. Gastric biopsy obtained by endoscopic means. Note the very small size and fragility of specimen. ×27.5.

Fig. 37. Biopsy depicted in previous figure. Diagnosis is made readily by histological means. ×435.

Fig. 38. Trolley and equipment for lavage techniques ensuring portability of procedures.

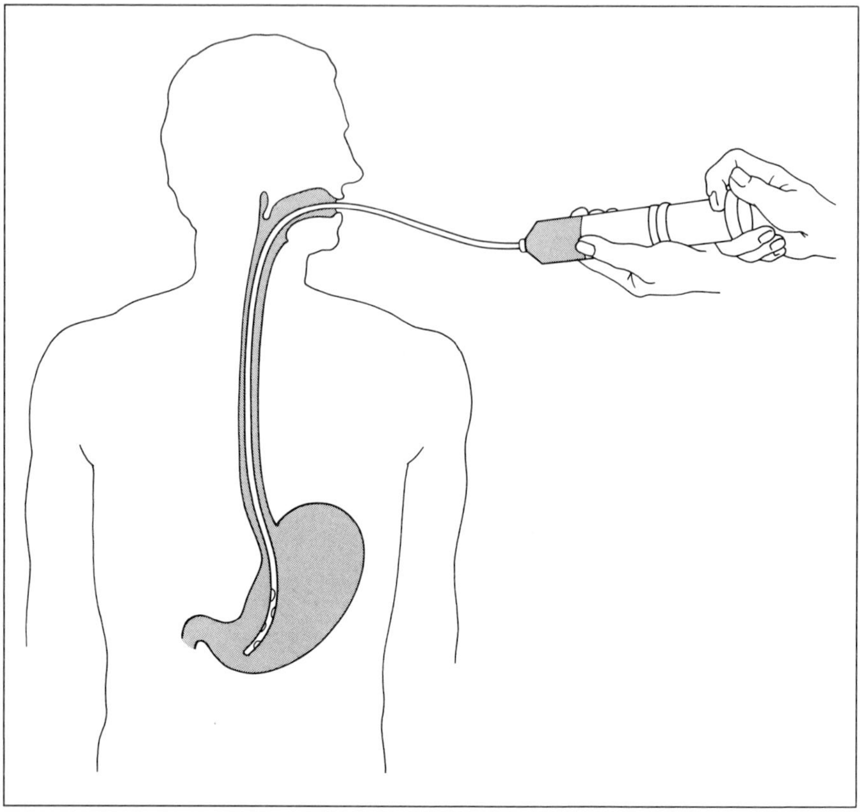

Fig. 39. Diagrammatic representation of gastric lavage technique.

lubricate the tube but usually moistening it with tap water is sufficient. A 100-ml glass syringe is attached to the proximal end of the tube, a special adaptor being used to ensure an airtight fit. The position of the tube is checked by blowing a small quantity of air into the stomach whilst listening with a stethoscope over the anterior abdominal wall (fig. 42). The fasting gastric juice is then withdrawn, its pH recorded, and the fluid discarded. A further check on the correct positioning of the tube can be made by testing the fasting juice with litmus paper (fig. 43). If a lesion is suspected at a particular site the tube is positioned accordingly, otherwise it is passed to about the 55-cm mark. Approximately 300 ml of isotonic saline is then forcefully injected into the stomach (fig. 44). This fluid is withdrawn and

40

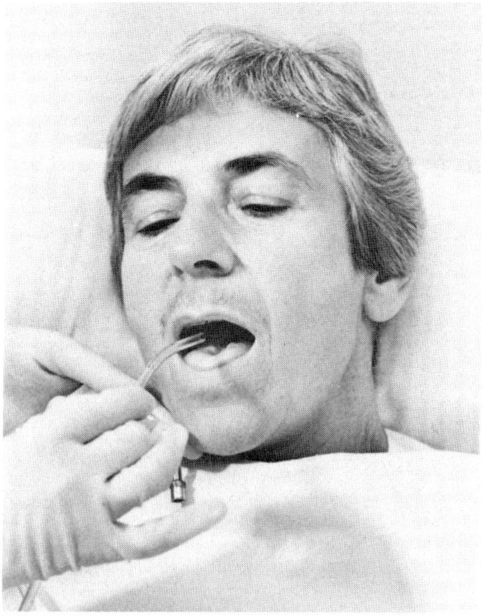
41

42
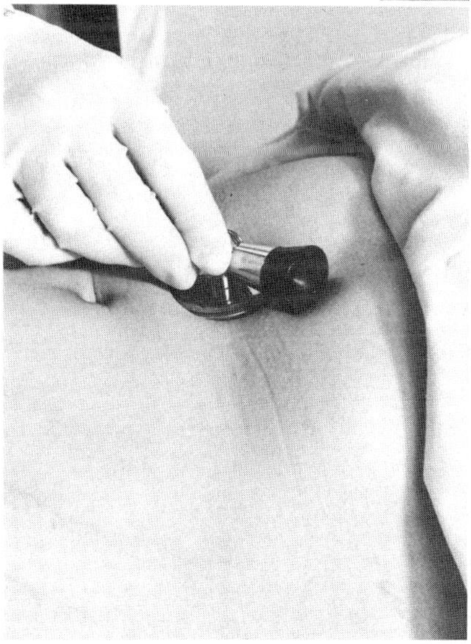

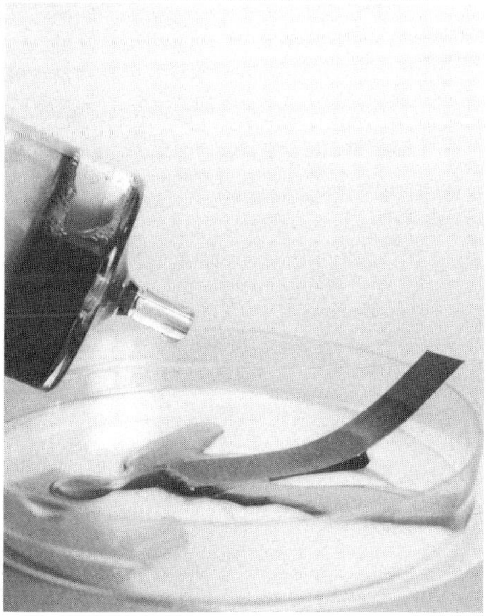
43

Fig. 40. Receptacle held by patient to encourage expectoration of all saliva, nasal and bronchial mucus.

Fig. 41. Passage of tube, preferably via mouth.

Fig. 42. Using stethoscope to check correct position of tube.

Fig. 43. Further check on correct position of tube using litmus paper.

re-injected, with as much force as possible, several times. Finally the fluid is aspirated and placed into 50-ml centrifuge tubes which are packed in ice to minimize cell degeneration. The procedure is repeated until approximately one litre of saline has been used. During the washes the patient is postured to improve access of the fluid to all parts of the gastric mucosa. Thus the wash is carried out with the patient seated, lying on his back and on his right and left sides. The upper abdomen is massaged and the stomach firmly palpated (fig. 45). Most of the injected saline can usually be recovered and this is also placed into centrifuge tubes packed in ice (fig. 46).

At the completion of thc wash the tube is withdrawn and the washings are centrifuged immediately. The tubes containing the washings are carefully balanced (fig. 47) and centrifugation is then carried out in a large capacity, refrigerated centrifuge (fig. 48). 5–10 min at 2,000 rpm is usually sufficient. The supernatant is then carefully decanted (fig. 49) and the tubes inverted to remove excess fluid (fig. 50). Depending on the consistency of the centrifuged deposit, a swab dipped in albumin (fig. 51) or a Pasteur pipette (fig. 52) is used to spread the deposit evenly on albuminized slides. The slides are placed in 95% ethyl alcohol for fixation (fig. 53) and subsequently stained by the Papanicolaou method (fig. 54). A good wash should yield 12–20 slides, each slide liberally coated with cytological material (fig. 55).

Occasionally the washings are heavily contaminated by food residue particularly if there is any impediment to gastric emptying such as may be caused by a neoplasm in the pyloric region. In these circumstances the wash may have to be abandoned and the patient placed on a fluid diet for several days prior to the test being repeated.

Very occasionally, if thick mucus appears to be interfering with the collection of material or its evaluation, consideration should be given to repeating the wash with a mucolytic agent added to the saline. The use of such an agent was first suggested by *Rosenthal and Traut* [1] who recommended the addition of buffered papain to the lavage fluid to facilitate the collection of cells from the gastric mucosa. This method was used subsequently by a number of workers but, although simple and rapid to apply, it did not always yield well preserved cells. This problem was attributed to the unstable nature of the enzyme mixture which decreases in activity with the passage of time at a rate dependent on many uncontrollable factors. Thus although cell preservation was sometimes excellent at times both mucus and cells were digested. Having experienced this problem *Rubin and Benditt* [2] carried out a series of comparative studies and as a result of these selected chymotrypsin as a suitable mucolytic agent for gastric lavage.

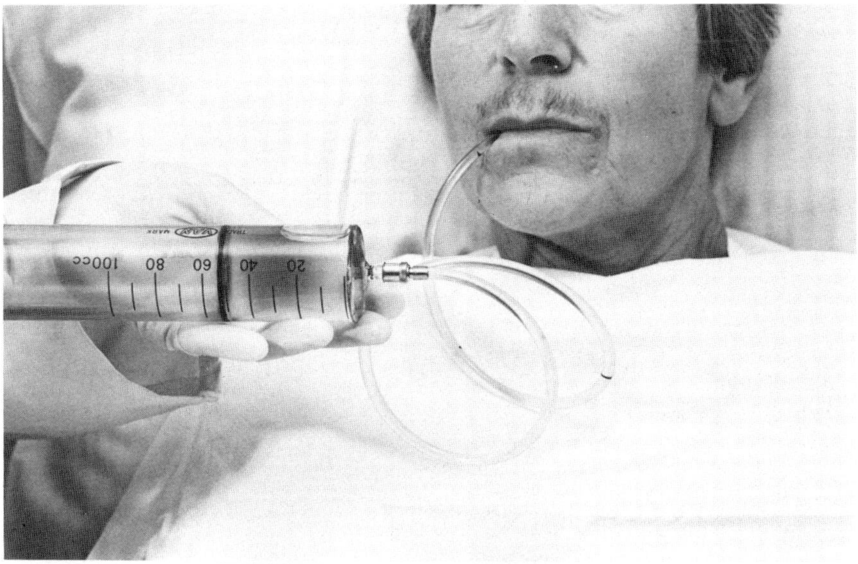

44

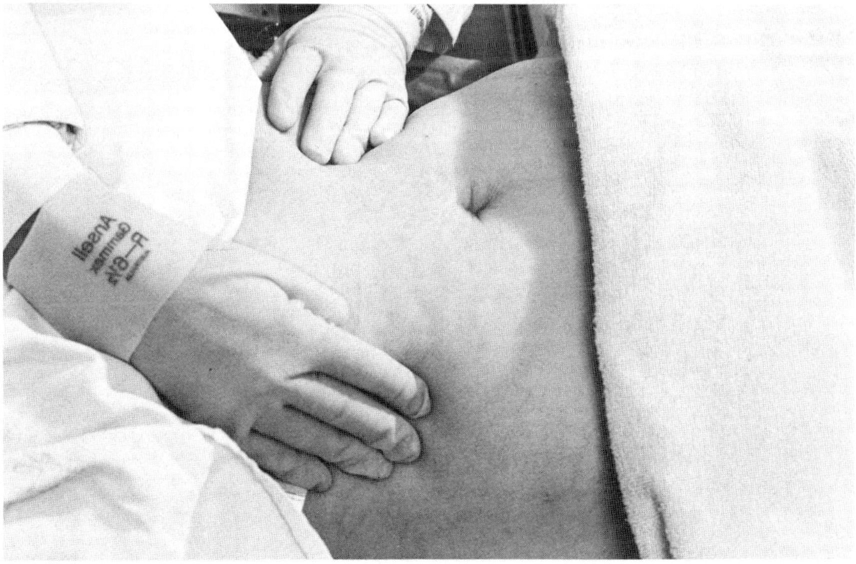

45

Fig. 44. Forceful injection of isotonic saline.
Fig. 45. Vigorous palpation of abdomen to ensure thorough washing of gastric mucosa.

46

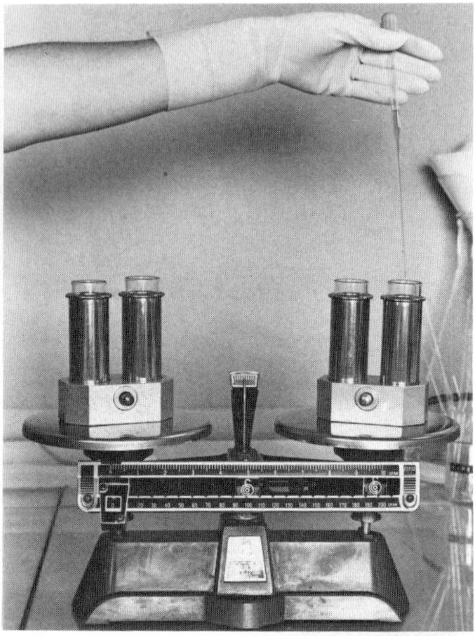

47

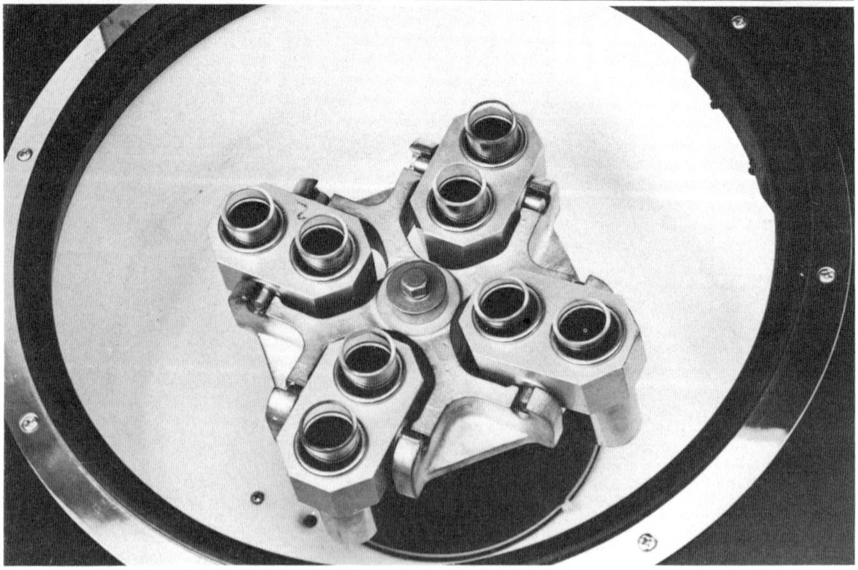

48

Fig. 46. Tubes containing gastric washings packed in ice to minimize cell degeneration.

Fig. 47. Careful balancing of centrifuge tubes.

Fig. 48. Use of large capacity, refrigerated, centrifuge to ensure optimal processing of fluid.

49

50
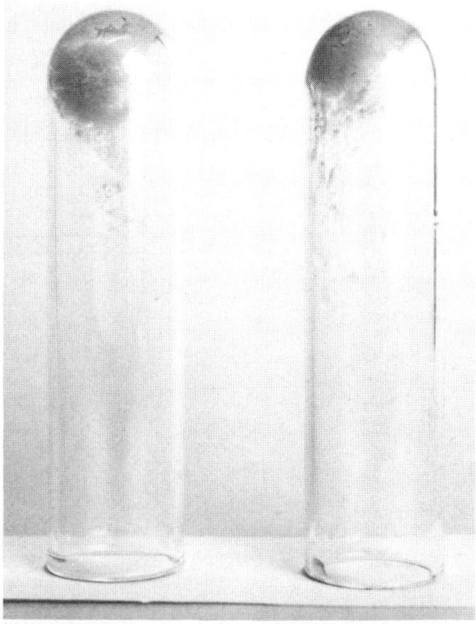

51
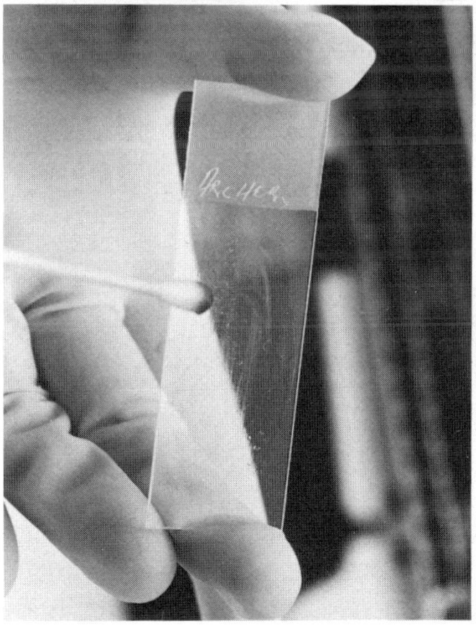

52
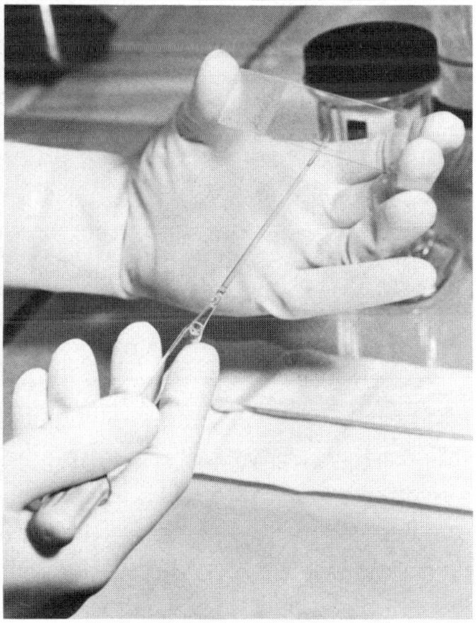

Fig. 49. Careful decanting of supernatant following centrifugation.

Fig. 50. Inversion of tubes to remove all excess fluid.

Fig. 51. Spreading centrifuged deposit on albuminized, pre-labelled slide using swab dipped in albumin.

Fig. 52. Preparation of specimen using Pasteur pipette.

These experiments also indicated the advisability of using an acetate buffer. They recommended preparing the enzyme lavage solution immediately prior to use by adding 7 mg of crystalline alpha-chymotrypsin to 500 ml of 0.1 *M* acetate buffer. They further recommended that the solution remain in the stomach for 10 min but no longer.

Subsequently chymotrypsin was used by a number of people, some variations on Rubin's technique being advocated. In 1964 *Yamada* [5] and *Yamada* et al. [6] reported two quite extensive studies of proteolytic enzyme lavage methods, the first a carefully controlled in vitro investigation of the relative merits of pepsin, trypsin and chymotrypsin at various concentrations, and the second a study of the use of trypsin and chymotrypsin in the clinical situation. On the basis of these studies they concluded that lavage of the stomach with trypsin and chymotrypsin solution (5 mg/ml in pH 5.6, 0.1 *N* acetate buffer) was a simple and reliable technique that had practical value for the detection of gastric malignancy including early or superficial cancers. These workers have continued to use this technique and in 1978 they reported [7] the examination of 420 patients by the method this examination resulting in the detection of 12 gastric cancers, each measuring less than 1.0 cm in maximum diameter. In this later study the concentration of the chymotrypsin had been increased, the authors now advocating a concentration of 10 mg/ml, and a radiopaque substance, 6% Urografin, was added to allow the washing to be carried out under fluoroscopic control. The authors stress that most of the lesions detected cytologically would not have been visible on X ray or on endoscopic examination. Similar successes have been reported by a number of other workers who have used the chymotrypsin technique.

Within our laboratory mucolytic agents are rarely used, a simple saline lavage usually being adequate. In the occasional case where tenacious mucus does appear to be a problem the following method is adopted:

Preparation is the same as for a normal saline lavage but at the commencement of the procedure the patient is given 7 mg of crystalline alpha-chymotrypsin dissolved in a glass of water to drink. 30 min later a chilled Levin tube is passed through the mouth and the residual gastric contents are aspirated and saved. The stomach is then vigorously washed for 3 min using 200 ml quantities of normal saline. These washings are retrieved and saved for processing.

Following this preliminary washing 500 ml of acetate buffer solution (13.6 sodium acetate and 0.6 ml glacial acetic acid in 1 litre of distilled water) containing 7 mg of alpha-chymotrypsin is instilled into the stomach.

53

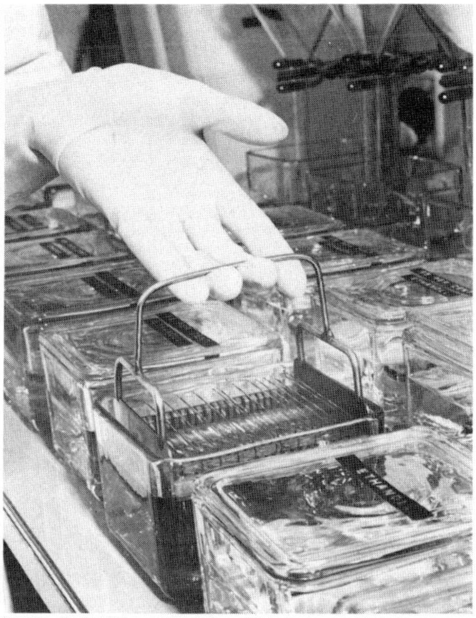

54

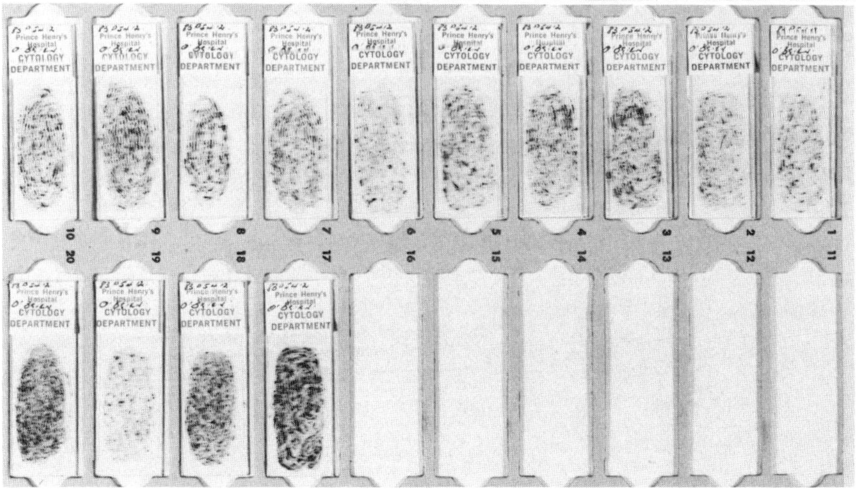

55

Fig. 53. Immediate fixation of cell preparation by immersion in ethyl alcohol.
Fig. 54. Subsequent staining by Papanicolaou method.
Fig. 55. Typical yield from well performed gastric lavage.

The patient is gently rolled from side to side and the chymotrypsin is immediately aspirated. All washings are then processed in the way described previously.

This technique sometimes results in gastric bleeding. The injection of 200–300 ml of iced water or saline at the conclusion of the test usually arrests this bleeding.

Esophageal Washing

Esophageal lavage differs somewhat in that an attempt is made to position the tip of the tube immediately distal to an area of suspicion or, if no such area has been localized, at the lower end of the esophagus. The patient is then asked to swallow the lavage fluid which is aspirated as it reaches the tip of the tube (fig. 56). Subsequently the esophagus is washed at multiple levels by progressively withdrawing the tube, injecting fluid and re-aspirating it at each level. Finally a modified gastric wash is carried out and both esophageal and gastric specimens are treated in the same way as previously.

Endoscopic Techniques

As indicated in an earlier chapter, the gastrofiberscope was first introduced by *Hirschowitz* and his colleagues in 1958. Since that time the instrument has undergone considerable modifications and improvements and shorter instruments have been specifically devised for esophageal sampling. Both the gastro-fiberscope and the esophago-fiberscope contain channels within their shafts through which biopsy forceps, a polyethylene tube for lavage purposes, or a flexible probe tipped with a nylon brush can be passed. The tips of both endoscopes are extremely mobile, flexion through an arc of 180° allowing any selected area of the mucosa, or any visible lesion, to be specifically sampled by whatever modality is being employed (fig. 57–61).

The patient requires to be prepared for this procedure, this preparation involving overnight fasting or for a minimum of 6 h. The attitude to premedication varies some operators giving valium or pethidine, with or without atropine, approximately 1 h prior to the procedure. Intravenous valium is usually given at the time of commencing the examination and this is almost always accompanied by a xylocaine throat spray. Adequate explanation and reassurance is essential and the patient should be observed for 2 h following completion of the procedure.

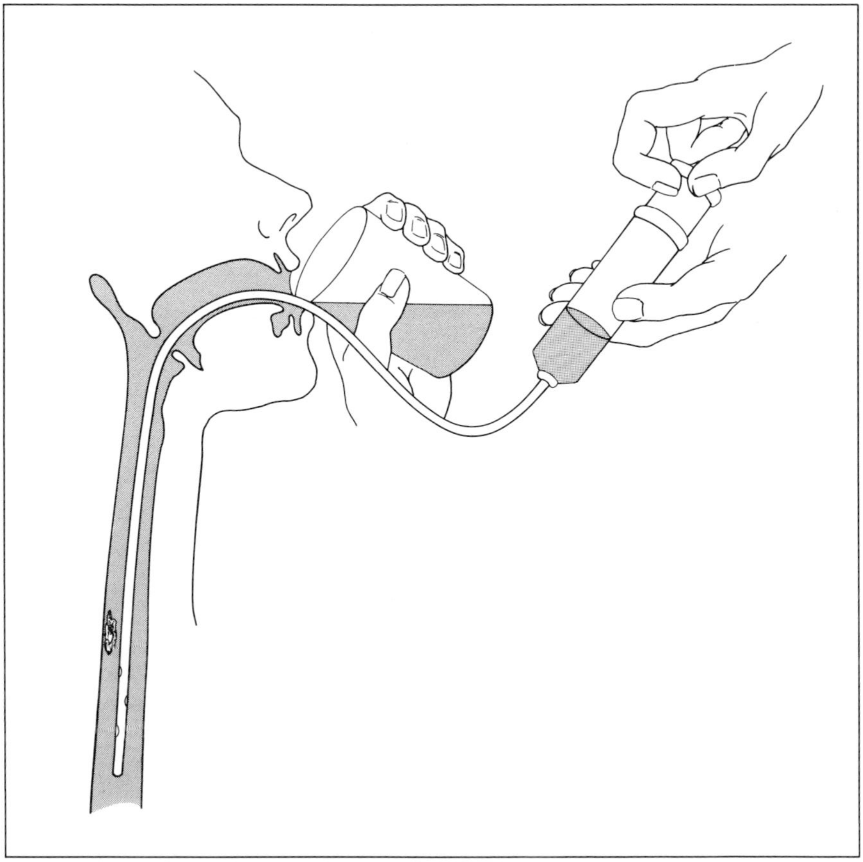

Fig. 56. Diagrammatic representation of esophageal lavage procedure.

Brushing

This is the preferred technique within our institution. The nylon brush is introduced by way of the appropriate channel in the endoscope. The area or lesion to be brushed is visualized (fig. 62) and the brush is then advanced and retracted several times to thoroughly brush the area. When an ulcerated lesion is being investigated in this way it is important to brush the edges of the ulcer as well as the floor as the latter may yield necrotic debris and inflammatory exudate only. When the brushing is completed the brush is withdrawn and smeared on to glass slides (fig. 63). The smears are immediately fixed.

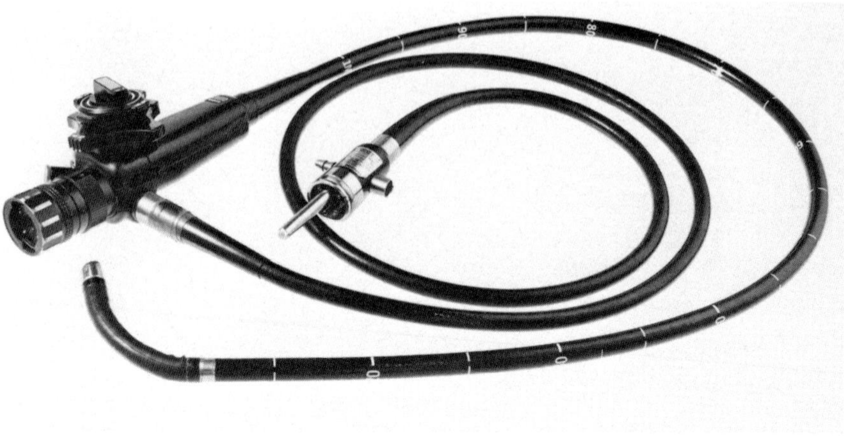

57

58

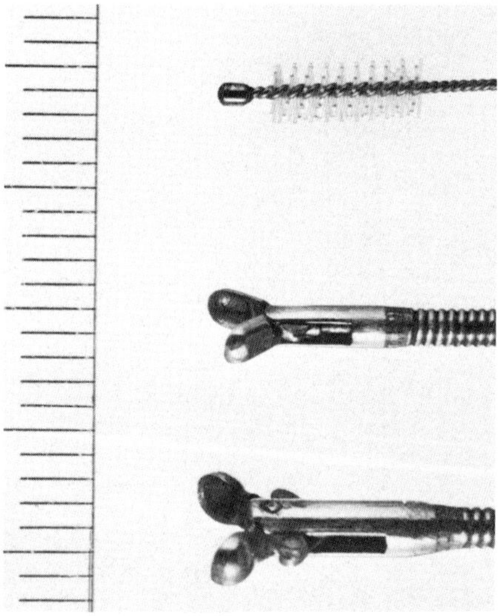

59

Fig. 57. Olympus gastric fiberoptic endoscope.
Fig. 58. Handle and eyepiece of endoscope.
Fig. 59. Nylon brush (upper) and biopsy forceps used in association with endoscope.

60

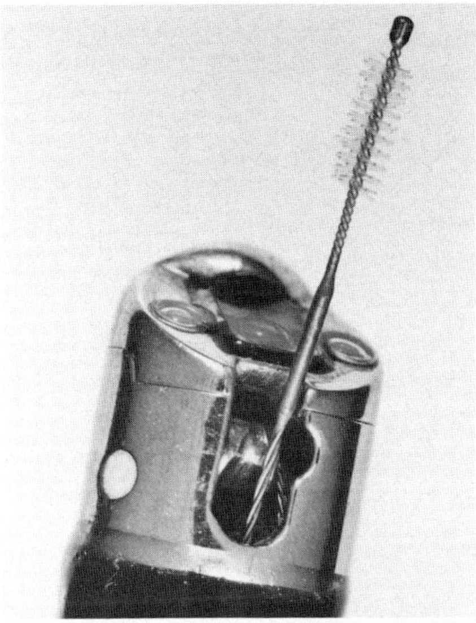

61

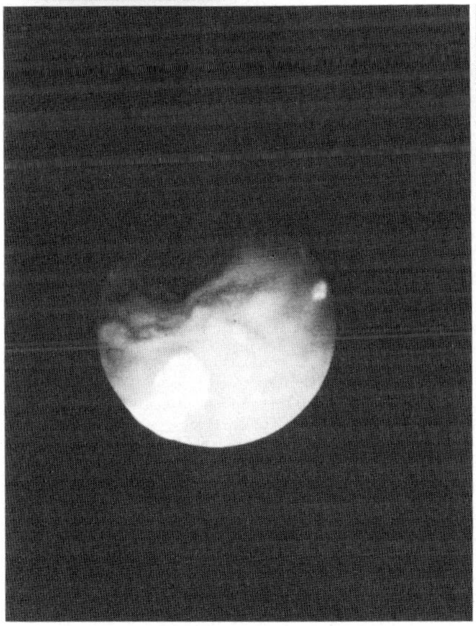

62

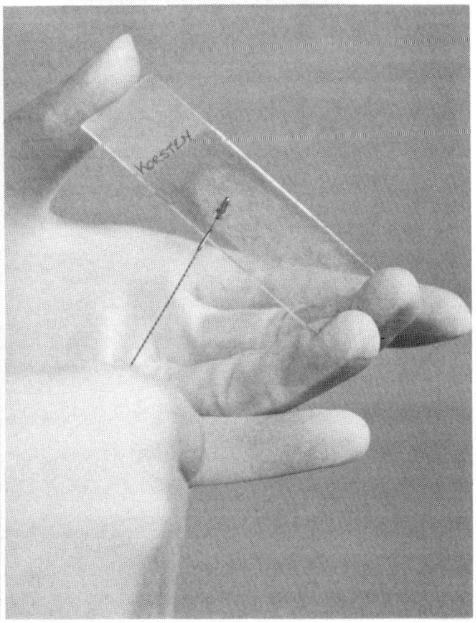

63

Fig. 60. Biopsy forceps protruding from tip of endoscope.
Fig. 61. Brush protruding from tip of endoscope.
Fig. 62. View through endoscope allowing brushing of lesion under direct vision.
Fig. 63. Immediate preparation of smear from material on brush.

Washing

A preliminary washing of the stomach is desirable before embarking on this procedure. This is achieved using a standard Levin-type gastric tube, aspirating the residual or fasting gastric juice, and then washing the stomach with a suitable fluid, such as physiological saline, until the aspirated washings become clear. The fiberoptic gastroscope is then inserted into the stomach, the lavage tube inserted through the channel in the shaft of the instrument, and all suspicious areas of the gastric mucosa are thoroughly washed. As soon as this direct lavage procedure is completed the gastroscope is withdrawn and the fluid is aspirated from the stomach by re-introducing a Levin-type gastric tube.

Table I. Staining procedures

	Solution	Time
1	Wash in running water	
2	Gill's haematoxylin	2 min
3	Wash in running water	5 min
4	Scott's tap water	1 min
5	Wash in running water	5 min
6	95% ethanol	20 dips
7	95% ethanol	20 dips
8	95% ethanol	20 dips
9	Orange G	2 min
10	95% ethanol	20 dips
11	95% ethanol	20 dips
12	95% ethanol	20 dips
13	EA	1–2 min
14	95% ethanol	20 dips
15	95% ethanol	20 dips
16	95% ethanol	20 dips
17	Absolute ethanol	20 dips + 1 min
18	Absolute ethanol	20 dips + 1 min
19	Absolute ethanol	20 dips
20	Xylene	20 dips
21	Xylene	1 min
22	Xylene	until coverslipped (at least 5 min)

Biopsy Techniques

The biopsy forceps are also introduced by way of a channel in the fiberoptic gastroscope and the suspicious areas sampled under direct vision. As the biopsies are small each individual tissue fragment is not as representative of the area under investigation as, for example, are thorough brushings. Hence multiple biopsies should be taken, each case yielding as many as 10 tissue fragments.

It is difficult to arbitrate on which is the method of choice as the success achieved with each is a function largely of the skill and experience of the endoscopist. At the outset the author would again emphasize his preference for the sectioning of biopsies rather than using them for imprint, smear, or squash preparations. The endoscopically directed lavage technique does have the disadvantage that it requires three separate intubations – a preparatory washing of the stomach, the passage of the endoscope, and the subsequent passage of the Levin tube for retrieval of the washings. These three procedures increase the total time spent on specimen collection and are also less acceptable to the patient than the single endoscopy for brushing. In addition, the washings carried out under direct vision are inevitably contaminated by other gastric contents thereby increasing the difficulties of screening and subsequent interpretation. Conversely it is believed by some operators that the material obtained by direct brushing is more difficult to interpret, the sheer bulk of material thus obtained, and the traumatic manner in which it is detached, leading to diagnostic problems. The interpretation of cells obtained by brushing procedures versus those by washing will be discussed below.

Specimen Preparation

Whatever technique of specimen collection is employed the end result is a greater or lesser number of slides liberally coated with material derived from the esophagus or stomach. Whilst some may prefer to allow this material to air dry and subsequently stain it by the May-Grünwald-Giemsa technique, wet fixation and Papanicolaou staining is recommended. This latter technique gives excellent and reproducible results and does allow a more direct correlation of the cytological findings with subsequent histological material – the latter being usually stained by the haematoxylin and eosin method. A variety of special stains may be used in special circumstances, e.g. mucin staining, staining for melanin etc. There is some merit, if sufficient material is available, in leaving one or two slides unstained, to use if special staining techniques are indicated by the initial evaluation.

Other staining methods, such as the polychromic staining of Starr and Bertalanffy's acridine orange fluorescence stain have been advocated as a possible basis for mass screening projects but the author has had no personal experience with these techniques. However, it would seem probable that diagnostic problems are more likely to occur in screening programmes than in routine clinical investigations and hence adherence to a standard preparation technique would appear highly desirable.

In summary, therefore, it is recommended that cytological specimens from both esophagus and stomach be fixed immediately, special care being taken to avoid air drying, and that the specimens so fixed be stained by the Papanicolaou technique. The staining procedures outlined in table I yield excellent results.

Specimen Evaluation

Detection

Although the initial screening procedures involve similar skills to those employed in other areas of diagnostic cytology gastro-esophageal cytology does require special experience on the part of the cytotechnologist. Obviously there is a need for a thorough knowledge of the structures and cells that may be encountered in specimens from these two organs both in the normal state and in association with the spectrum of disease processes. In addition, the cytotechnologist must appreciate the elusiveness of the cells shed from some malignant processes. For example, as will be pointed out subsequently, the anaplastic carcinoma of the stomach usually sheds relatively few cells often in association with a considerable amount of necrotic debris and inflammatory exudate. The carcinoma cells are small and relatively inconspicuous and may easily be overlooked by the technologist inexperienced in the evaluation of gastric material (fig. 64).

Before systematically screening each slide a rapid and random examination is desirable. This enables the technologist to obtain an overall impression of both the epithelial cells present and also the background of inflammatory exudate, blood, contaminants, such as meat fibres and vegetable cells, and any other non-cellular material which may be present such as barium. In addition, as will be discussed in the appropriate chapter, specimens from patients with a malignant lymphoma often have a curious pale blue background. Although the nature of this is uncertain, its presence should alert the technologist to search most diligently for small single malignant cells which are often missed by the inexperienced observer or mistaken for lymphocytes and ignored. Indeed an examination of lymphocytes under high power may be necessary to ensure that they are normal lymphocytes, such as may frequently accompany gastritis or peptic ulceration, and do not have criteria indicative of a malignant lymphoma.

After making this preliminary assessment the specimens must be screened systematically to ensure that all cells on every slide are examined. It is preferable to screen in a vertical manner (fig. 65a) and all abnormal cells should be marked for further evaluation. The method of marking varies considerably amongst different laboratories. The use of rings to mark the slide is undesirable as they tend to obscure too much cellular detail. Small dots placed to one side of the abnormal cells is preferred and it is strongly recommended that white marking ink is used. It is surprising how many laboratories continue to use a black or dark coloured marking ink which is quite invisible against the black background of the microscope stage. Where a group of cells is considered to be of particular diagnostic significance the relevant dot may be highlighted by a small line or arrow (fig. 65b)

Interpretation

The final evaluation depends on a detailed knowledge of the disease processes that may be encountered and their cytological manifestations. However, in making this evaluation it is essential for the cytotechnologist and the cytopathologist to be aware of the way in which the specimen was collected and to be familiar with the very significant differences between cells obtained by direct brushing and those by washing – particularly by the blind lavage techniques.

Usually direct brushings yield large sheets of cells or tissue fragments (fig. 66) in contrast to the single cells or small groups of cells that are commonly seen in lavage specimens. In addition, in the latter specimens, the cells tend to round up in the fluid medium – a phenomenon of surface tension. They are thus seen as three dimensional structures or 'balls' of cells (fig. 67) rather than the flatter sheets of cells seen in brushings. These features are particularly relevant to the problem of differentiating between the regenerative changes that may accompany chronic gastritis or a chronic peptic ulcer, and those changes indicating malignancy. In lavage specimens the tendency for tissue showing regenerative activity only to appear as large sheets (fig. 68) contrasts with the small groups of cells so typical of a carcinoma. This diagnostic aid may be lost in brushing specimens although the poor cohesion of malignant cells is usually manifest even in material collected by the brushing techniques (fig. 69).

There are also differences in the appearances of the individual cells collected by the two methods. The cytoplasm of cells in a gastric brushing is usually more abundant and has an uneven outline, whereas in washings the cells tend to round up, the cytoplasm appears less, and the cell outline is smoother. There is some disagreement as to variations in nuclear detail. Thus *Shida and Ishioka* [3] state that the cells collected by the lavage technique have an extremely abnormal chromatin pattern with uneven condensation and irregularity of chromatin arrangement, thickening of nuclear borders and prominent nucleoli. This contrasts with the features seen in direct smears where, as a rule, the cell pattern appears less malignant. In our experience the comparative features of specimens collected by the two techniques are very variable. Whilst it is true that in some cases the nuclear malignant criteria are accentuated in lavage specimens, in other cases the reverse is true (fig. 70).

Finally, drying artefact is more likely to be seen in the specimen collected by the brushing technique than by lavage methods. This artefact is particularly noticeable at the periphery of the smear where drying commences (fig. 71). To avoid this problem it may be necessary to have a cytotechnologist attend the endoscopy to personally supervise the immediate fixation of the smear.

However, it is important to note that, regardless of the method of collection and preparation, the basic principles of diagnosis of malignancy are always applicable as is the recognition of differentiation characteristics.

Malignant Criteria

It is important to note from the outset that in both esophageal and gastric carcinoma cells malignant criteria are usually present and clearly recognizable. Indeed it is stressed that identification of such criteria is essential if false diagnoses of malignancy are to be avoided. As in all areas of diagnostic cytology there is no absolute criterion of malignancy – there is no single cytologic change which, if present, is diagnostic of malignancy, nor is there any single cytologic change, the absence of which excludes malignan-

64

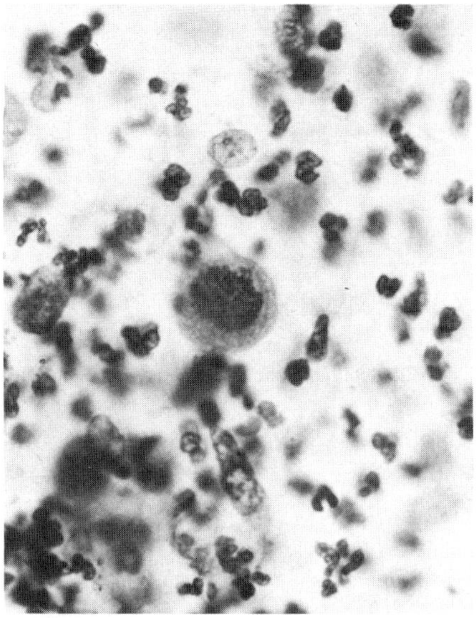

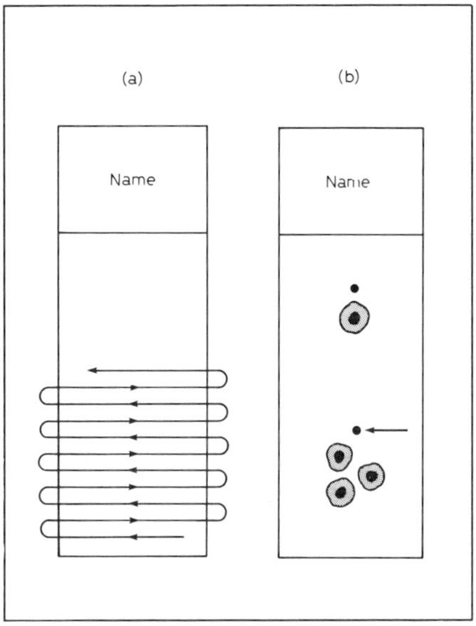

65

66

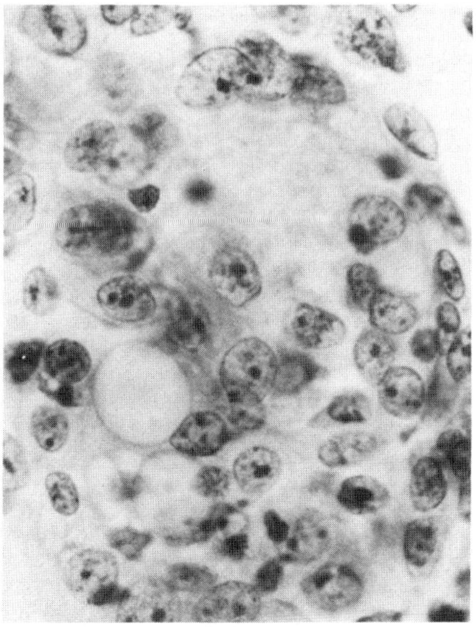

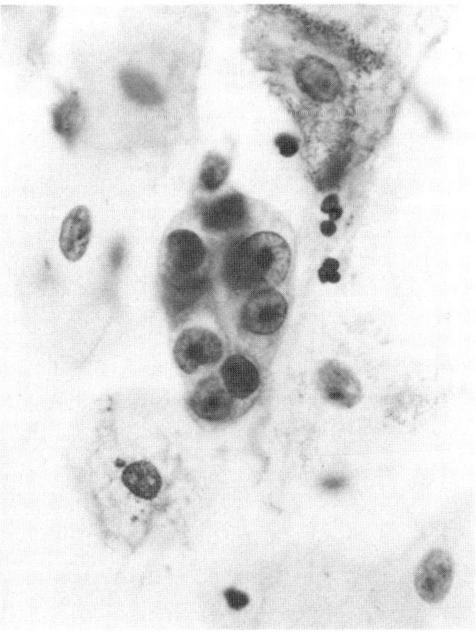

67

Fig. 64. Undifferentiated carcinoma cell partially obscured by inflammatory exudate. × 545.

Fig. 65. Method of screening (*a*) and marking (*b*) slides.

Fig. 66. Large sheet of cells characteristic of gastric brushing. Cells are from adenocarcinoma – note central acinar structure. × 545.

Fig. 67. Three dimensional structure characteristic of lavage techniques. Cells are from a papillary adenocarcinoma. × 545.

cy. Nevertheless esophageal and gastric cancers are usually characterized by a number of malignant criteria present in numerous cells. The following criteria are the most useful:

Nuclear Abnormalities

The chromatin pattern is usually altered sometimes showing a coarse granularity but, more frequently, large chromatin clumps are evident. These clumps have an irregular, jagged outline and usually clearly defined borders (fig. 72).

In association with the clumping of the chromatin material clear areas within the nucleus may be prominent.

A nucleolus is usually present, is frequently enlarged, and may be irregular in outline. Multiple nucleoli are quite common (fig. 73). It will be stressed, however, that nucleoli in gastric cancer cells may be quite uniform in outline and, conversely, large, very irregular nucleoli may be seen in cells showing regenerative activity in association with gastritis or peptic ulceration.

The nuclear membrane is usually irregular in outline (fig. 74) and may show prominent notching or projections (fig. 75).

Mitotic figures are seen frequently. They may be normal but abnormal forms are common (fig. 76).

The nuclear-cytoplasmic ratio is very variable, many gastric cancer cells having relatively abundant cytoplasm. However, a high N/C ratio may be a valuable diagnostic criterion, particularly in single cells, and is a characteristic feature of an undifferentiated carcinoma (fig. 77).

Cell Clumps and Tissue Fragments

As will be emphasized subsequently, gastric adenocarcinoma is most commonly manifested cytologically by small sheets or clumps of malignant cells. Whilst the individual cells usually show good malignant criteria a study of the relationship of one cell to another within a group is also of considerable value in making a diagnosis.

There may be considerable variation of individual cells within a group with respect to their nuclear size and shape and their chromatin pattern (fig. 78).

There may be considerable crowding of nuclei within a group, with pronounced nuclear moulding (fig. 79) whilst a loss of polarity of the cells is common. This latter feature of cellular disorganization or disarray is a most

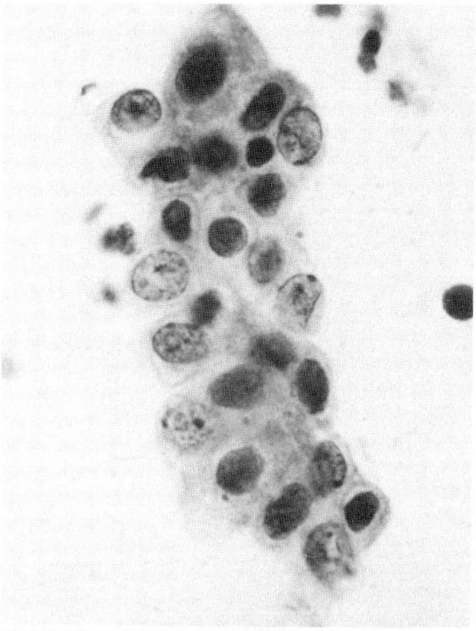

68

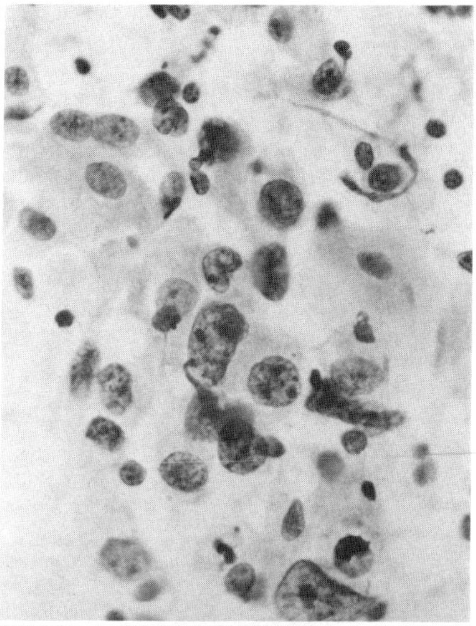

69

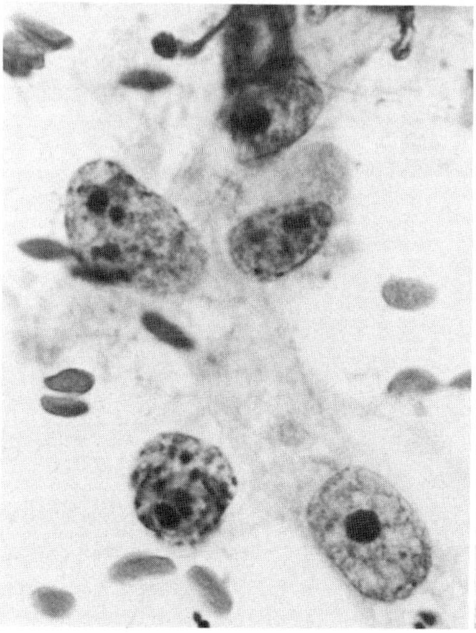

70

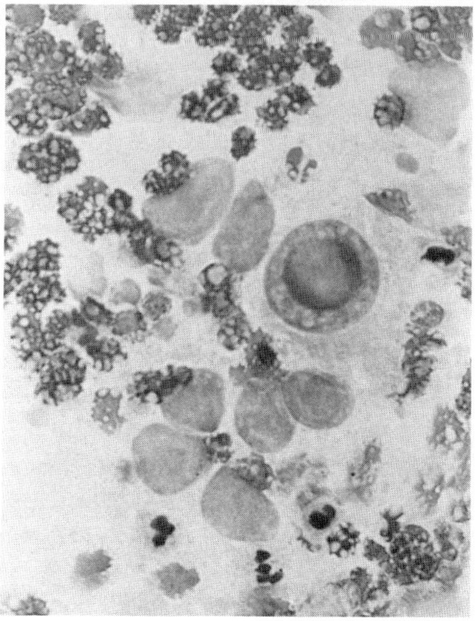

71

Fig. 68. Cohesion of non-malignant cells seen particularly in lavage specimens. Cells are from edge of peptic ulcer. × 630.

Fig. 69. Poor cohesion of malignant cells evident in gastric brushings. Cells are from poorly differentiated adenocarcinoma. × 790.

Fig. 70. Prominent malignant criteria in cells obtained by brushing technique.

Fig. 71. Drying artefact commonly seen at the periphery of brushing specimens.

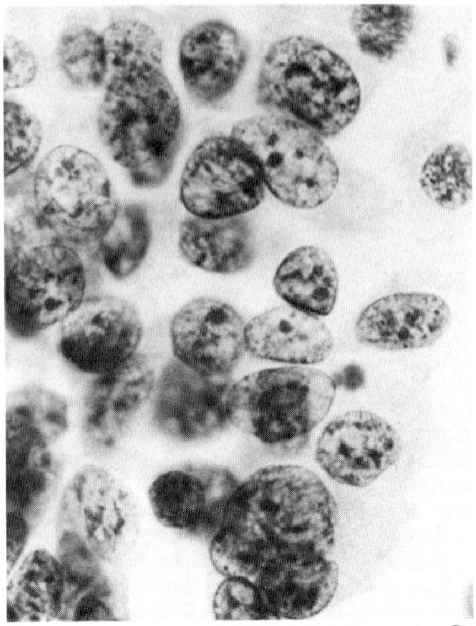

72

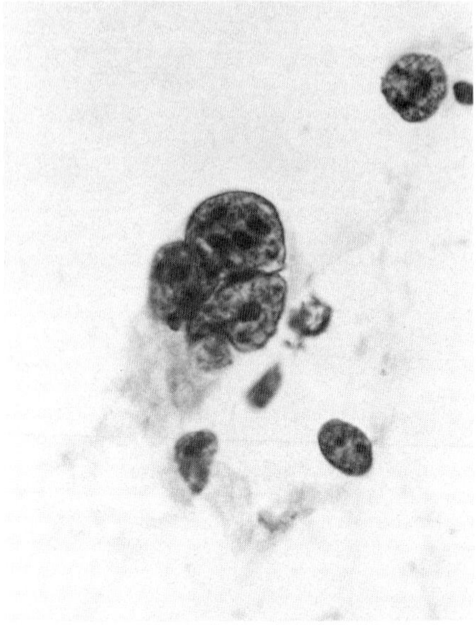

73

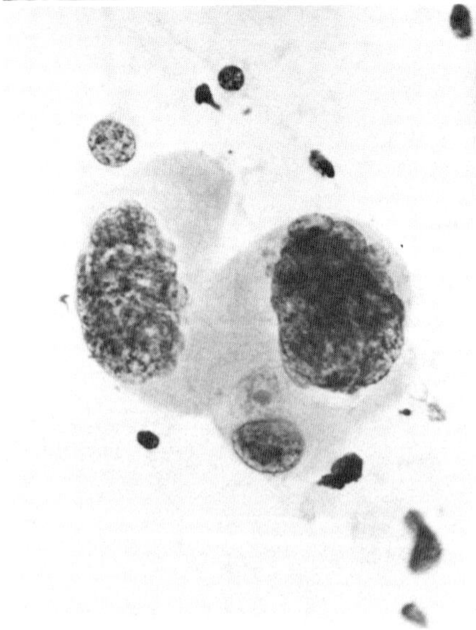

74

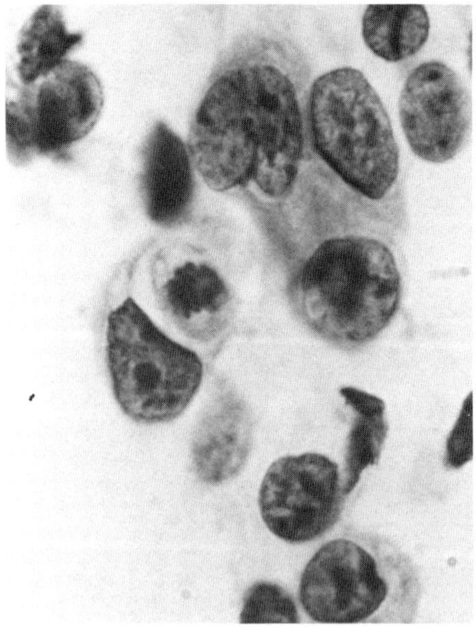

75

Fig. 72. Irregular chromatin clumps in cells from undifferentiated carcinoma of esophagus. × 545.

Fig. 73. Prominent and multiple nucleoli in cells from very poorly differentiated adenocarcinoma of stomach. × 545.

Fig. 74. Marked irregularity of nuclear membranes in cells from poorly differentiated adenocarcinoma of esophagus. × 545.

Fig. 75. Notches in and projections of nuclear membranes in cells from adenocarcinoma of stomach. × 545.

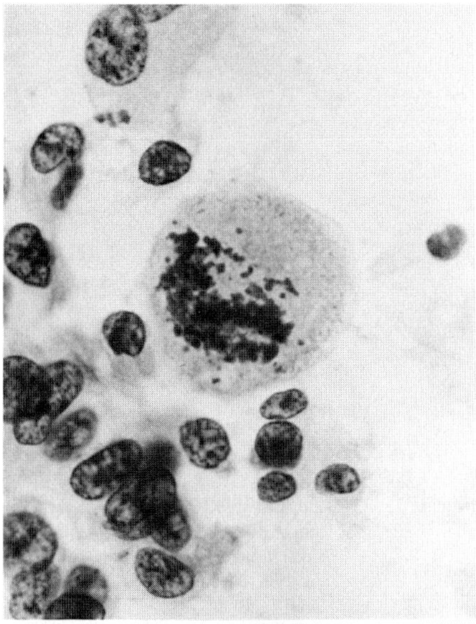

76

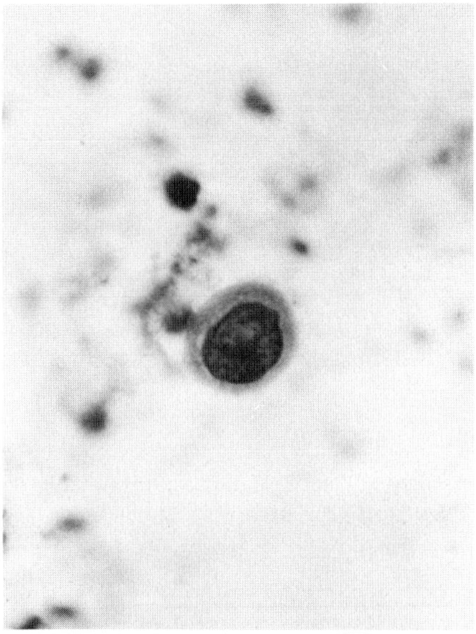

77

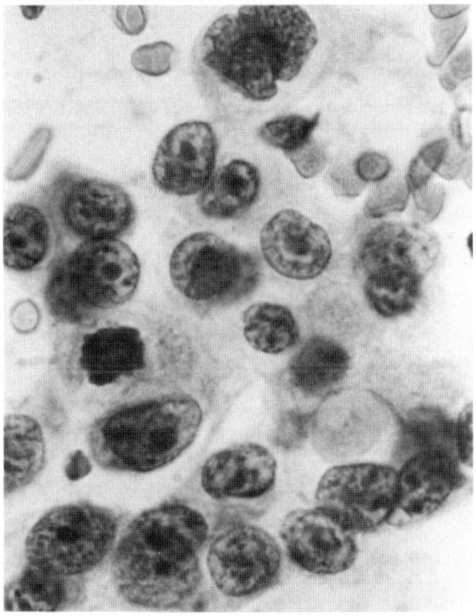

78

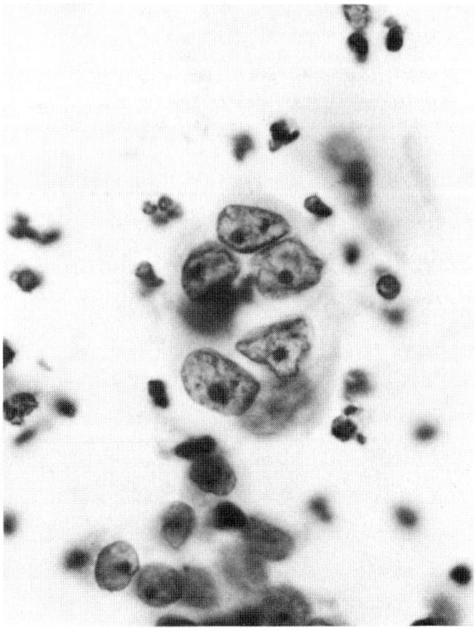

79

Fig. 76. Abnormal mitotic figure in cells from undifferentiated carcinoma of esophagus. × 545.

Fig. 77. High nuclear-cytoplasmic ratio in cell from undifferentiated carcinoma of stomach. × 545.

Fig. 78. Variation in size and shape of nuclei in cells from poorly differentiated adenocarcinoma of stomach. × 860.

Fig. 79. Nuclear moulding and loss of polarity in cells from poorly differentiated adenocarcinoma of stomach. × 545.

valuable one and, as will be emphasized, may indicate the possibility of a surface or intramucosal carcinoma.

The presence of tissue structures such as acini and papillae may be a diagnostic aid. However, as we are dealing with a glandular epithelium, the cells comprising these tissue structures must show the abnormalities of carcinoma.

Differentiation Characteristics

Although the primary role of cytology is to make a diagnosis of malignancy the determination of the type of neoplasm that is present may be of considerable assistance in the further investigation and management of the patient. The characteristics of differentiation, which are seen mostly within the cytoplasm, should only be applied to cells that have already been diagnosed as being malignant on the basis of the malignant criteria outlined above. The most useful indications of differentiation in gastro-esophageal cytology are as follows:

Squamous carcinoma cells tend to be shed singly. Their cytoplasm is often keratinized appearing highly refractile or 'glassy'. The cell border is sharp (fig. 80).

Characteristic cell groups may be seen. These may take the form of malignant 'pearls' or, alternatively, a 'cell-in-cell' appearance may be evident (fig. 81). This phenomenon, sometimes referred to as 'cannibalism', is due usually to one cell lying in a depression in the cytoplasm of another.

In contrast, cells from an adenocarcinoma are usually seen in groups, sheets, or tissue fragments. They often have abundant, delicate, finely vacuolated or 'foamy' cytoplasm (fig. 82). Since the cells are essentially secretory in nature cytoplasmic vacuoles are often prominent (fig. 83). These secretory vacuoles must be distinguished from degenerative vacuoles that may occur in any type of cell.

Malignant criteria and differentiation characteristics will be considered in more detail in relation to the specific neoplasms that occur in both esophagus and stomach.

References

1 Rosenthal, M.; Traut, H.G.: The mucolytic action of papain for cell concentration in the diagnosis of gastric cancer. Cancer *4:* 147–149 (1951).

2 Rubin, C.E.; Benditt, E.P.: A simplified technique using chymotrypsin lavage for the cytological diagnosis of gastric cancer. Cancer *8:* 1137–1141 (1955).

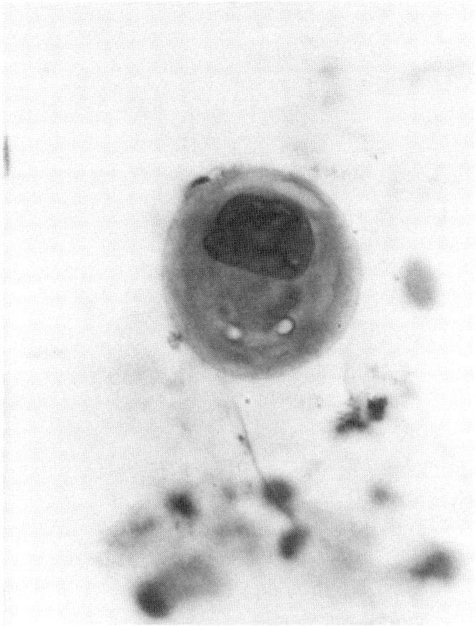
80

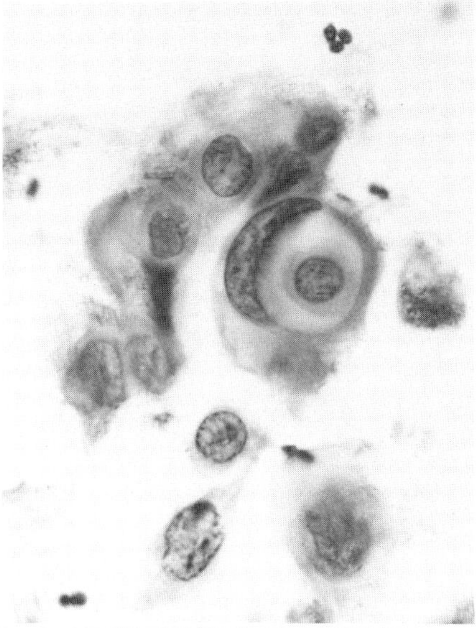
81

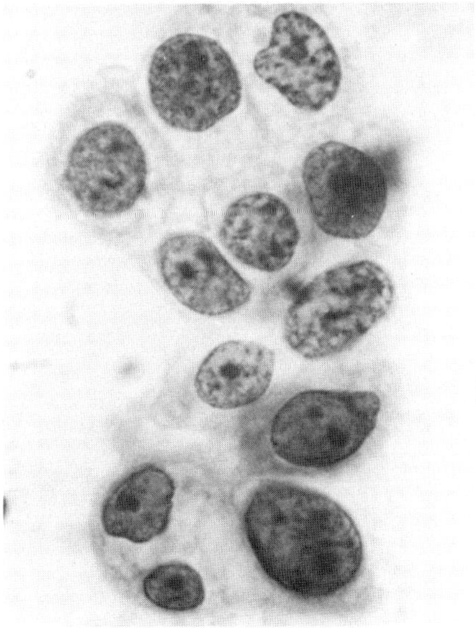
82

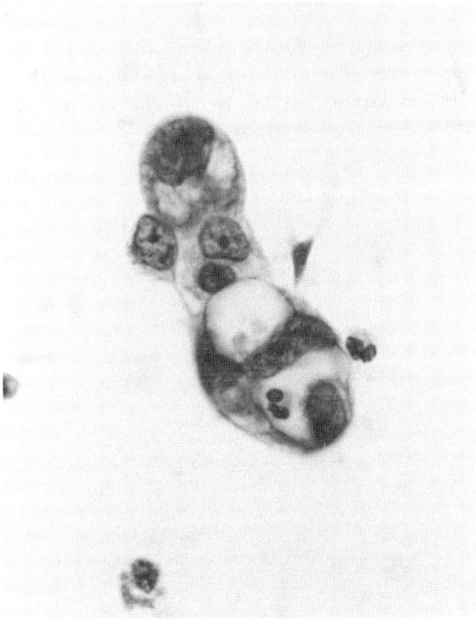
83

Fig. 80. Squamous carcinoma cell showing refractile keratinized cytoplasm and sharp cell border. × 545.

Fig. 81. 'Cell-in-cell' appearance in squamous cell carcinoma. × 435.

Fig. 82. Cells from adenocarcinoma of stomach with abundant finely vacuolated cytoplasm. × 860.

Fig. 83. Cells from adenocarcinoma of stomach showing prominent cytoplasmic secretory vacuoles. × 545.

3 Shida, S.; Ishioka, K.: Gastric cytology: its evaluation for the diagnosis of early gastric cancer; in Wied, Koss, Reagan, Compendium on diagnostic cytology; 4th ed. (Tutorials of Cytology, Chicago 1976).

4 Takeda, M.: Atlas of diagnostic gastrointestinal cytology (Igaku-Shoin, New York 1983).

5 Yamada, T.: Basic study of the proteolytic lavage method in gastric diagnosis, especially in the comparative analysis of the exfoliative tendency of malignant and benign gastric epithelial cells. Acta cytol. *8:* 19–26 (1964).

6 Yamada, T.; Matsumoto, S.; Sankawa, H.; Seino, Y.: Clinical evaluation of proteolytic enzyme lavage method in the gastric cytodiagnosis, especially in the detection of early cancer of the stomach. Acta cytol. *8:* 27–33 (1964).

7 Yamada, T.; Murohisa, B.; Yoshihiro, M., et al.: Point, minute and small cancers of the stomach at the early developmental stage detected by improved chymotrypsin lavage method for diagnostic cytology. Acta cytol. *22:* 460–469 (1978).

4. Esophageal Cytology – Normal and Abnormal

Esophageal cytology is, for the most part, relatively straightforward. The normal cytology is extremely simple whilst abnormal cytology is concerned mostly with the distinction between inflammatory conditions and carcinoma, the latter being almost exclusively squamous cell carcinoma and adenocarcinoma.

Normal Esophageal Cytology

The epithelial cell component of both washings and brushings of the normal esophagus is composed of squamous cells of superficial and intermediate cell type. The superficial cells are polygonal with abundant eosinophilic cytoplasm and a single pyknotic nucleus (fig. 84). The cytoplasm may contain some keratohyalin granules although actual keratinization does not normally occur within the esophagus. Sometimes the superficial cells present as small fragments or epithelial 'pearls' (fig. 85). The cells of the intermediate layers of the esophagus are also polygonal and also have abundant cytoplasm but with a green or greenish-blue staining reaction. The nucleus is larger than that of a superficial cell, round or, more commonly, oval with a vesicular chromatin pattern (fig. 86). Cells from the deeper layers, the parabasal cells, may occasionally be seen in a normal brushing, presumably reflecting the vigour of that procedure. More often, however, their presence indicates an undue friability of the mucosa associated with inflammation or, more probably, an ulceration of the mucosa with exposure of the deeper layers. The cells are round or oval in shape, have a relatively narrow rim of thick cytoplasm which stains blue or blue-grey, and a round or oval, rather hyperchromatic nucleus with a vesicular chromatin pattern. These cells may be seen in association with cells transitional between parabasal and intermediate cells – sometimes referred to as 'deep intermediate cells' (fig. 87).

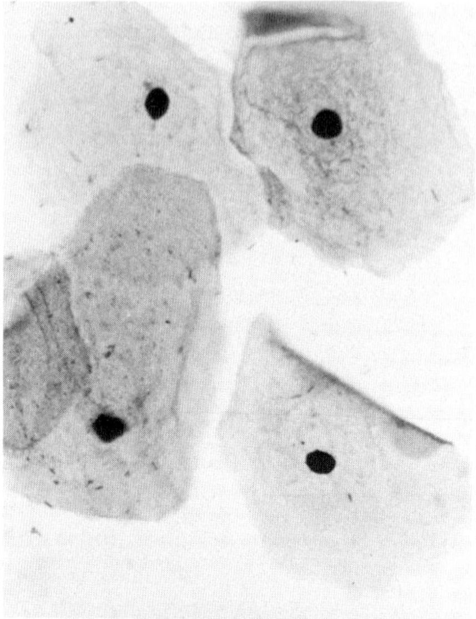

84

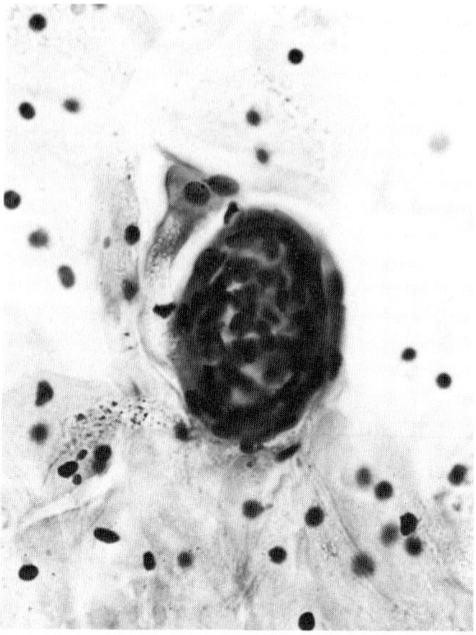

85

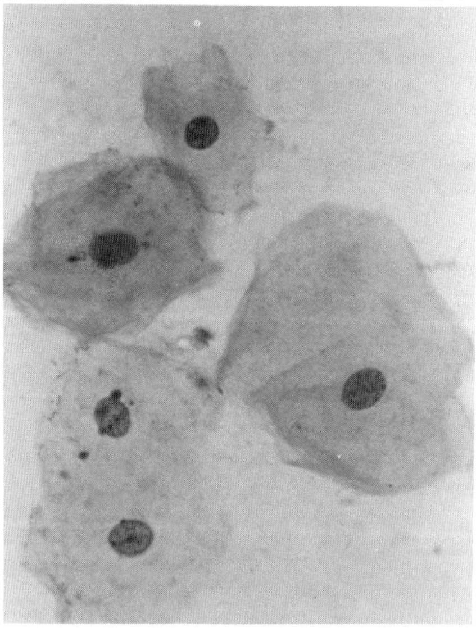

86

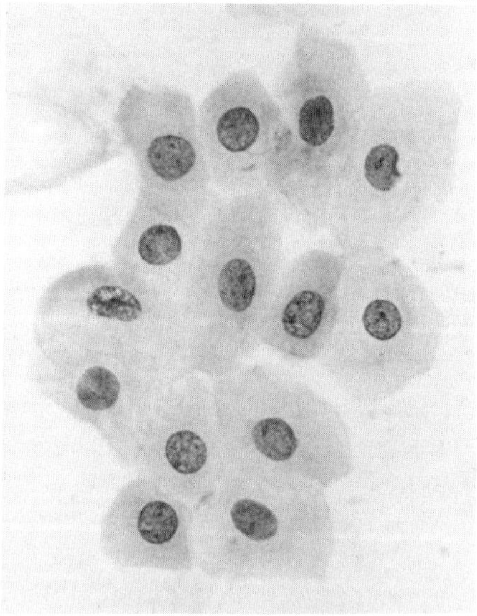

87

Fig. 84. Normal superficial squamous cells from esophagus. Note pyknotic nuclei and abundant transparent cytoplasm. × 545.

Fig. 85. Normal superficial squamous cells from esophagus including group presenting as epithelial 'pearl' formation. × 345.

Fig. 86. Normal intermediate squamous cells from esophagus. Note vesicular nucleus and abundant cytoplasm. × 545.

Fig. 87. A mixture of parabasal and 'deep intermediate cells' from esophageal mucosa. Nuclei are vesicular and cytoplasm is thick and densely staining. × 545.

88
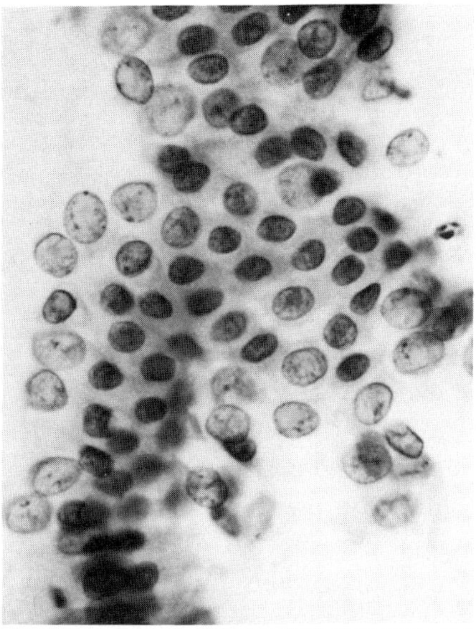

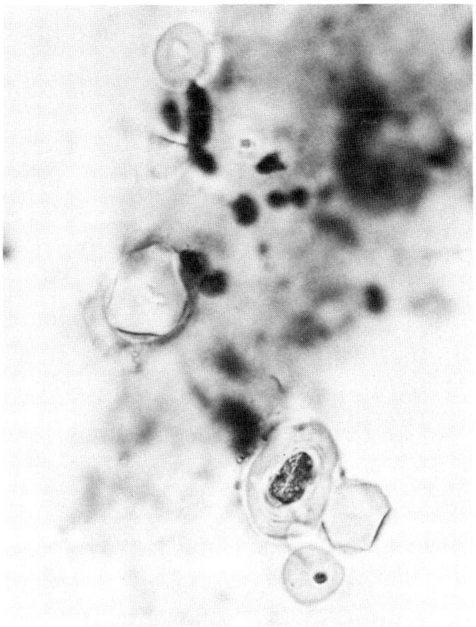
89

90
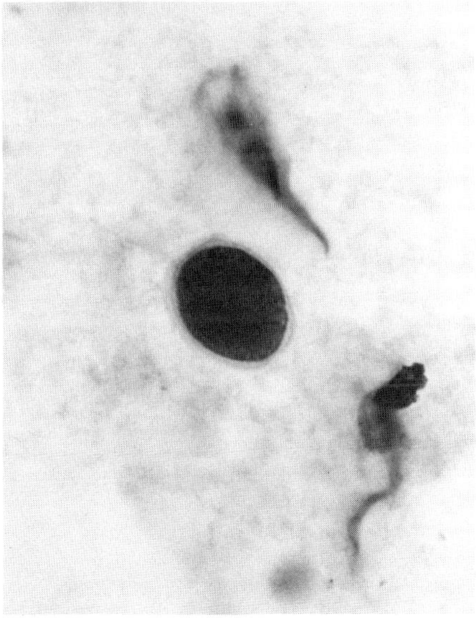

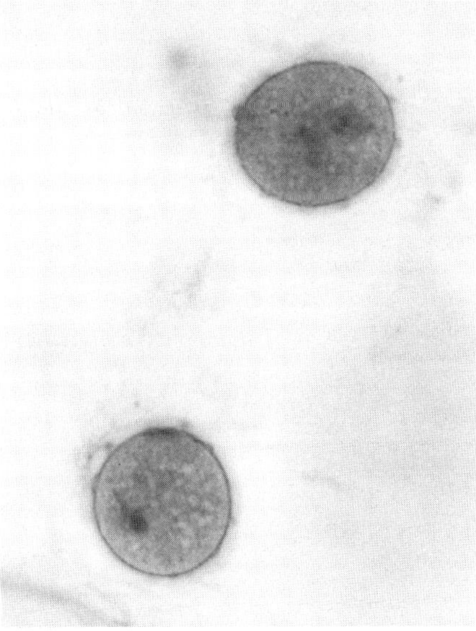
91

Fig. 88. Gastric mucosal cells in esophageal brushings. Cells were brushed from site proximal to gastro-esophageal junction and hence indicate area of metaplasia. × 545.

Fig. 89. Talcum crystals in esophageal brushings – presumably derived from operator's gloves. × 545.

Fig. 90. Pollen grain in esophageal washing – probably a contaminant during specimen preparation. Note resemblance to squamous carcinoma cell. × 545.

Fig. 91. Pollen grains in esophageal washing showing a more clearly recognizable structure than that depicted in previous figure. × 860.

In addition to these normal cellular components other structures may be found within the esophagus particularly if there is any delay at the cardio-esophageal junction. It is not uncommon to find swallowed respiratory elements in both esophageal washings and brushings, particularly the former. Both ciliated respiratory columnar cells and dust-containing macrophages may be demonstrated. Gastric columnar epithelial cells may also be seen (fig. 88). These may indicate merely that the gastric cardia has been sampled but, alternatively, they may be indicative of a metaplasia or heterotopia.

A variety of contaminants may be seen also. Talcum crystals (fig. 89), usually derived from the operator's gloves, are common as are pollen grains. The latter, which may have been swallowed or, more commonly, are the result of contamination of the specimen during preparation, can be of importance as they may have a superficial resemblance to a squamous carcinoma cell (fig. 90). Closer examination usually reveals their characteristic structure (fig. 91).

Non-Neoplastic Abnormalities

Infective Esophagitis

Micro-organisms are often seen in the esophageal specimen as a contaminant, relatively common examples being the fungi *Candida* and *Aspergillus.* However, many different viruses, bacteria and fungi can, on occasions, cause an infection of the esophagus. Hence the possibility of such infection should always be borne in mind, particularly if consistent with the patient's clinical status. Perhaps the commonest causes now of primary infection of the esophagus are *Candida albicans* or *Monilia albicans* and the herpes simplex virus.

Esophagitis Due to Candida albicans *(Monilial Esophagitis)*

Fungal infections have been noted with increasing frequency in recent years. The vast majority of these infections are caused by *Candida albicans* although other *Candida* species are occasionally associated with esophagitis

---

Fig. 92. Esophagus showing extensive mucosal ulceration and plaque formation due to infection by *Candida albicans.* Autopsy specimen.

Fig. 93. Prominent mucosal plaques in esophagus infected by *Candida albicans.* Autopsy specimen.

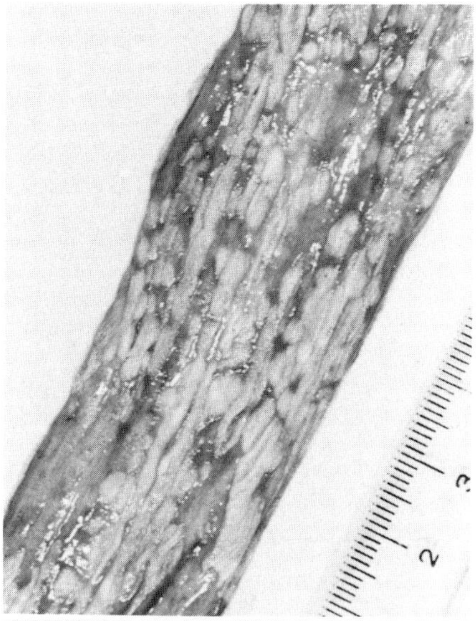

92

93

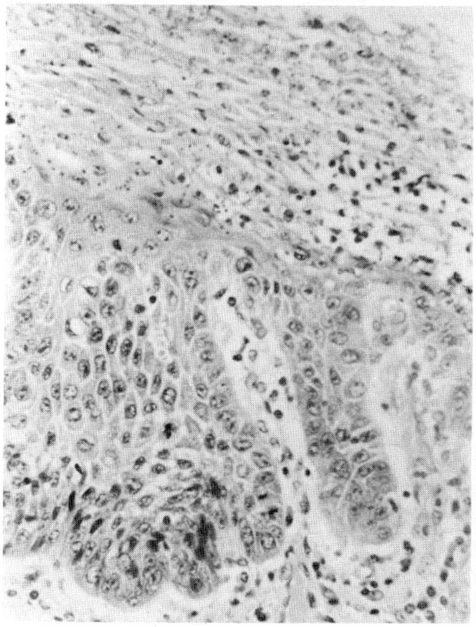

94

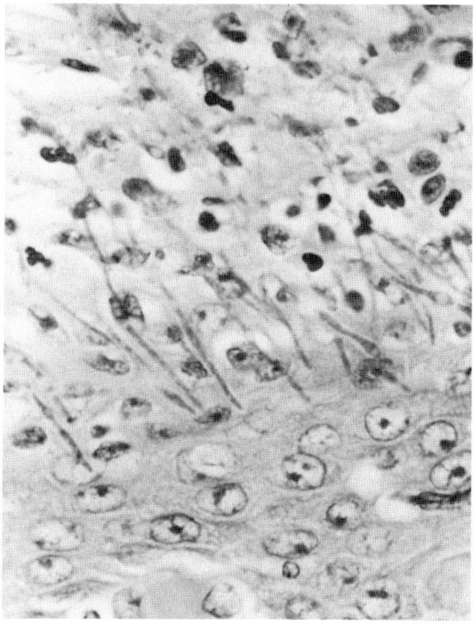

95

Fig. 94. Esophagitis due to *Candida albicans.* Partial loss of surface epithelium with thick layer of necrotic debris, inflammatory cells and desquamated epithelial cells. × 173.

Fig. 95. Fungal elements clearly recognizable in surface exudate and within superficial portion of surviving epithelium. Monilial esophagitis. × 435.

[23]. *Candida* can cause significant inflammatory lesions in any part of the gastro-intestinal tract but the esophagus is the organ most frequently involved. In several autopsy series approximately 50% of the alimentary tract disease was present in the esophagus [17, 50].

Candida is an ubiquitous yeast that is present frequently in the intestinal tract of healthy individuals. Infection of the esophagus may follow antibiotic therapy which facilitates the proliferation of the organism. Poor esophageal drainage due to achalasia, strictures or neoplasms also favour such infection [29]. Currently the commonest predisposing factor to serious fungal esophagitis is immunosuppression particularly if this is associated with antibiotic therapy. Very occasionally acute monilial esophagitis may occur in an otherwise healthy person [9].

The patients usually present with dysphagia, odynophagia and pain, the latter being most severe in the upper retrosternal region. The radiological appearances are characteristic a barium swallow showing a shaggy irregular outline of the barium column. There is a loss of normal mucosal folds and the mucosa may have a 'cobblestone' appearance [26].

Endoscopically the disease is manifested by the appearance of white plaques on a reddened mucosa. The endoscopic appearances have been classified into four grades [29] ranging from small white plaques on a hyperaemic mucosa to nodular elevated plaques with mucosal ulceration. Most specimens subject to direct macroscopic examination are derived from autopsies. Here both ulceration and plaque formation may be extensive (fig. 92). The latter may be particularly prominent (fig. 93). Histologically there may be partial or complete ulceration of the squamous epithelium the eroded surface being covered by a mass of necrotic debris.

If the deeper portion of the mucosa survives the squamous epithelial cells show considerable abnormalities, these abnormalities being due to coexisting degenerative and regenerative activity. The abnormal cells are shed frequently into the surface exudate (fig. 94). Mycelia are often prominent within this surface exudate and also within the superficial epithelium (fig. 95).

Fig. 96. Cells from case of monilial esophagitis. Note inflammatory cells and squamous epithelial cells. The latter show degenerative changes with nuclear enlargement and prominent cytoplasmic vacuolation. × 545.

Fig. 97. Squamous epithelial cells from case of monilial esophagitis showing regenerative and degenerative changes, the former predominating. Note nuclear enlargement and irregularity, chromatin granularity and prominent nucleoli. × 545.

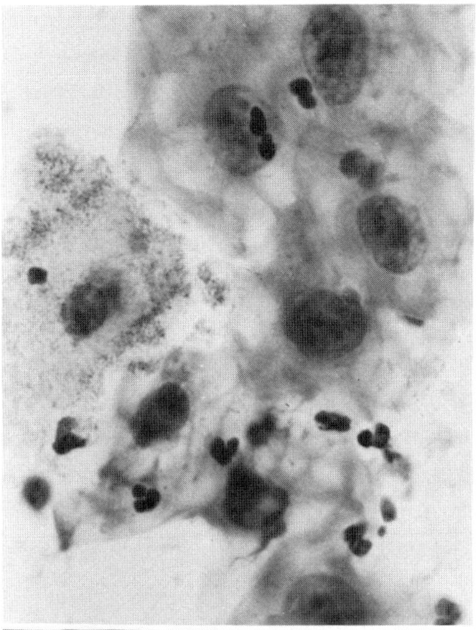

96

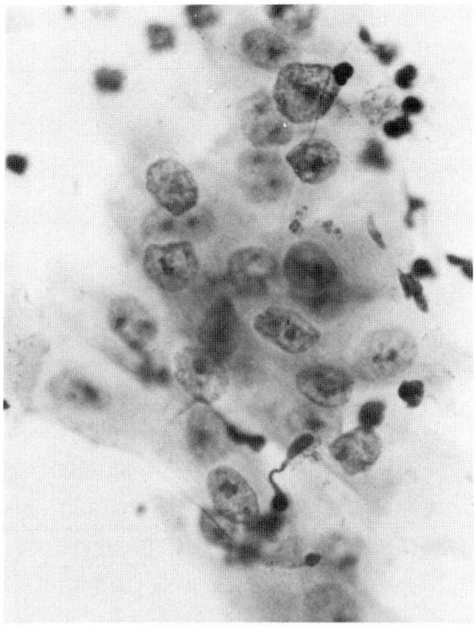

97

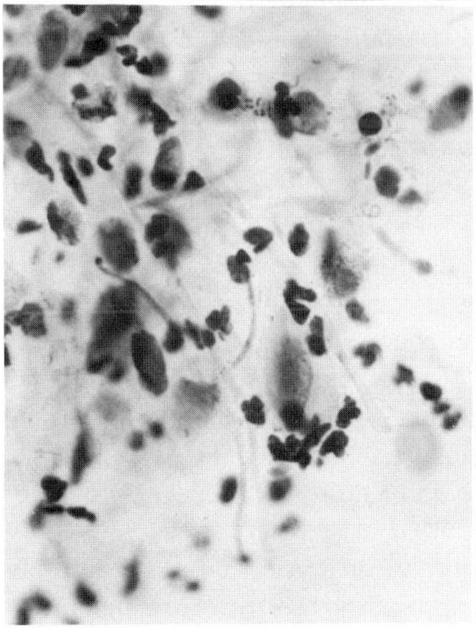

98

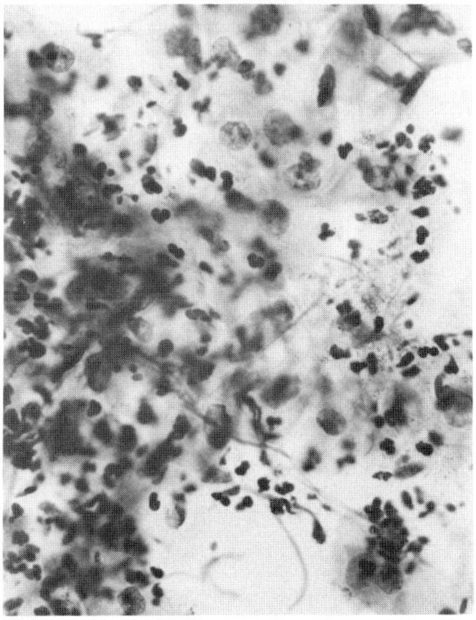

99

Fig. 98. Surface exudate from case of monilial esophagitis. Note inflammatory cells, degenerate squamous epithelial cells, and degenerate but clearly recognizable fungal elements. × 545.

Fig. 99. Surface exudate from case of monilial esophagitis. Features are similar to those seen in previous figure. Fungal elements are again clearly visible as are numerous inflammatory cells. × 345.

The surface material is collected readily in the cytological specimen. Polymorphonuclear leucocytes are usually present, indicative of an acute inflammatory reaction, whilst both degenerative (fig. 96) and regenerative (fig. 97) changes in epithelial cells are also prominent. The organism is usually seen within the necrotic debris, inflammatory cells and abnormal epithelial cells (fig. 98, 99).

Herpes simplex Virus Esophagitis

Esophagitis due to infection by the herpes simplex virus has also increased in incidence in recent years. In 1943 *Pearce and Dagradi* [48] described four cases of esophagitis associated with ulceration in which the cells at the ulcer margins displayed intranuclear inclusion bodies. Subsequently herpes simplex virus was isolated from the ulcers of similar cases and in 1955 *Berg* [7] emphasized the importance of esophageal herpes as a complication of cancer therapy. *Buss and Scharyj* [10] reviewed their autopsy materials for a period of 10 years and found 56 patients with histological evidence of herpes virus infection involving non-genital viscera. Herpes virus infections were suspected clinically in only 7 of these patients with 40 showing, on review, no clinical signs or symptoms of herpetic infection. The viscus most commonly involved by herpetic infection was the esophagus, ulcers of that organ being present in 50 cases. *Rosen and Hajdu* [52] also reviewed a large series of autopsy reports and found visceral lesions consistent with either herpes simplex or varicella zoster infection in 31 patients in 6,694 autopsied patients. Of the 31 patients herpetic esophageal lesions were present in 25. An antemortem diagnosis, based on skin lesions, had been made in 10 cases but more widespread disease was not usually considered. These authors also noted that the herpetic infection may predispose to the establishment of monilial esophagitis – an association that has been noted by others [40]. The organism is now recognized as one of the common causes of esophageal infection and a number of reports of diagnosis by cytological means have been reported in recent years [30, 31, 54].

The factors that appear to predispose to such infections are trauma to the esophagus, such as may occur with nasogastric intubation, malignancy –

Fig. 100. Autopsy specimen of esophagus showing extensive mucosal ulceration and plaque formation due to herpes simplex viral (HSV) infection.

Fig. 101. Section of esophagus showing earliest change in squamous epithelium due to HSV infection. $\times 69$.

100

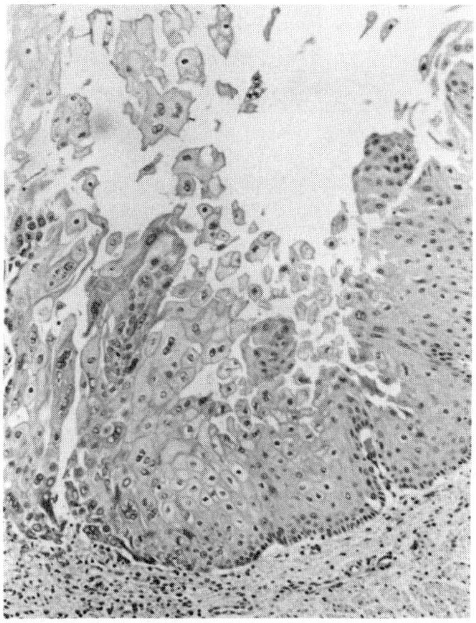

101

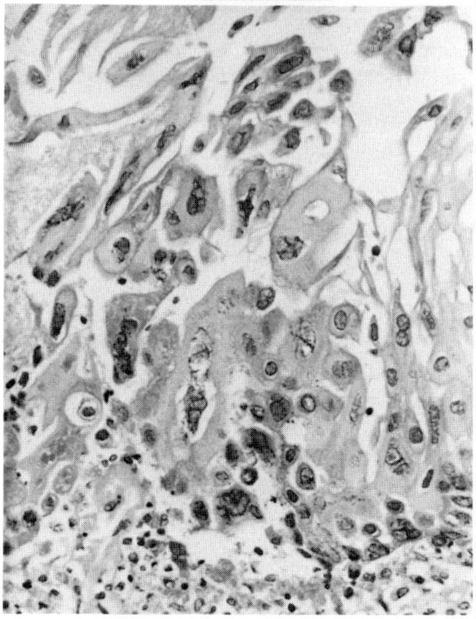

102

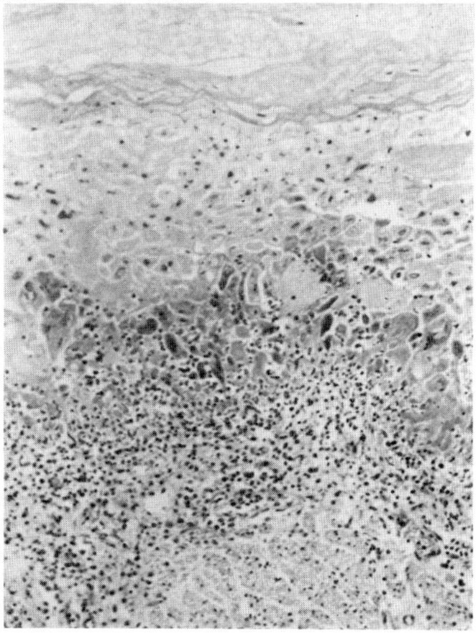

103

Fig. 102. Section of esophagus showing more advanced lesion due to HSV infection. Mucosal squamous cells show characteristic viral changes and mucosa is beginning to separate from underlying connective tissues. × 173.

Fig. 103. Section of esophagus showing mucosal necrosis, viral changes in deeper epithelial cells, and dense inflammatory cell infiltrate at site of separation of mucosa from connective tissues. × 69.

particularly lymphomas and leukaemias – chemotherapy, mediastinal irradiation, immunosuppressive therapy and immunodeficiency states. However, it has recently been appreciated that herpetic esophagitis can attack apparently normal healthy individuals [12, 18, 44, 47, 59]. In most patients infection probably occurs by contiguous spread of virus from the oropharynx to the esophagus. In others it may result from re-activation of herpes simplex virus from latently infected sensory ganglion cells.

The earliest demonstrable lesion is a vesicle which rapidly enlarges to produce an ulcer. In severe cases the ulcers become confluent over a segment of esophagus to produce an extensive area that is denuded of mucosa. Often the confluent lesions are seen in the lower third of the esophagus, multiple punched out ulcers being seen in the more proximal portions. The ulcers may be accompanied by raised mucosal plaques (fig. 100) the appearances then resembling those of monilial infection.

Histologically the earliest lesion is seen in the squamous epithelium the surface epithelial cells showing changes characteristic of the viral infection (fig. 101). The infected epithelium separates from the underlying connective tissues (fig. 102) and a dense accumulation of inflammatory cells is seen at this site of separation (fig. 103). The epithelium, which is now necrotic, sloughs away to leave an ulcer the floor of which is covered with fibrin, debris, and inflammatory cells (fig. 104). The cells in the infected epithelium during the formation of the ulcer, and adjacent to the ulcer when formed, show marked abnormalities (fig. 105). These abnormalities are clearly manifest in the cytological specimen collected from the infected esophagus.

The cells may be mononuclear but more commonly are multinucleate (fig. 106). The most characteristic feature is the margination of the nuclear chromatin material the bulk of the nucleus having a striking refractile or 'ground-glass' appearance (fig. 107). Within the multinucleate cells the enlarged nuclei show conspicuous moulding (fig. 108, 109). Intranuclear inclusions are characteristic. These may be extremely small but much more commonly are large, eosinophilic and surrounded by a clear halo (fig. 110, 111).

Fig. 104. Esophageal ulcer due to HSV infection. Floor of ulcer is covered with fibrinous exudate, cell debris and inflammatory cells. × 69.

Fig. 105. Cells at periphery of esophageal ulcer showing characteristic changes due to herpes simplex viral infection. × 275.

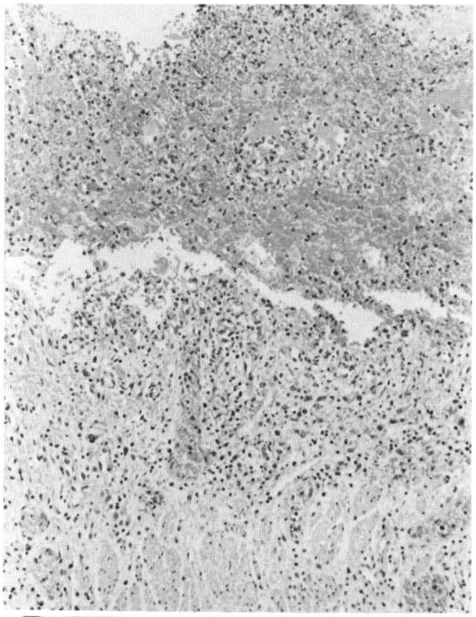

104

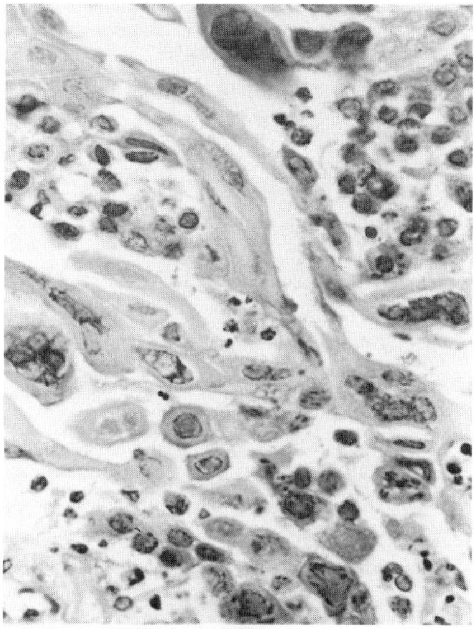

105

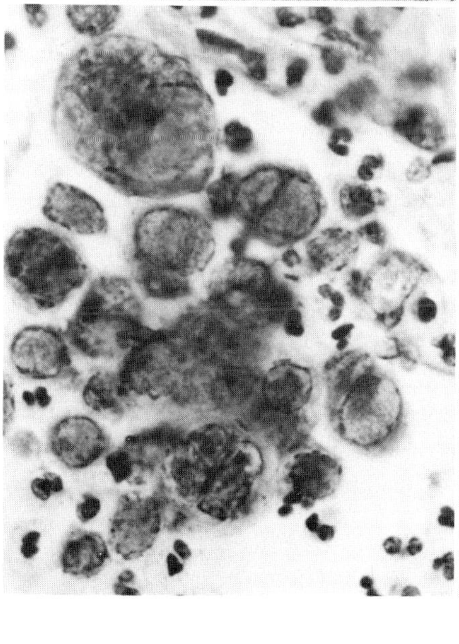

106

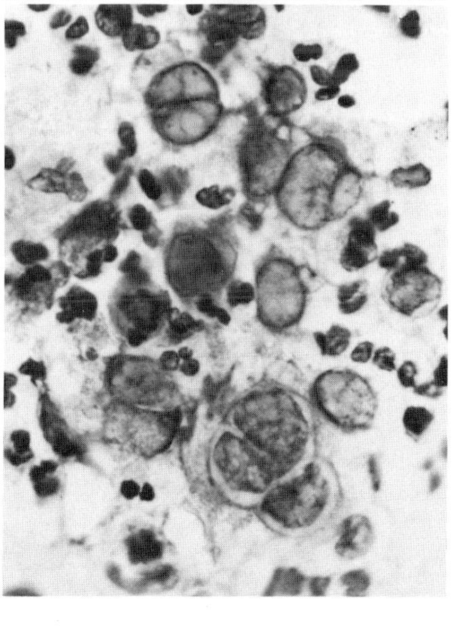

107

Fig. 106. Cells from case of HSV esophagitis. Note enlarged nuclei which show margination of chromatin and have characteristic 'ground-glass' appearance. × 545.

Fig. 107. Cells from case of HSV esophagitis with features similar to those seen in previous figure. Binuclear cells are prominent. × 545.

Aspergillus *Esophagitis*

As already indicated, the fungus *Aspergillus* (fig. 112) is a relatively common contaminant of esophageal specimens as indeed it is in specimens from other body sites. However, as also emphasized, the possibility of a true infection must always be considered. The esophagus depicted in figure 113, for example, was from a 27-year old man who had received intensive chemotherapy for acute myeloblastic leukaemia. He ultimately died of an intercurrent infection and at postmortem examination his esophageal mucosa was noted to be markedly haemorrhagic with multiple small ulcerations. Histological examination showed masses of *Aspergillus* throughout the esophageal mucosa (fig. 114) some of which appeared to be invading and destroying the intramural blood vessels (fig. 115). No cytological investigations had been carried out prior to death.

Chronic Granulomatous Esophagitis

Both tuberculous [14, 25] and syphilitic [60] infections of the esophagus have been reported but both conditions are so rare as to be of no practical importance to the diagnostic cytologist. Other granulomatous conditions of the esophagus such as Crohn's disease [33] and sarcoidosis, both of which are of obscure etiology, can also be dismissed as pathological curiosities.

Non-Infective Esophagitis

Non-infective acute and chronic esophagitis is a relatively common condition and may be due to a variety of causes. Amongst these causes the most important to the diagnostic cytologist are achalasia (cardiospasm) and reflux esophagitis – the former because of its clinical similarity to esophageal cancer and the latter because of its frequent occurrence.

Achalasia (Cardiospasm)

As already discussed normal esophageal motility is characterized by peristaltic mechanisms in the wall of the esophagus associated with sphincteric mechanisms at both proximal and distal ends of the tubular structure. In this way swallowed material is transported into the stomach whilst gastric content is prevented from entering the esophagus. Achalasia is a disease of unknown etiology characterized by an absence of peristalsis in the body

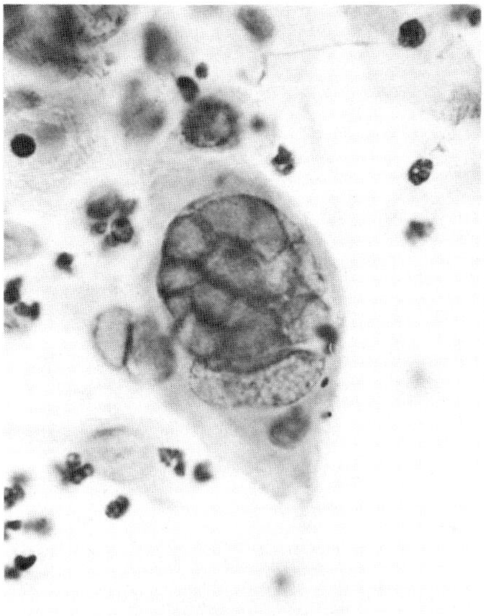

108

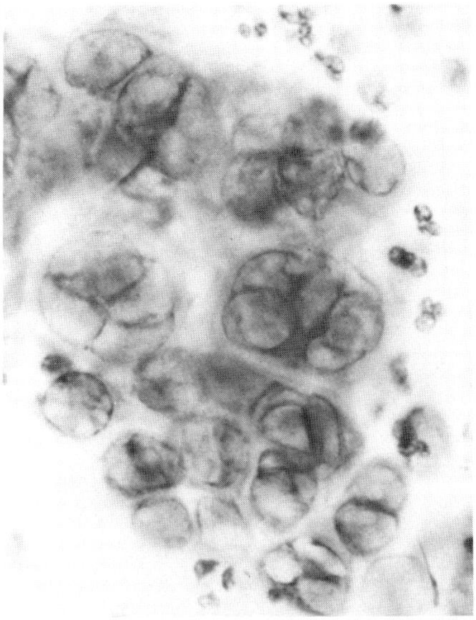

109

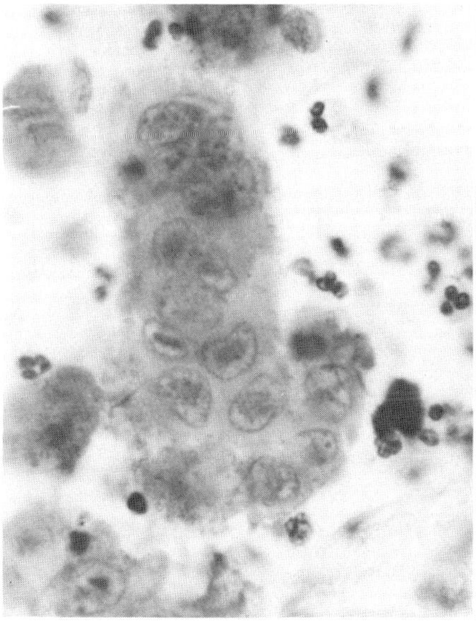

110

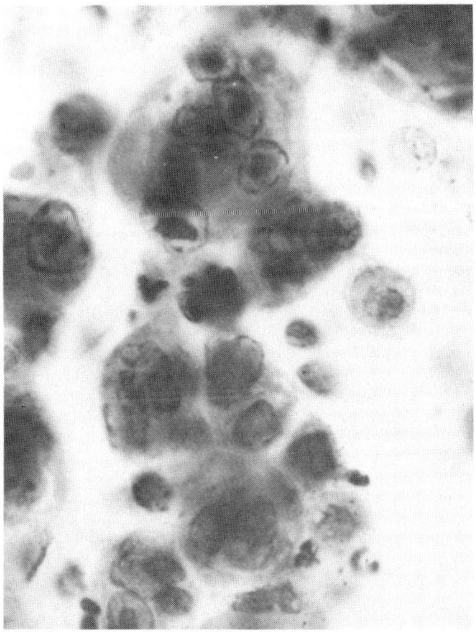

111

Fig. 108. Typical herpes simplex viral changes in squamous cells. Multinucleation is prominent as is moulding of nuclei within multinucleate cell. × 545.

Fig. 109. Squamous cells showing characteristic changes of HSV infection. Multinucleation and nuclear moulding are again prominent. × 545.

Fig. 110, 111. Cells from case of herpes simplex viral esophagitis showing prominent nuclear inclusions. Inclusion bodies are large, irregular in shape, eosinophilic in the routine Papanicolaou stain, and frequently surrounded by a clear area. × 545.

of the stomach and a failure of the gastro-esophageal sphincter to relax normally in response to swallowing. Loss of the propulsive peristaltic contractions, together with the defective sphincter mechanism, results in stasis of food in a progressively dilating esophagus. The obstruction, and consequent esophageal dilatation, are clearly demonstrable radiologically (fig. 116).

Dysphagia for liquids and solids is the most prominent symptom and is present in nearly all patients. In addition, regurgitation of swallowed food and fluid, substernal pain and nocturnal coughing spells are common.

Inflammatory changes in the mucosa, submucosa and muscle coats are variable and are probably due largely to the stagnation of esophageal content. Ulceration of the mucosa of the distal esophagus is common and this is frequently followed by, and associated with, an intense epithelial regenerative process. The cytological manifestations of these processes will be described below. Neoplasms occur more commonly than usual in the esophagus of patients with achalasia [3, 64] and associated conditions.

Reflux Esophagitis

Reflux of gastric content into the lower esophagus may occur only occasionally, it may be intermittent, or it may be continuous. It is commonly associated with hiatus hernia and with increased intra-abdominal pressure particularly due to over-eating. The major abnormality would appear to be an incompetence of the cardio-esophageal sphincter mechanism, presumably as a result of neuromuscular dysfunction, but the precise cause is unknown. The competence of the sphincter depends primarily on the strength of its circular muscle component. It could be that this muscle tissue is abnormal or, alternatively, that there is a disturbance of its innervation. Finally, there is some evidence that gastro-intestinal neuropeptides may have some role in controlling the competence of the cardio-esophageal sphincter and the sphincter malfunction could be due to an abnormality of these substances.

Regardless of the cause, the incompetence of the sphincter mechanisms results in the abnormal exposure of the mucosa of the lower esophagus to

---

Fig. 112. Aspergillus fumigatus from esophageal washings. No clinical evidence of esophagitis and hence fungus presumably a contaminant. ×205.

Fig. 113. Esophagus from autopsy on patient dying following treatment for leukaemia. Note haemorrhagic ulcerated mucosal surface.

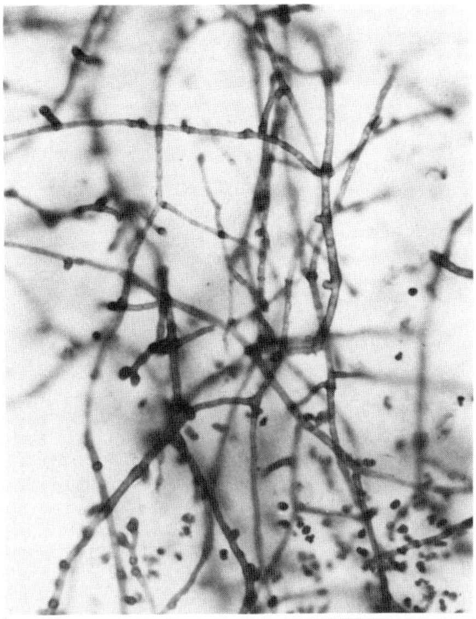

112

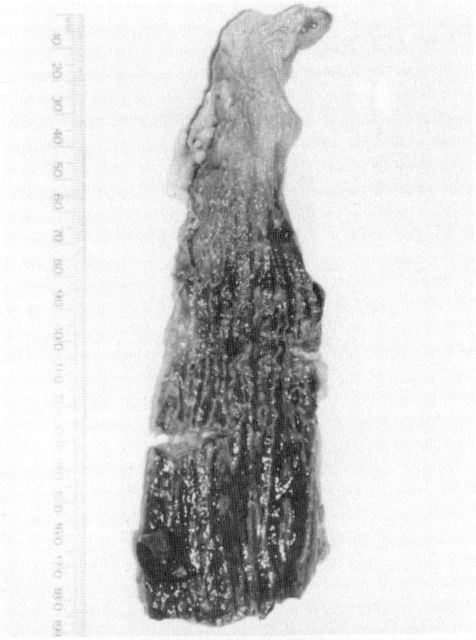

113

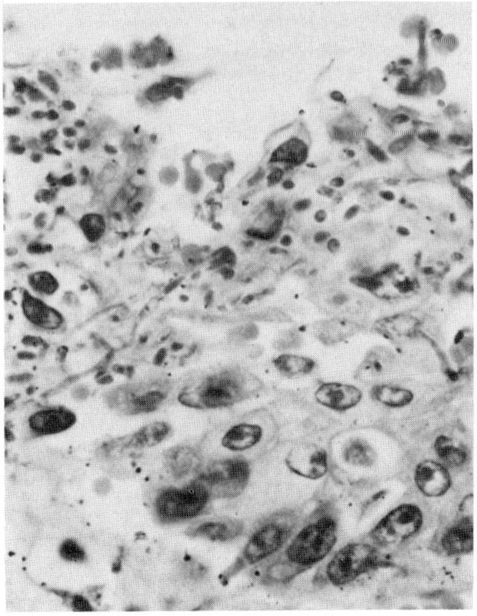

114

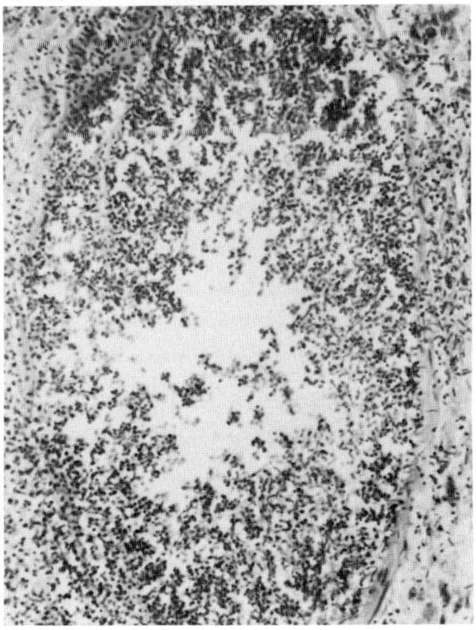

115

Fig. 114. Section of esophagus depicted in previous figure. Note desquamated degenerate surface epithelial cells, inflammatory cells and fungal elements. × 545.

Fig. 115. Section of esophagus showing dense colonies of fungus within submucosal blood vessels with destruction of vessel wall. × 173.

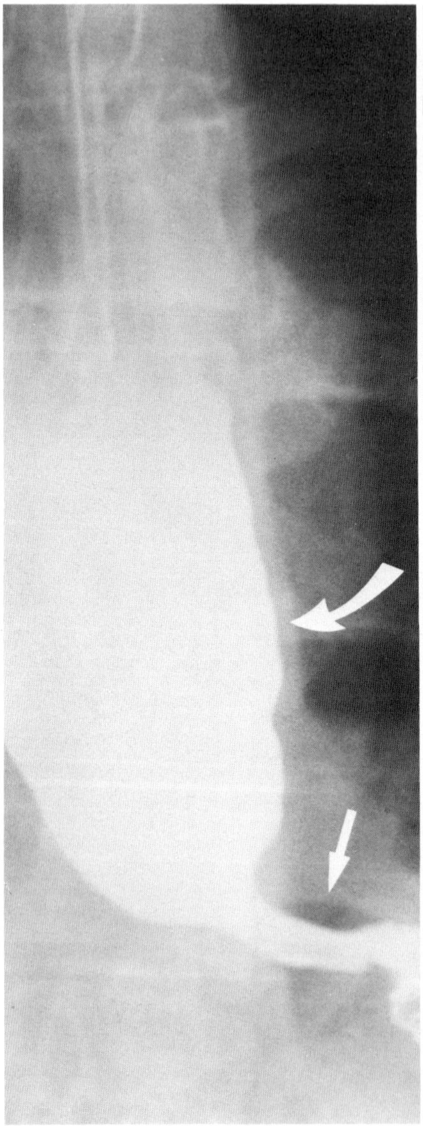

116

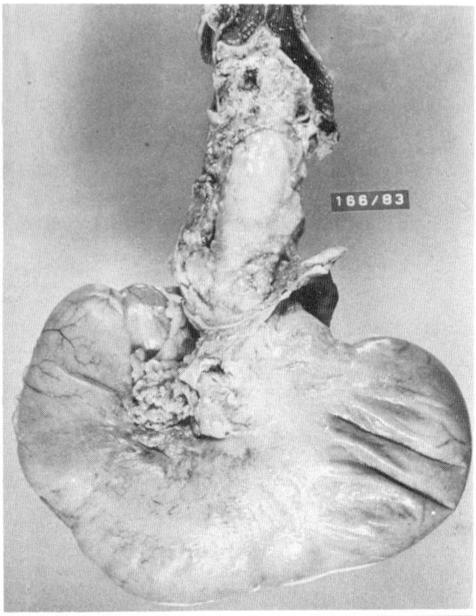

117

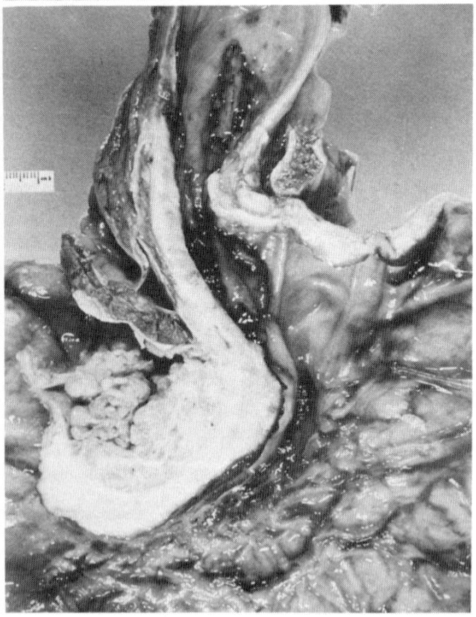

118

Fig. 116. Barium swallow in patient with achalasia of esophagus. The straight arrow indicates the typical narrowing of the esophagus immediately above the diaphragm whilst the curved arrow shows the associated dilatation of the esophagus proximally.

Fig. 117, 118. Specimen of esophagus and stomach showing carcinoma arising in hiatus hernia. Photographs courtesy of Professor *P. Bhathal,* Royal Melbourne Hospital.

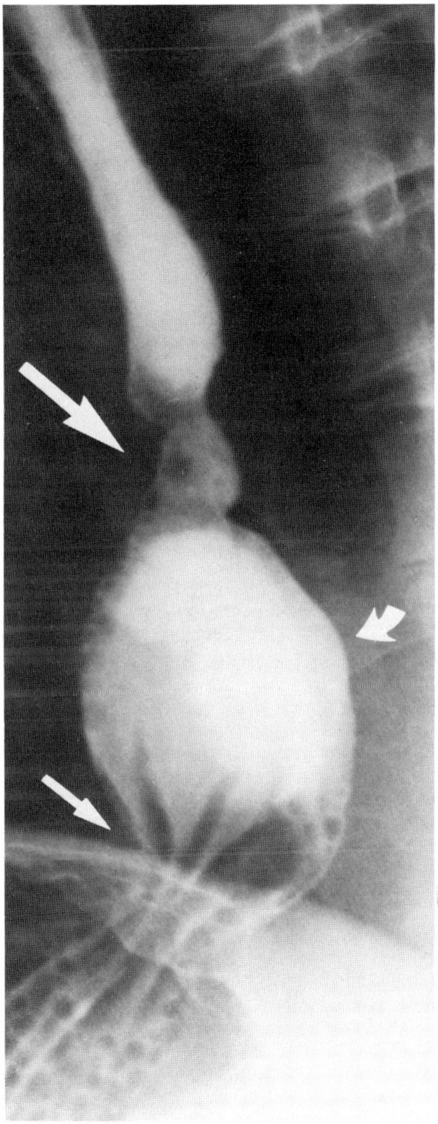

119

120

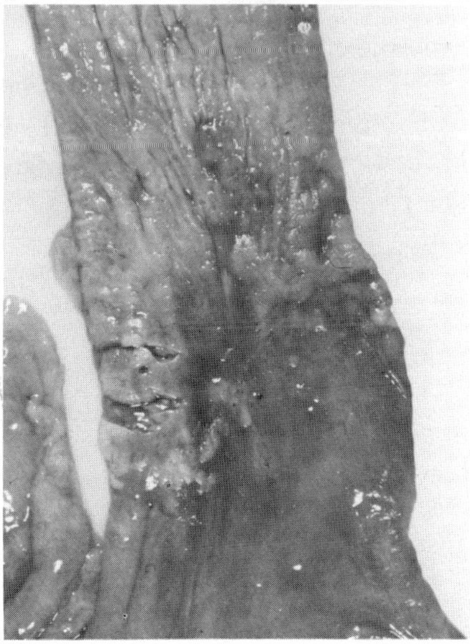

121

Fig. 119. Barium swallow showing large hiatus hernia with stricture formation proximally.

Fig. 120. Peptic ulcer developing in hiatus hernia with subsequent perforation.

Fig. 121. Ulceration of esophageal squamous mucosa due to prolonged esophageal reflux.

gastric content. This exposure leads to a number of complications particularly if the reflux is associated with a hiatus hernia. Thus a recognized [46], although unusual, complication of such a hernia is the development of a carcinoma within the mucosa of the herniated portion of stomach (fig. 117, 118). Much more commonly peptic ulceration of the herniated gastric mucosa, or of metaplastic gastric epithelium in the lower esophagus, may occur. These ulcers are subject to the same complications as peptic ulcers elsewhere namely, haemorrhage, stricture formation (fig. 119) and perforation (fig. 120). Both the peptic ulceration and the malignant transformation will, of course, have their own characteristic cytological manifestations and these will be described in considerable detail below.

The squamous epithelium of the lower esophagus appears to have a modest resistance to the corrosive effects of hydrochloric acid, pepsin, and bile acids but repeated and prolonged episodes of reflux lead ultimately to mucosal damage (fig. 121). Initially there is a dilatation of mucosal capillaries, a thickening of the basal layers of the epithelium, and an increased prominence of the submucosal papillae [4, 24, 34]. Inflammatory cells collect within the upper submucosa and also infiltrate diffusely the more superficial layers of the surface epithelium (fig. 122). Persistent or more severe reflux causes progression of the mucosal injury with increased desquamation of the mucosal cells (fig. 123) and ultimately ulceration. During this process the surface of the esophagus is covered by a mass of necrotic debris, inflammatory cells and desquamated epithelial cells (fig. 124). The subsequent epithelial responses depend, to some extent, on the control of the reflux. Thus if some control is possible the breach in the surface epithelium may be healed by regenerative activity in the adjacent squamous mucosa the end result being replacement of the lost epithelium by squamous epithelium of the same sort as that which was previously destroyed. During this regenerative process considerable cytological abnormalities may be apparent and these will be described below. If reflux persists unabated repair may take place by the formation of a metaplastic epithelium

---

Fig. 122. Early changes in reflux esophagitis. Note infiltration of surface epithelium, and also of subepithelial connective tissues, by inflammatory cells. × 173.

Fig. 123. Partial loss of surface epithelium due to esophageal reflux. Necrotic superficial layers of the mucosa are densely infiltrated by polymorphonuclear leucocytes. × 205.

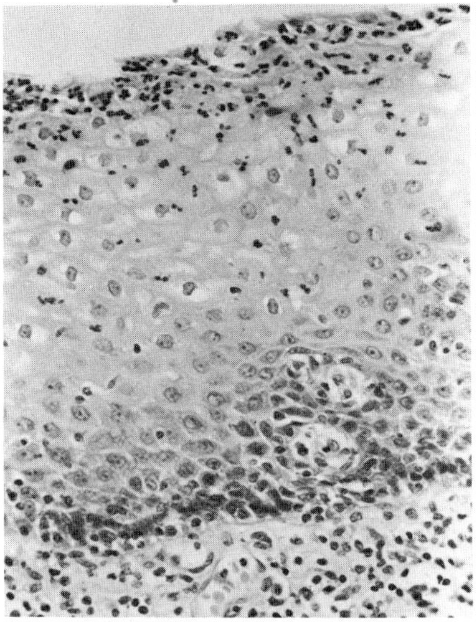

122

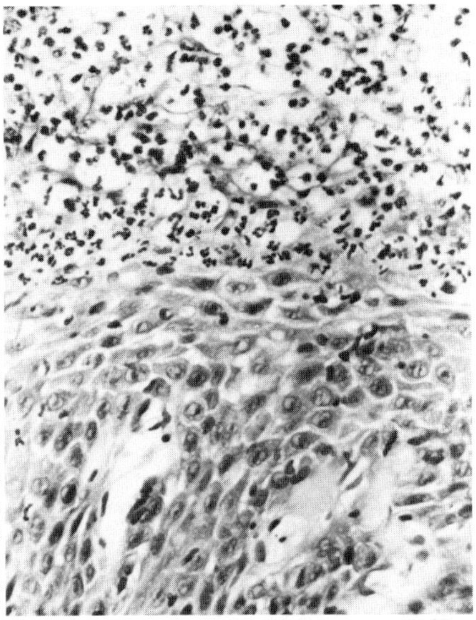

123

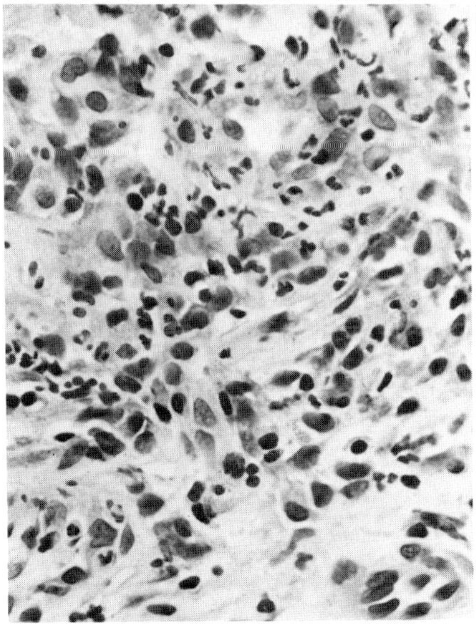

124

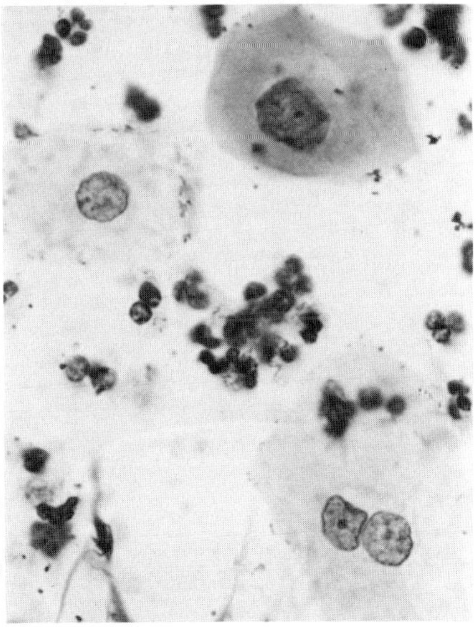

125

Fig. 124. Superficial layers of esophageal mucosa damaged by prolonged esophageal reflux. Note desquamated degenerate squamous cells and numerous inflammatory cells. ×345.

Fig. 125. Abnormal squamous cells in case of reflux esophagitis. Nuclear enlargement is prominent as is cytoplasmic degeneration. Many inflammatory cells are present. ×545.

consisting of gastric-type mucosal cells or, in some cases, a mixture of gastric and intestinal mucosal glands. This metaplastic process may result in the so-called Barrett's esophagus also to be described below.

Cytological Features of Non-Infective Esophagitis

When the squamous epithelium is involved in the inflammatory process the washings or brushings will reveal squamous cells of all three layers, the parabasal cells being shed frequently. There is often enlargement of the squamous cell nuclei (fig. 125) whilst clumping of the nuclear chromatin material and prominent nucleoli may be seen (fig. 126). These latter changes reflect the regenerative activity associated with tissue destruction and attempts at healing. Inflammatory cells will also be present these usually admixed with the squamous epithelial cells. Polymorphonuclear leucocytes predominate in the acute inflammatory process although some lymphocytes, plasma cells and histiocytes are usually present also. Finally many of the squamous epithelial cells will show a variety of degenerative changes such as cytoplasmic vacuolation and fragmentation, and nuclear degeneration (fig. 127).

Barrett's Esophagus

As stated by *Barrett* himself, columnar lined esophagus 'is a condition denied by some, misunderstood by others, and ignored by the majority' [2]. Initially regarded as a congenital disorder and indeed referred to as 'congenitally short esophagus' the term Barrett's esophagus, as currently used, refers to a metaplastic process in which a variable segment of the normal squamous epithelium above the physiological sphincter is replaced by a

---

Fig. 126. Desquamated esophageal squamous cells in case of reflux esophagitis. Regenerative and degenerative changes are evident, the former predominating. Note enlargement of nuclei, chromatin clumping and prominent nucleoli. × 545.

Fig. 127. Desquamated esophageal squamous cells in case of reflux esophagitis. Regenerative and degenerative changes are again evident with prominent cytoplasmic vacuolation. × 545.

Fig. 128. Section from case of Barrett's esophagus. Normal esophageal squamous epithelium is apparent superiorly but distally the squamous epithelium has been replaced by metaplastic glandular epithelium predominately of intestinal type. × 69.

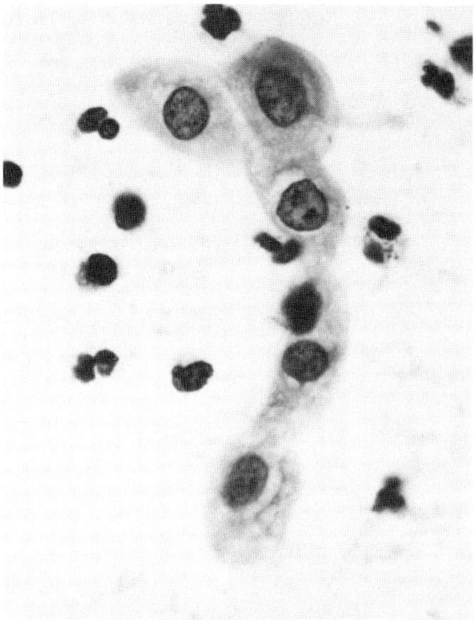

126

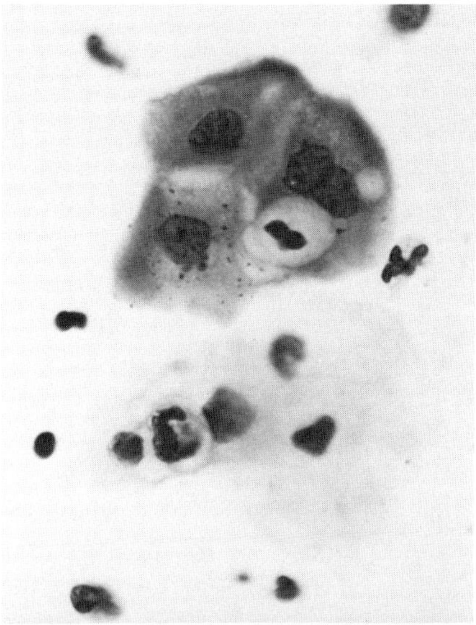

127

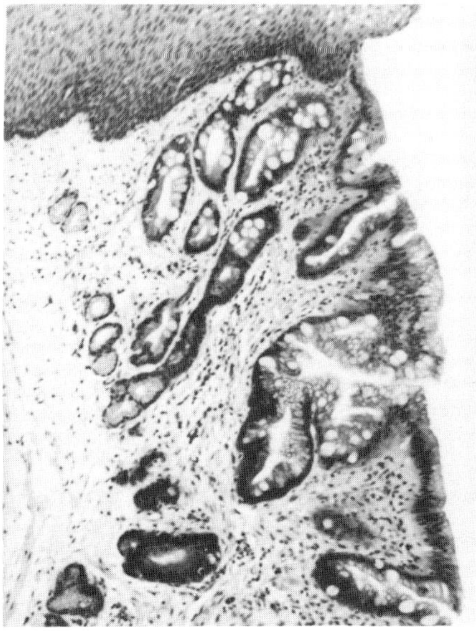

128

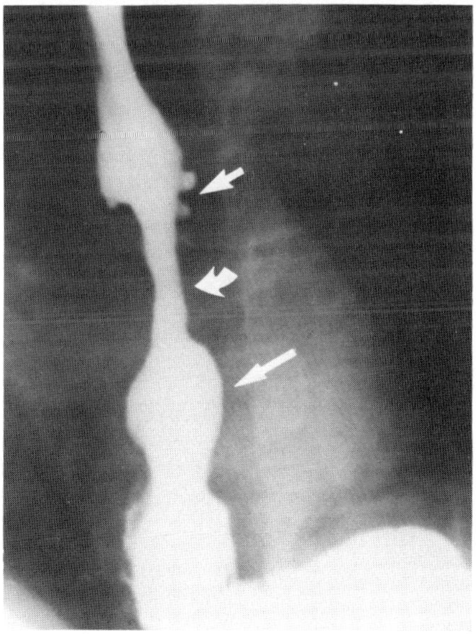

129

Fig. 129. Barium swallow in patient with Barrett's esophagus. The lower straight arrow indicates a prominent hiatus hernia. There is a stricture of the esophagus, as indicated by the curved arrow, and superiorly three small ulcer craters, filled with contrast medium, are evident.

complex mixture of columnar epithelium [45] resembling that found in the gastric and small bowel mucosa (fig. 128). At the upper edge of the metaplastic columnar epithelium there is usually a zone of ulceration with fibrosis and often stricture formation. There is now abundant evidence that the so-called Barrett's esophagus is due to reflux of gastric content, and also of bile-containing small bowel content into the lower esophagus. There is probably an initial ulceration of the squamous epithelium followed by a regeneration under conditions that lead to the formation of a mucosa composed of the various types of glandular epithelium described. Most patients with Barrett's esophagus have a clinical history of reflux esophagitis characterized by regurgitation, substernal discomfort or pain, usually referred to as 'heartburn', and dysphagia. A stricture may be demonstrable radiologically (fig. 129) and this may suggest the diagnosis of esophageal cancer.

The condition is of importance to the diagnostic cytologist for two reasons. Firstly the diagnosis of Barrett's esophagus may be indicated in esophageal washings or brushings by the presence of mucus-producing columnar cells including goblet cells (fig. 130). Secondly there is now clear evidence that the metaplastic epithelium of Barrett's esophagus is frequently the site of dysplasia and adenocarcinoma (fig. 131). The incidence figures for adenocarcinoma associated with metaplasia vary considerably depending on the criteria used for the acceptance of the diagnosis. Thus *Haggitt* et al. [21] suggested that 86% of esophageal adenocarcinomas arise in a background of Barrett's metaplasia. However, many authors exclude any case that involves the gastric cardia, ignoring other features such as the presence of early neoplastic lesions or of foci of columnar metaplasia in the adjacent esophagus. On this basis Barrett's metaplasia probably accounts for approximately 8.5% of esophageal adenocarcinomas [11].

In summary, Barrett's esophagus is of importance to the gastro-esophageal cytologist because of its cytological manifestations in the uncomplicated case and because of the undoubted development of adenocarcinoma in a significant number of cases. In the latter event the cytological features

---

Fig. 130. Mucus producing columnar cells in esophageal brushing from patient with Barrett's esophagus. × 345.

Fig. 131. Adenocarcinoma arising in Barrett's esophagus. Proximal to the carcinoma a clear line of demarcation between the esophageal squamous epithelium and the metaplastic glandular epithelium can be seen. Photograph courtesy of Professor *P. Bhathal,* Royal Melbourne Hospital.

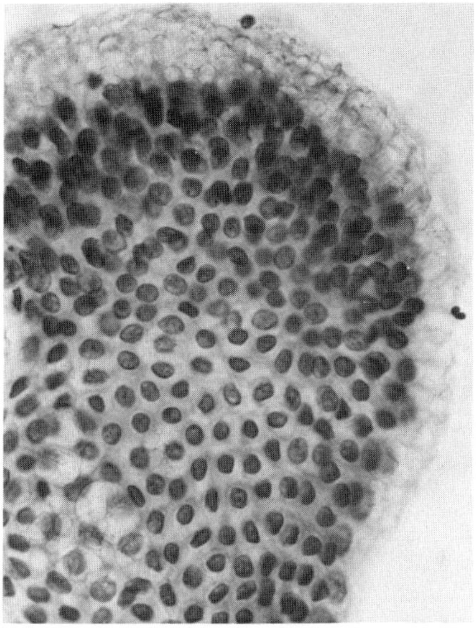

130

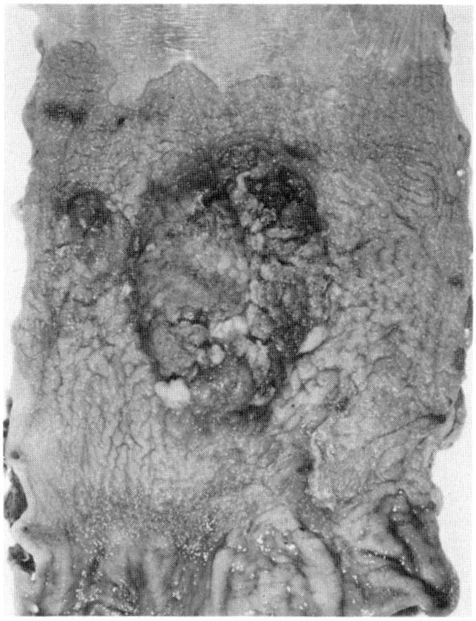

131

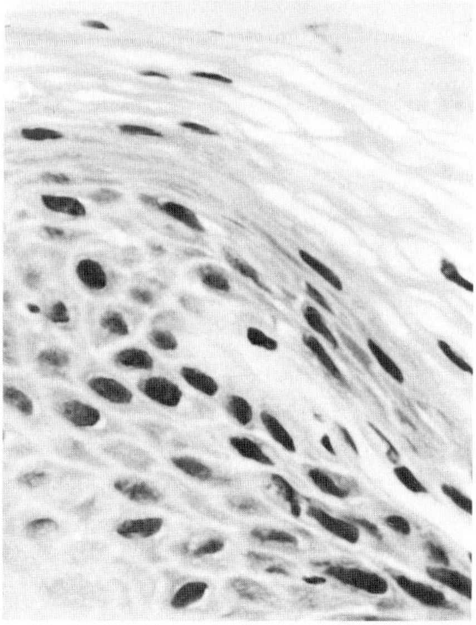

132

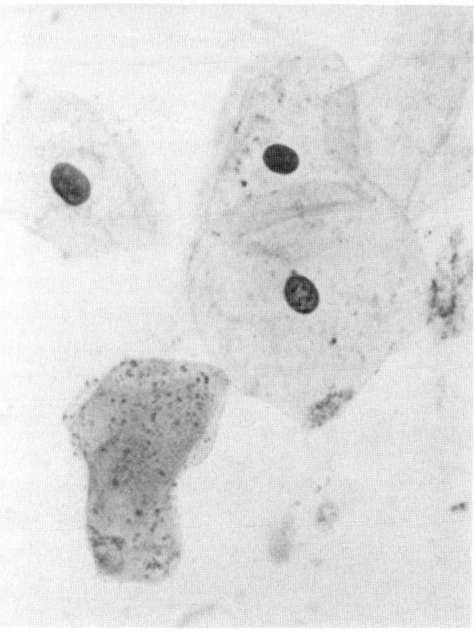

133

Fig. 132. Focus of esophageal leukoplakia. Section shows surface hyperkeratinization and moderate dysplasia of the deeper layers of the epithelium. × 545.

Fig. 133. Cells in esophageal brushings from patient with leukoplakia. Note the anucleate squames in the upper right and lower left corners of the photograph. × 545.

are those of any esophageal adenocarcinoma. Indeed it has been suggested [6] that the proper surveillance for malignant change in patients with a columnar lined lower esophagus should include regular cytological and histological studies.

Miscellaneous Non-Neoplastic Abnormalities

Leukoplakia

Although some authorities state that leukoplakia does not occur in the esophagus [41] others regard it as being relatively common [1]. The problem is largely one of semantics. Thus the word leukoplakia, meaning literally 'white patch', may be used to describe the clinical or macroscopic appearances of a mucosal surface the 'white patch' being due to a variety of causes including simple hyperkeratinization. To the pathologist, however, the diagnosis of leukoplakia implies, in addition to surface hyperkeratosis, atypia or dysplasia of the underlying squamous epithelium. It is this connotation that probably leads to the belief by some [55] that esophageal leukoplakia is a precancerous lesion.

A more common cause of white patches on the esophageal mucosa is the condition known as glycogenic acanthosis. This is a lesion characterized by focal hyperplasia of epithelial cells containing abundant cytoplasmic glycogen. The mucosal plaques are white, round, usually discrete, smooth lesions. They may be found throughout the esophagus but are most common in the distal third and are usually multiple. Hyperkeratosis, cellular atypia, and dysplasia are absent. The aetiology of the condition is obscure but it does not appear to have the malignant potential of leukoplakia. *Bender* et al. [5] believe that leukoplakia of the esophagus has not been convincingly described as an entity and that most cases probably represent glycogenic acanthosis. In 1978 *Herschman* et al. [22] described a case of 'esophageal leukoplakia' stating that 'to date this apparently benign condition has not been reported'. The lesion thus described consisted of two discrete, slightly raised, firm white plaques in the distal esophagus. Esophageal

---

Fig. 134. Squamous cells showing marked cell enlargement as a result of irradiation. × 205.

Fig. 135. Enlarged squamous cells with prominent cytoplasmic vacuolation as a result of irradiation. × 545.

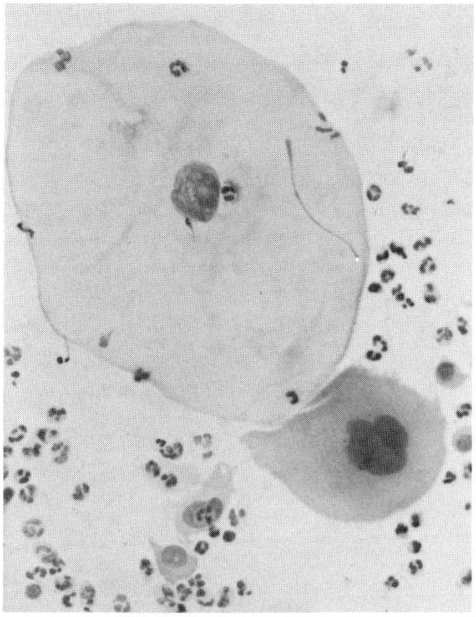

134

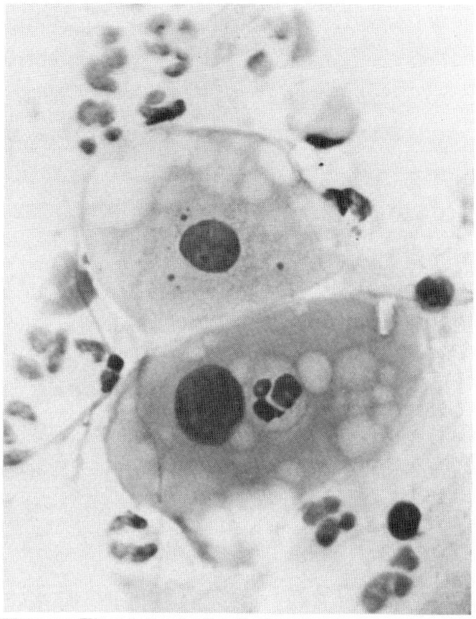

135

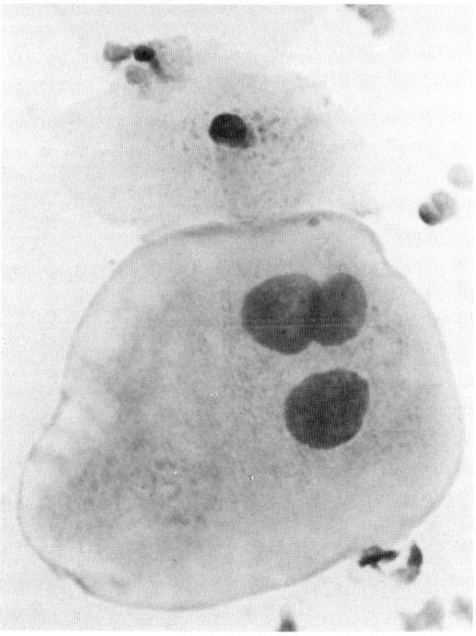

136

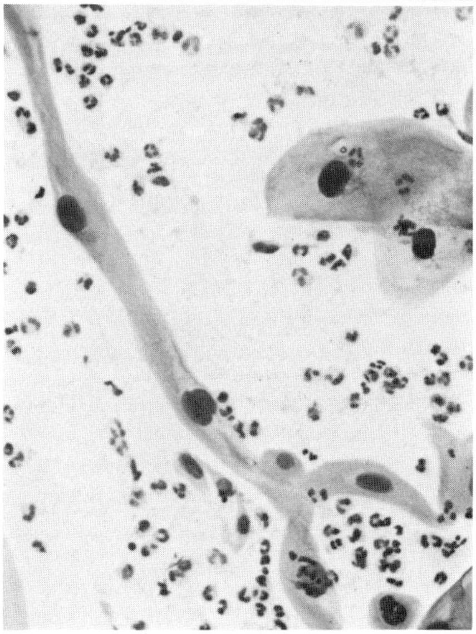

137

Fig. 136. Cellular enlargement and multinucleation due to irradiation. × 545.

Fig. 137. Bizarre shaped squamous cells in esophageal brushings from patient with history of previous irradiation. × 205.

brushings yielded 'abundant keratinized cells'. Biopsies revealed a surface layer of parallel horizontally oriented eosinophilic cornified epithelium containing occasional pyknotic nuclei. No vacuolation of cells was noted and there was no evidence of cell atypia or dysplasia. The authors concluded that additional studies of a larger number of cases are necessary to determine the malignant potential of the condition.

Nevertheless histological lesions resembling 'true' leukoplakia are occasionally seen in the esophagus (fig. 132). The areas of leukoplakia will yield abundant superficial type cells with eosinophilic or orangeophilic cytoplasm. The cells may be anucleate or contain pyknotic nuclei (fig. 133). In blind lavage specimens these cells cannot be distinguished from cells originating in areas of leukoplakia of the buccal mucosa. However, in brushing specimens their site of origin can, of course, be documented.

Radiation Change

The esophagus is quite frequently exposed to radiation in the course of radiotherapy to the lungs or mediastinum for malignant disease or indeed radiotherapy may be used in the treatment of esophageal cancer. In this event the esophageal squamous cells will show changes identical to those seen in the cervix and vagina following radiotherapy to those organs. Most characteristic is cell enlargement, cytoplasmic vacuolation, multinucleation and bizarre cell shapes (fig. 134–137). These appearances may cause concern if the cytologist is not aware of the history of irradiation.

In a recent paper, *O'Morchoe* et al. [42] described the cytological changes seen in esophageal brush specimens from patients who did not have either gastric or esophageal cancer but who had received a variety of chemotherapeutic agents. 10 patients received chemotherapy alone, and 9 received combined chemotherapy and radiation therapy. Of the 10 patients, who had received chemotherapy alone, 3 showed evidence of moderate to severe epithelial atypia, which was not seen in the combined treatment group or the control group. The cytologic atypias in the 3 patients consisted of variation in nuclear size with crowding and overlapping, an increased nuclear-cytoplasmic ratio and multiple nucleoli, many of which varied in size and shape. There was also evidence of cell death and keratinization. Although these patients had not been irradiated, other striking changes were similar to those seen following radiation therapy.

The authors comment that the cytologic changes reported by them pose a serious diagnostic problem as they may mimic malignant transformation. We have not had the opportunity to study similar cases but the abnormalities illustrated in this paper would suggest that the authors' comment is a very valid one. With the increasing use of cytotoxic drugs there is no doubt that the diagnostic problem will be encountered more frequently.

An interesting sidelight, and one that is relevant to an earlier section in this chapter, is the comment that evidence of infection by herpes simplex alone or herpes simplex associated with *Candida* was seen in 32% of the treated patients and in 4% of the control group.

Pemphigus vulgaris

Pemphigus vulgaris is an uncommon disease which affects the skin and is manifested by the formation of successive crops of vesicles or bullae. In chronic pemphigus, oral lesions develop in most cases at some time during the course of the disease but in 50% of cases of acute pemphigus vulgaris the lesions appear first in the oral cavity [38]. The mucous membranes of the mouth, pharynx, conjunctiva, nasal mucosa and vagina are often involved, but the esophagus only occasionally. More frequently the esophagus is involved in the related 'blistering diseases' such as epidermolysis bullosa and bullous pemphigoid, particularly the so-called benign mucous membrane variant of the latter [19, 35].

The cytological manifestations of pemphigus vulgaris have been described in considerable detail by *Medak* et al. [38]. The most characteristic feature is the presence of squamous cells with prominent bar or bullet-shaped nucleoli (fig. 138). The cells may show a curious concentration of the chromatin material at one side of the nucleus (fig. 139). A cell-in-cell arrangement may be seen the cell group being similar to that of an epithelial 'pearl' (fig. 140). Finally multinucleated cells are sometimes prominent (fig. 141).

Benign Neoplasms

Benign esophageal tumours are rare and in nearly all published reports there is a lack of histological description and illustration. From the few reliable reports available the majority are non-epithelial and consist of lipomas, leiomyomas, hemangiomas and lymphangiomas. Squamous papillomas have been described some of which appear genuine but the majority of which are probably inflammatory lesions with reactive squamous hyperplasia of the mucosa. Occasionally simple adenomas arise from glands within the esophagus.

It is doubtful if cytology has any role to play in the diagnosis of these lesions and, in the author's experience, their presence has no cytological implications.

Malignant Neoplasms

A variety of malignant neoplasms may occur within the esophagus. The relevant publication of the World Health Organization [43], for example, includes the following types:

Malignant Epithelial Tumours
- Squamous cell carcinoma
- Adenocarcinoma
- Adenoid cystic carcinoma
- Mucoepidermoid carcinoma
- Adenosquamous carcinoma
- Undifferentiated carcinoma

Malignant Non-Epithelial Tumours
- Leiomyosarcoma

Miscellaneous Tumours
- Carcinosarcoma
- Malignant melanoma

Despite this relatively large number of malignant esophageal neoplasms it is important to stress at the outset that only two occur with sufficient frequency to be of major concern to the diagnostic cytologist. These are squamous cell carcinoma and adenocarcinoma, the former predominating. However, the cytological manifestations of the less common types are occasionally seen and are usually sufficiently characteristic to enable a diagnosis to be made.

Squamous Cell Carcinoma

The figures given for the relative incidence of the various histological types of esophageal cancer vary considerably depending upon whether or not the author includes post-cricoid neoplasms and carcinomas of the cardia in their series. *Morson and Dawson* [41] state that 60–70% of malignant

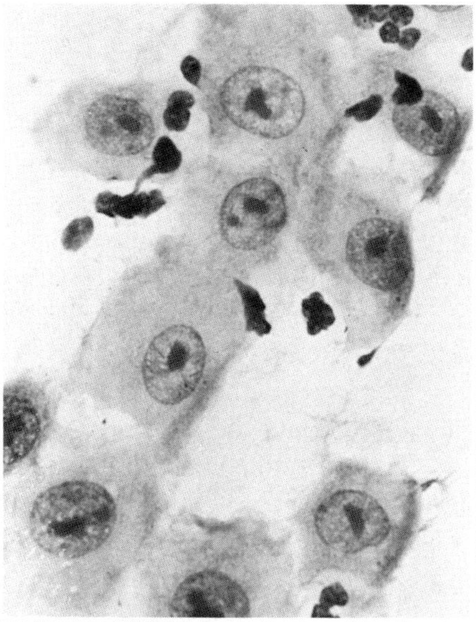

138

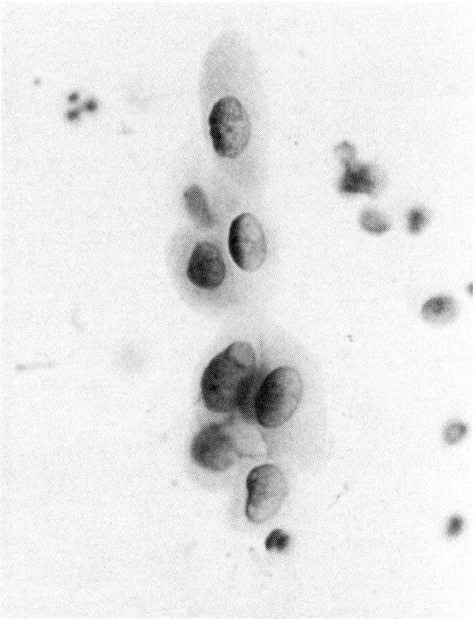

139

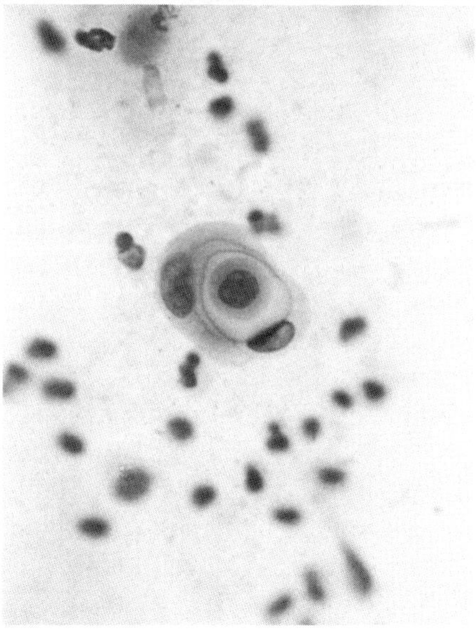

140

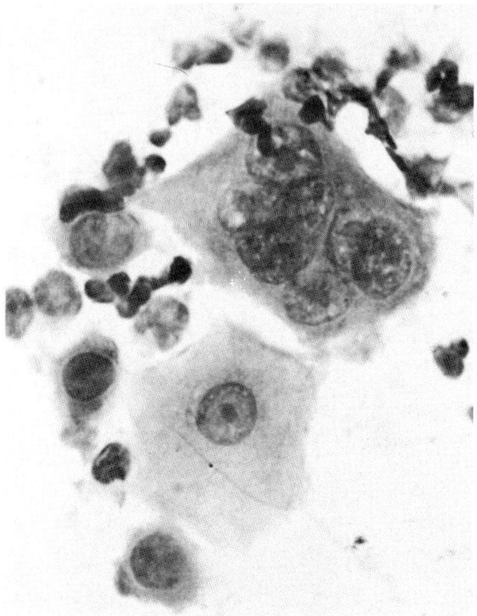

141

Fig. 138. Squamous cells from case of pemphigus vulgaris showing characteristic bar or bullet-shaped nuclei. × 545.

Fig. 139. Squamous cells from case of pemphigus vulgaris showing concentration of chromatin material at one side of nucleus. × 545.

Fig. 140. Cell-in-cell arrangement of squamous cells from case of pemphigus vulgaris. × 545.

Fig. 141. Multinucleate cell from case of pemphigus vulgaris. × 545.

tumours of the esophagus are squamous cell carcinomas whilst both *Ashley* [1] and *Ming* [39] believe that almost 95% of esophageal cancers are of this type. These neoplasms may arise in any part of the esophagus but occur most commonly in the middle and lower thirds. Although most cases show a rather mixed growth pattern it is useful to divide them into three types. Thus, according to *Ming* about 60% are fungating in type, the neoplasm forming a large mass within the lumen. The surface of this mass is usually ulcerated or exhibits numerous polypoid projections (fig. 142). Approximately 25% of cases are ulcerative in type. The edges of the ulcer are usually raised or nodular, the crater is of variable depth, and the floor is covered with necrotic, often haemorrhagic, tissue. An infiltrating carcinoma is seen in about 15% of cases with extensive intramural tumour growth and often superficial ulceration of the overlying mucosa (fig. 143). Rarely this intramural spread may be very extensive the appearances being analogous to the so-called 'leather-bottle' stomach. The radiological features frequently reflect the growth pattern present (fig. 144).

Histological Features

The squamous cell carcinoma of the esophagus, as in other parts of the body, is graded as well, moderately well, and poorly differentiated. The well differentiated carcinomas usually have abundant keratin which may be seen as broad irregular bands which intersect the malignant epithelial cells (fig. 145). More frequently, perhaps, the keratin is seen as central masses within islands of carcinoma tissue (fig. 146). The nuclei tend to be fairly uniform in size and shape and intercellular bridges or 'prickle-processes' are prominent (fig. 147). At the other end of the spectrum evidence of differentiation is barely discernible the appearances in most areas being those of an undifferentiated carcinoma (fig. 148). This type of very poorly differentiated squamous cell, or large-cell undifferentiated carcinoma, must be distinguished from the much less common small-cell undifferentiated or 'oat-cell' carcinoma to be discussed below. Between these two extremes of differentiation are seen a range of histological patterns, keratin formation being less conspicuous but the squamous nature of the lesion being indicated by the mosaic arrangement of the tumour cells (fig. 149, 150). Not uncommonly within the larger islands of carcinoma tissue central degeneration occurs with fragmentation of cytoplasm and increased hyperchromasia or pyknosis of nuclei (fig. 151). These degenerative changes are reflected in the cytological specimen obtained by direct brushing of the lesion (fig. 152).

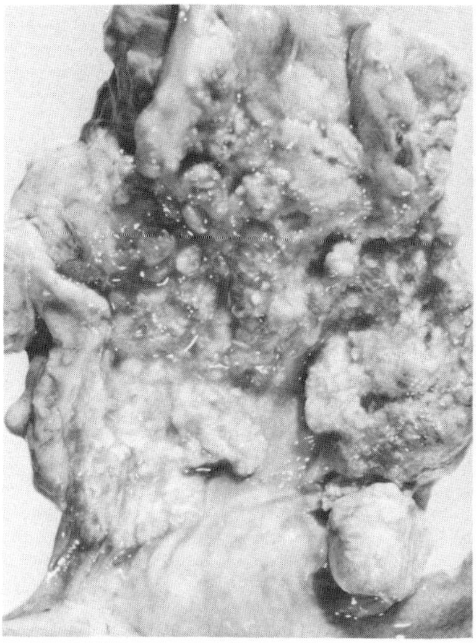

142

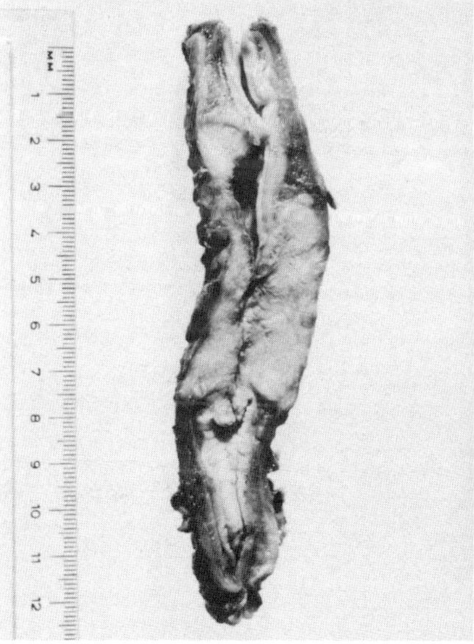

143

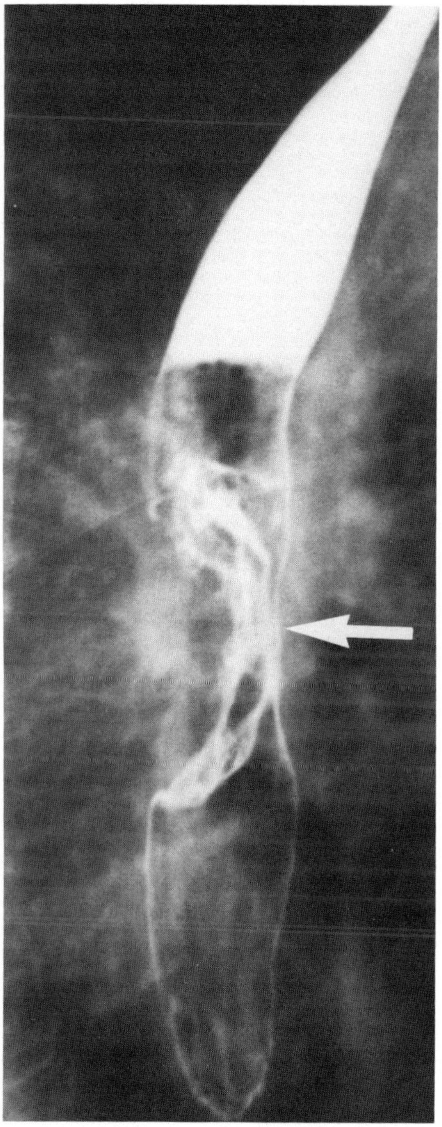

144

Fig. 142. Fungating carcinoma of distal esophagus with polypoid ulcerated surface.

Fig. 143. Extensive intramural spread of esophageal carcinoma with marked narrowing of the lumen.

Fig. 144. X ray of esophagus showing evidence of a fungating infiltrating carcinoma in the mid thoracic region. The lumen of the esophagus appears to be almost totally obliterated as indicated by the arrow.

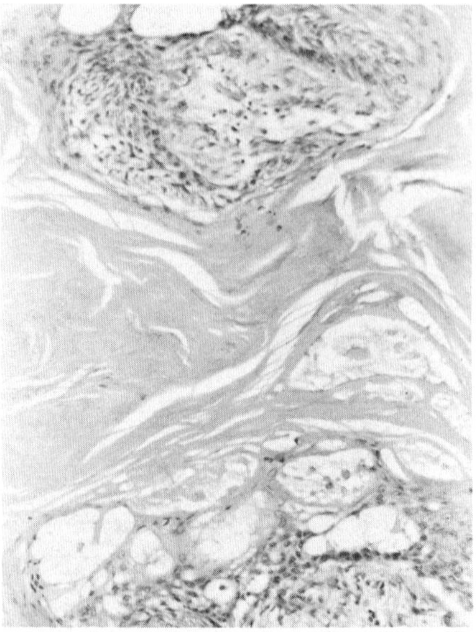

145

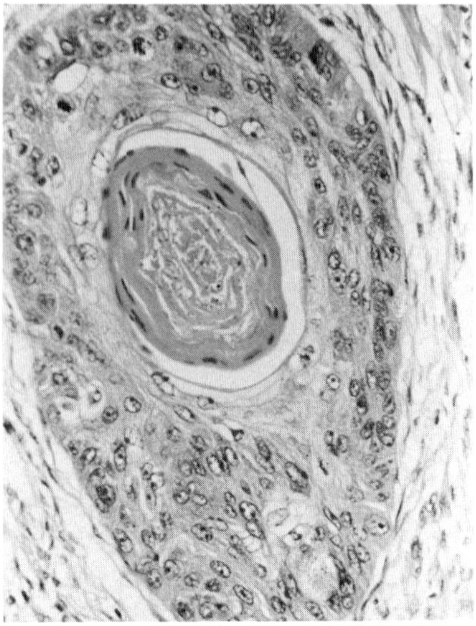

146

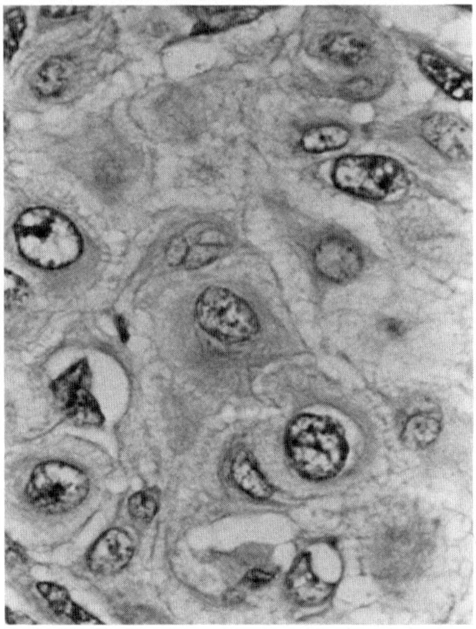

147

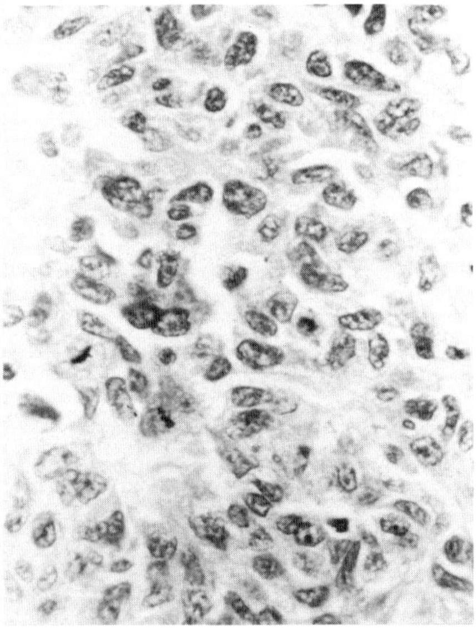

148

Fig. 145. Well differentiated squamous cell carcinoma of esophagus. Islands of carcinoma tissue are intersected by irregular bands of keratin. × 86.

Fig. 146. Well differentiated squamous cell carcinoma of esophagus. Island of carcinoma tissue with central mass of keratin. × 173.

Fig. 147. Well differentiated carcinoma of esophagus showing prominent intercellular bridges or 'prickle processes'. × 545.

Fig. 148. Very poorly differentiated squamous cell, or large-cell undifferentiated carcinoma of esophagus. × 345.

149

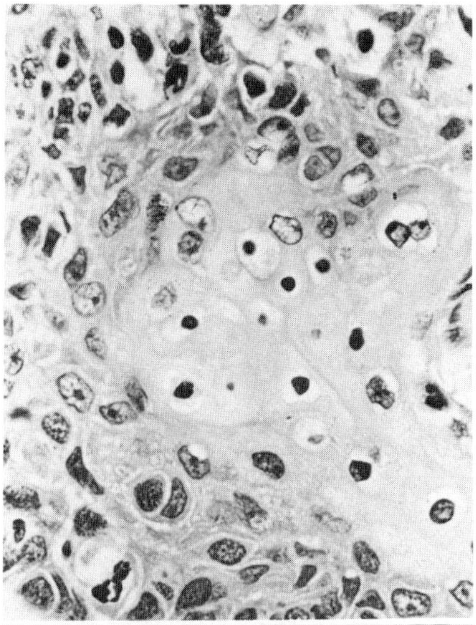

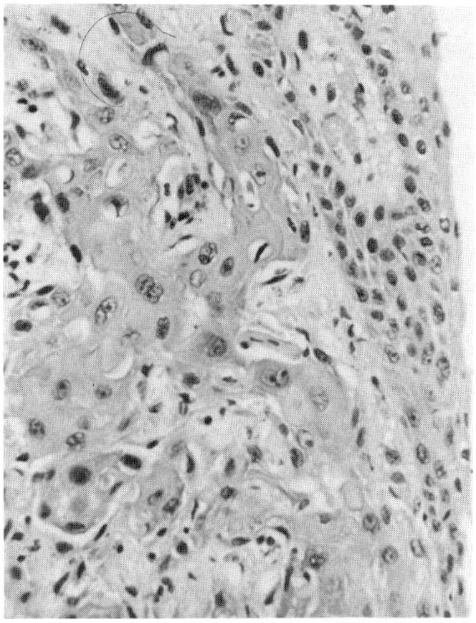

150

151

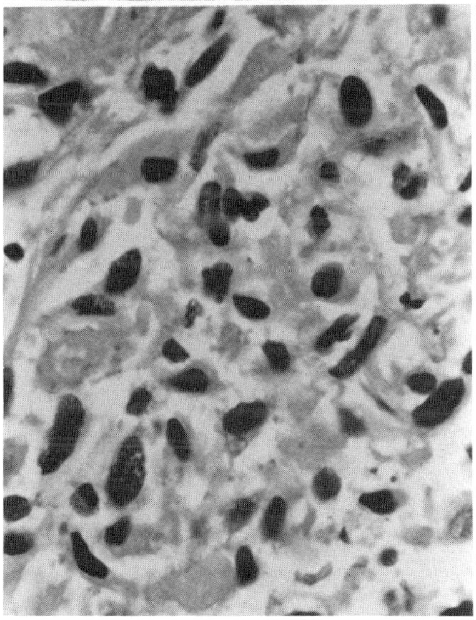

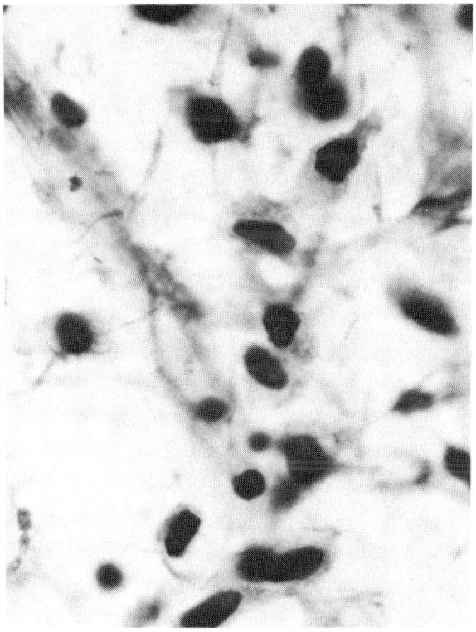

152

Fig. 149. Moderately well differentiated squamous cell carcinoma of esophagus with typical mosaic arrangement of tumour cells. × 345.

Fig. 150. Moderately well differentiated squamous cell carcinoma of esophagus with mosaic arrangement of tumour cells and focal keratin formation. × 173.

Fig. 151. Section of well differentiated squamous cell carcinoma of esophagus with evidence of tissue degeneration and consequent hyperchromasia of nuclei. × 545.

Fig. 152. Esophageal brushing from patient with well differentiated squamous cell carcinoma showing features similar to those depicted in previous figure. × 545.

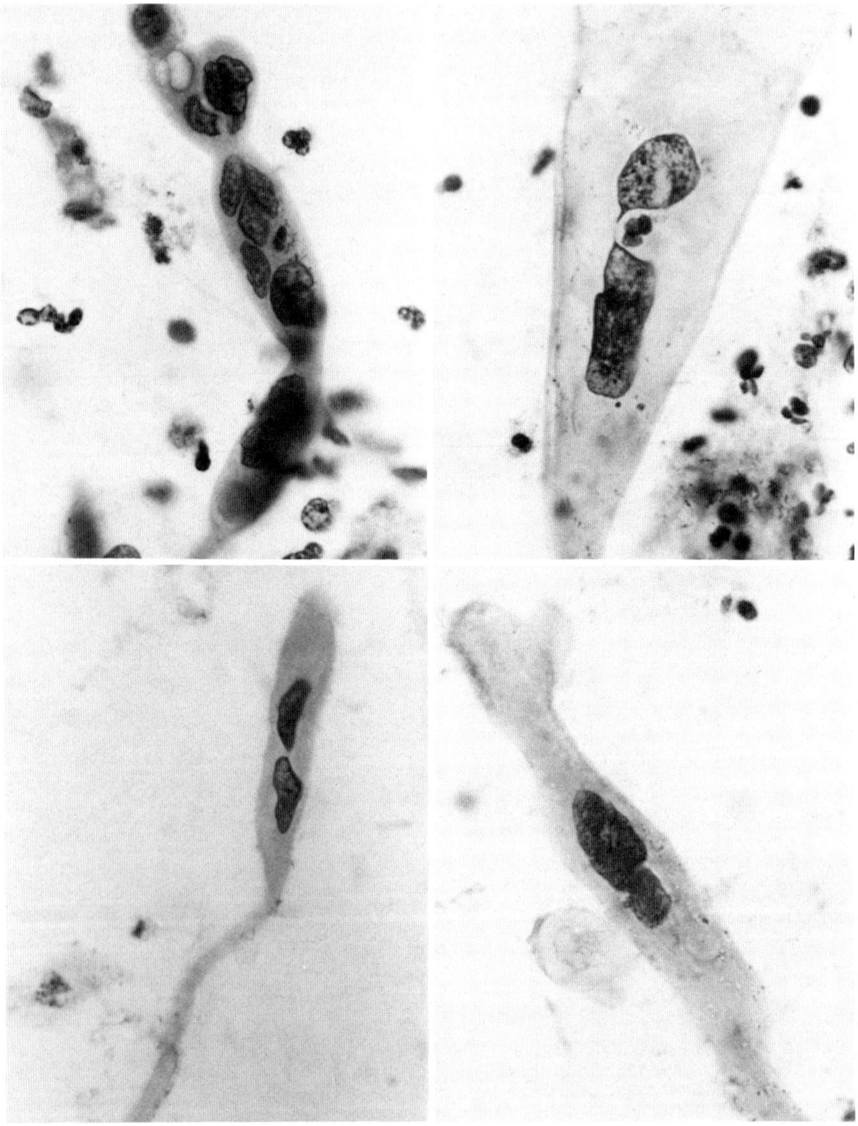

153 154 155 156

Fig. 153–156. Bizarre appearances of cells from well differentiated squamous cell carcinoma of esophagus. Note the malignant criteria evident in most nuclei. All photographs × 545.

Fig. 157. Single cell from squamous cell carcinoma of esophagus. Note the irregularity of the nuclear membrane, the chromatin clumping, and the cleared areas within the nucleus. × 545.

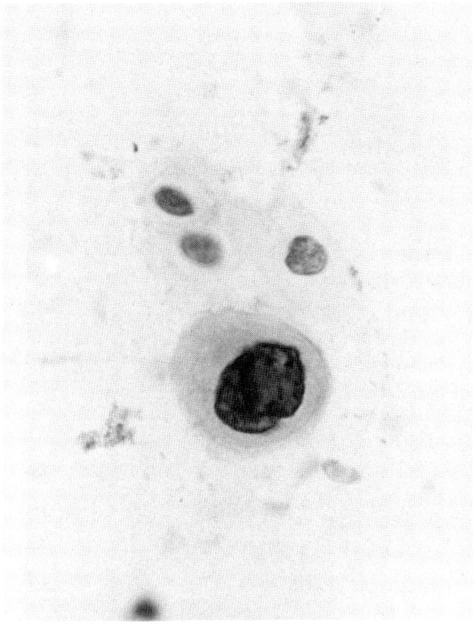

157

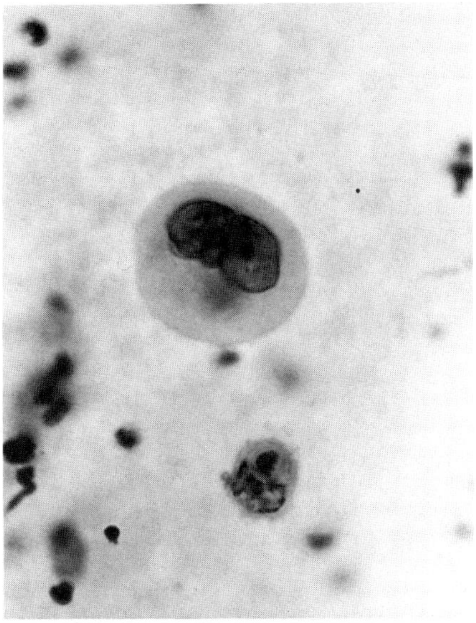

158

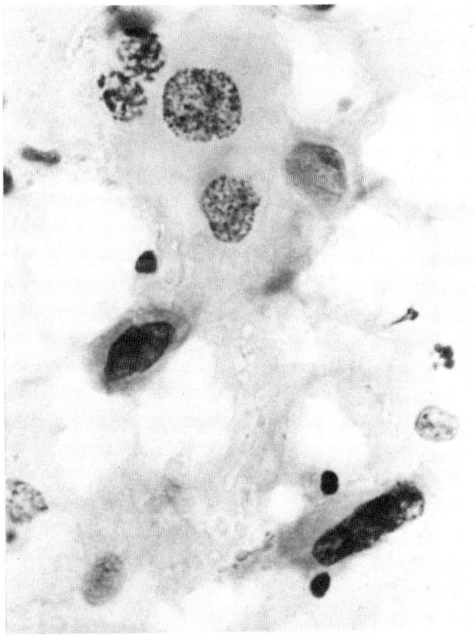

159

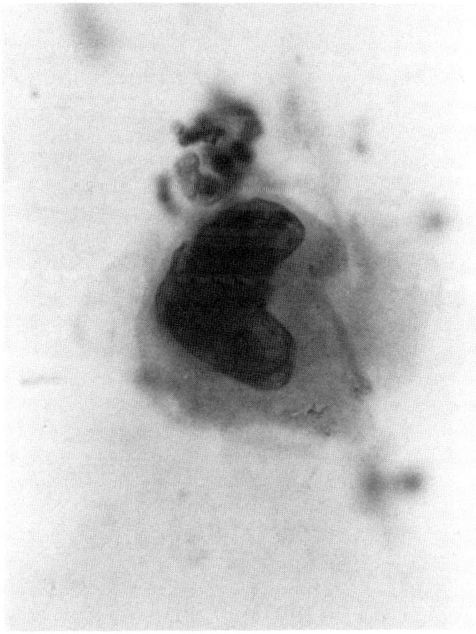

160

Fig. 158. Binucleate cell from well differentiated squamous cell carcinoma of esophagus. Note the irregularity of the nuclear membranes, the chromatin clumping, and the prominent nucleolus within one nucleus. × 545.

Fig. 159. Cells from squamous cell carcinoma of esophagus. Degenerative changes have produced prominent karyorrhexis in the upper cells and pyknosis of the nuclei in the two lower cells, the latter tending to obscure malignant criteria. × 545.

Fig. 160. Single large cell from squamous cell carcinoma of esophagus. Nuclear pyknosis obscures malignant criteria. × 545.

Cytological Manifestations

Examination of brushings or washings from a well differentiated squamous cell carcinoma reveals cells which show considerable variation in size and shape and which are frequently quite bizarre in appearance (fig. 153–156). Nuclear malignant criteria are usually clearly recognizable (fig. 157, 158) although increased hyperchromasia of the nucleus may tend to obscure the finer details of nuclear structure (fig. 159, 160). Evidence of squamous differentiation may be prominent with obvious cytoplasmic keratinization (fig. 161), spindle cells (fig. 162) and malignant epithelial 'pearls' (fig. 163). Less well differentiated forms may require careful examination to determine their squamous origin. The 'cell-in-cell' arrangement in the lower part of figure 164 is helpful as is the similar structure near the centre of figure 165. In this latter case degenerative vacuolation is confusing as it is in the case depicted in figure 166. Here the mosaic pattern and the suggestion of occasional intercellular bridges indicates the nature of the neoplasm. Finally undifferentiated or anaplastic carcinomas are seen. These yield cells which vary in size and shape, being round, oval or polygonal, usually have good nuclear malignant criteria and rather scanty nondescript cytoplasm (fig. 167, 168).

Adenocarcinoma

The true incidence of adenocarcinoma of the esophagus is difficult to establish. The inclusion of primary lesions of the stomach which invade the lower esophagus accounts, in part at least, for the lack of agreement as to the real incidence. Data from cancer registries show the incidence to be about 2% for the upper and middle thirds of the esophagus and this is undoubtedly a more reliable figure than can be derived for the lower third [8].

Adenocarcinomas involving the esophagus are usually located near the cardio-esophageal junction. Indeed some actually arise within the mucosa of the gastric cardia and, by upward extension, present as esophageal neoplasms. The macroscopic appearances are variable but in general resemble

---

Fig. 161. Cell from well differentiated squamous cell carcinoma of esophagus showing cytoplasmic keratinization. The latter has a refractile, brightly eosinophilic or orange appearance in routine Papanicolaou stain. × 545.

Fig. 162. Group of spindle cells from squamous cell carcinoma of esophagus. × 545.

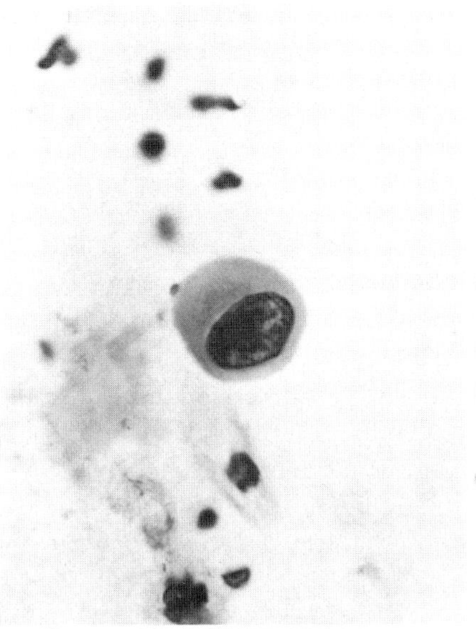

161

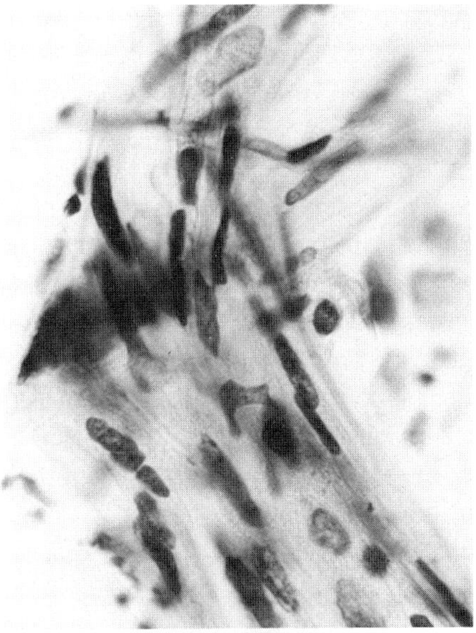

162

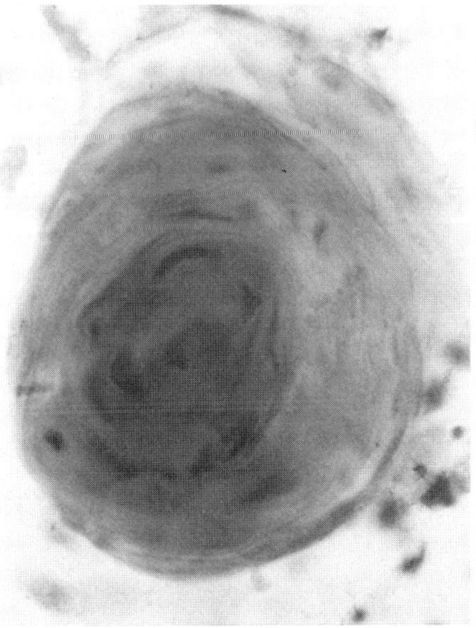

163

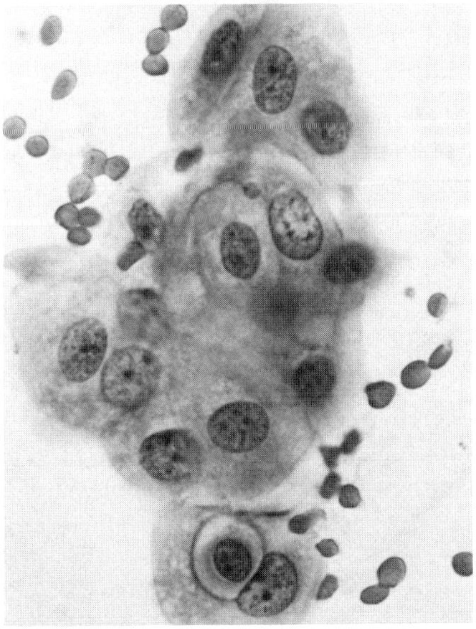

164

Fig. 163. Malignant epithelial 'pearl' from well differentiated squamous cell carcinoma of esophagus. × 545.

Fig. 164. Group of cells from squamous cell carcinoma of esophagus showing characteristic 'cell-in-cell' arrangement in lower part of photograph. × 545.

those already described for the squamous cell carcinoma. Thus ulcerating (fig. 169), fungating or polypoid (fig. 170) and infiltrating types are seen. The true primary adenocarcinomas in the lower esophagus arise within gastric type mucosa, either that which occurs naturally within the esophagus or within metaplastic gastric type epithelium [13, 49, 57, 58]. The occurrence of adenocarcinoma in Barrett's esophagus has already been described and illustrated in the discussion of the sequelae of esophageal reflux. A further example of this phenomenon is seen in figure 171. In this case the carcinoma is seen arising from metaplastic gastric epithelium the latter showing prominent intestinal features. When the tumour arises elsewhere in the esophagus the tissue of origin is somewhat controversial. Although the esophagus contains submucosal mucous glands, as already described, carcinoma arising from these is extremely rare. It is generally stated that most originate in metaplastic gastric type epithelium and indeed remnants of this are often found in the vicinity of the tumour. However, it is not uncommon to see the malignant glandular structures apparently 'budding off' from the basal layers of the squamous epithelium (fig. 172). It could be postulated that these cells are bipotential and that, on undergoing malignant transformation, are differentiating along a glandular pathway instead of the squamous one followed formerly. Whilst highly speculative, and based on morphologic features alone, the frequent observation of this phenomenon suggests the need for further study.

Histological Features

Regardless of the mode of formation the lesions are essentially adenocarcinomas and again all degrees of differentiation may be seen. Papillary formations may be prominent (fig. 173, 174) or, alternatively, the lesion may be predominantly tubular or acinar in type. The latter may be well differentiated, with well formed glandular spaces lined by fairly uniform cells (fig. 175) or it may be poorly differentiated, the acini being small and irregular and the surrounding cells pleomorphic (fig. 176).

Fig. 165. Cells from poorly differentiated squamous cell carcinoma of esophagus. Cell-in-cell arrangement near centre of photograph indicates squamous differentiation although degenerative vacuolation of cytoplasm may falsely suggest glandular pattern. × 545.

Fig. 166. Group of cells from poorly differentiated squamous cell carcinoma. Degenerative cytoplasmic vacuolation is confusing but mosaic pattern and suggestion of 'prickle-processes' aids in recognition of squamous structure. × 545.

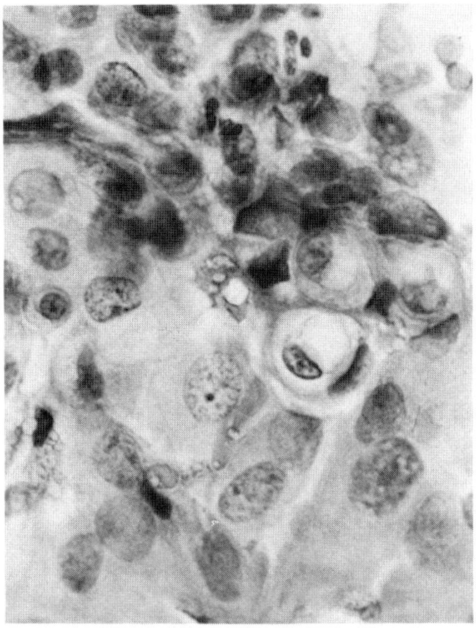

165

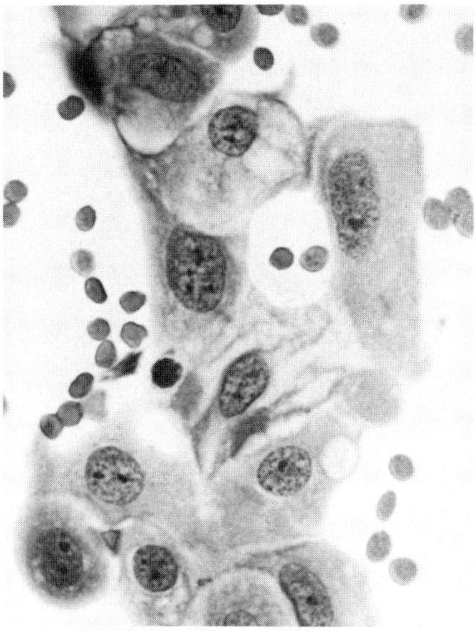

166

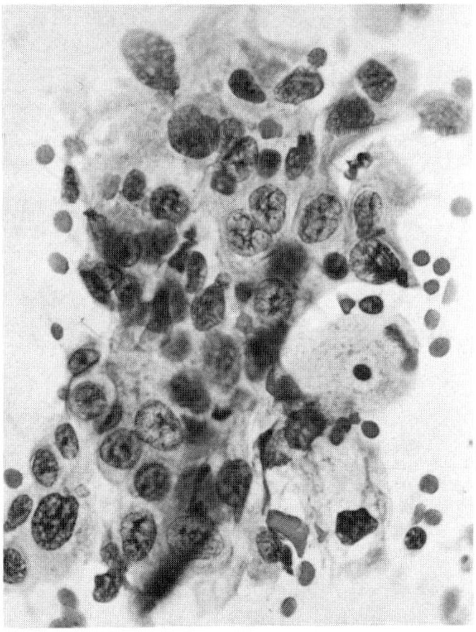

167

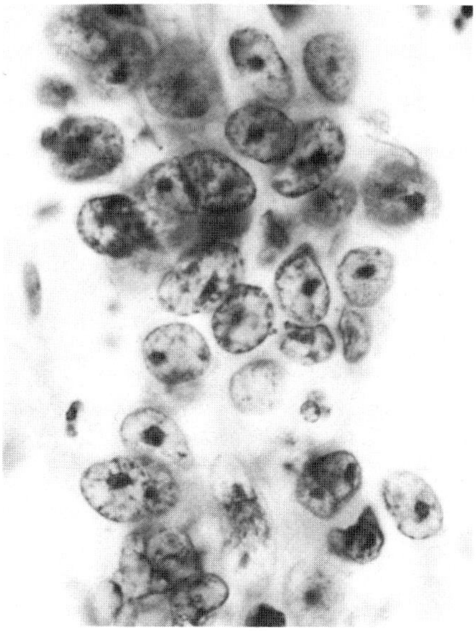

168

Fig. 167. Undifferentiated carcinoma of esophagus. Cells have poorly defined cell borders and no evidence of cytoplasmic specialization although suggestion of cell-in-cell arrangement in lower left of photograph. × 545.

Fig. 168. Cells from undifferentiated carcinoma of esophagus. Cytoplasm is poorly defined but good malignant criteria are evident in nuclei. × 545.

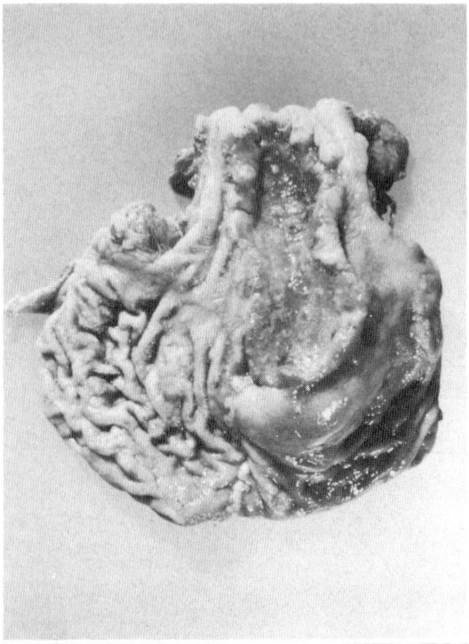

169

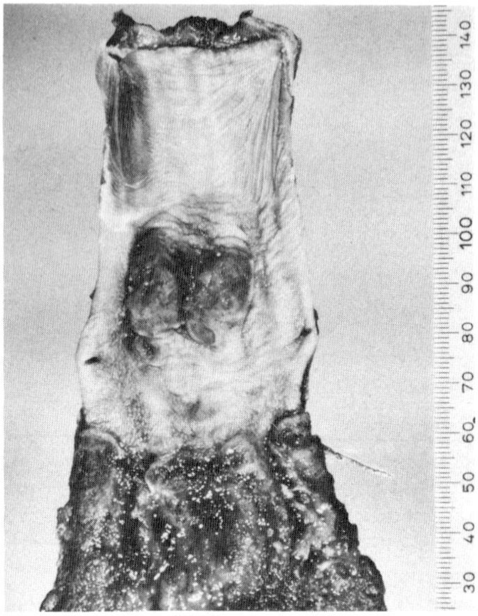

170

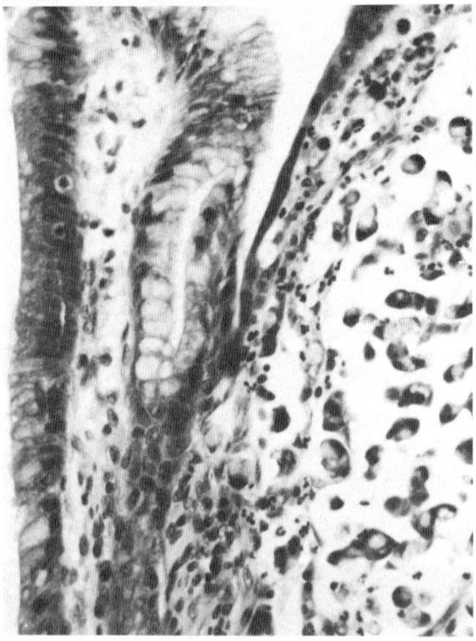

171

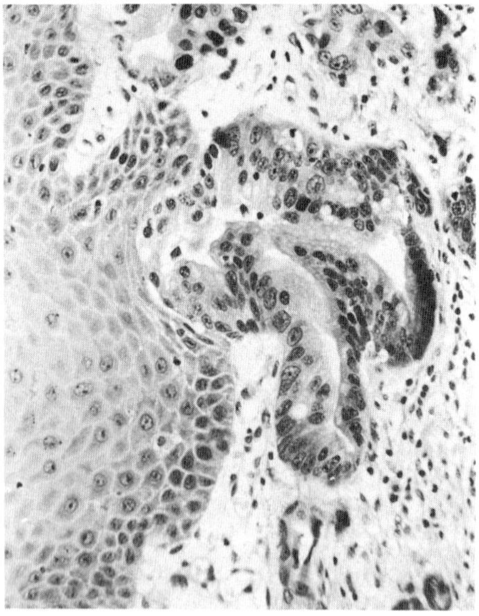

172

Fig. 169. Ulcerating adenocarcinoma of esophagus extending into cardia of stomach.

Fig. 170. Fungating or polypoid adenocarcinoma arising in lower esophagus.

Fig. 171. Signet-ring cell adenocarcinoma arising in metaplastic glandular epithelium of Barrett's esophagus. × 173.

Fig. 172. Esophageal adenocarcinoma apparently arising from basal layer of esophageal squamous epithelium. × 140.

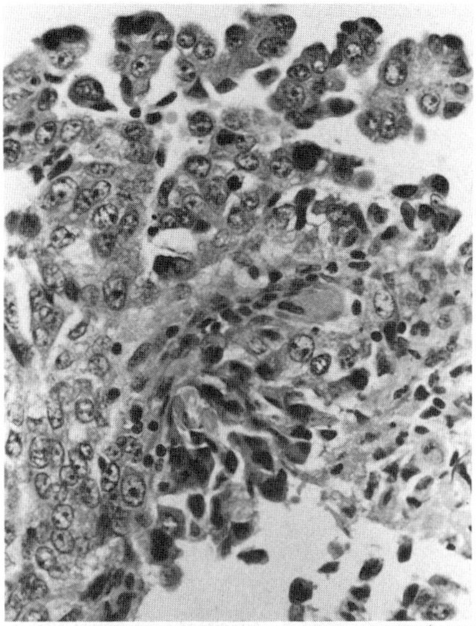

173

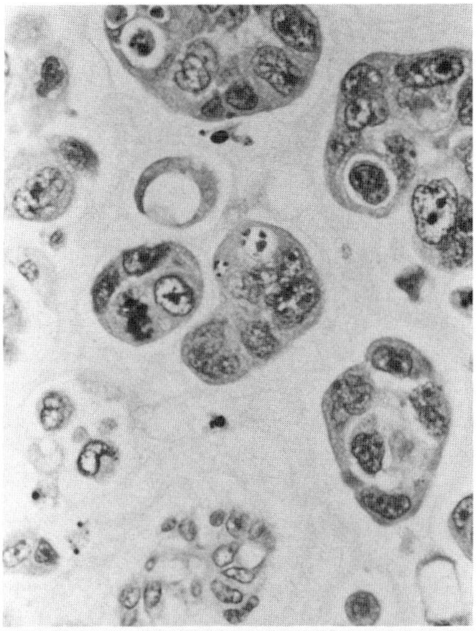

174

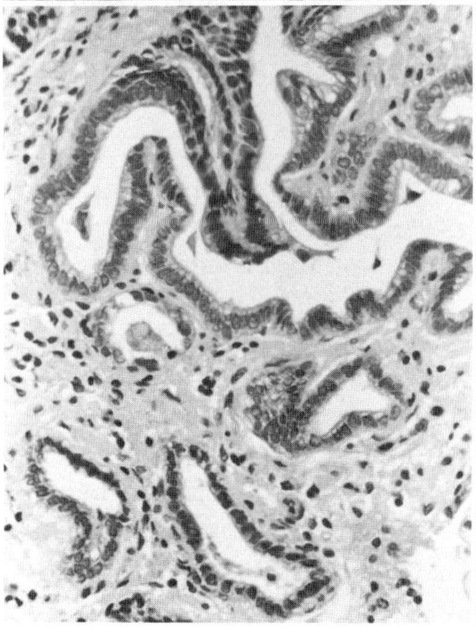

175

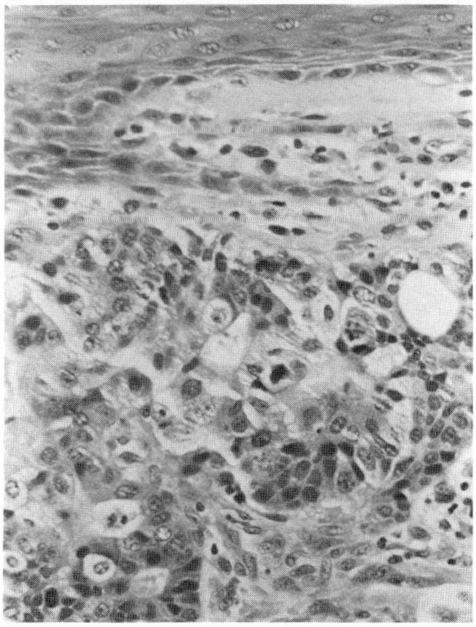

176

Fig. 173. Surface of papillary adenocarcinoma of esophagus. × 220.

Fig. 174. Section of papillary adenocarcinoma of esophagus showing prominent discrete papillary fragments. × 345.

Fig. 175. Well differentiated tubular or acinar adenocarcinoma of esophagus. × 205.

Fig. 176. Poorly differentiated tubular or acinar adenocarcinoma of esophagus. Note very small irregular acini in islands of carcinoma tissue. × 173.

Cytological Manifestations

The cytological appearances in both brushings and washings of the esophagus relect clearly these variations in histological pattern. Papillary fragments may be seen (fig. 176, 177a, b) as may acinar structures (fig. 178, 179). The less well differentiated forms are distinguished by cells which show good malignant nuclear criteria and cytoplasmic vacuolation (fig. 180–182) the latter being less conspicuous in the poorly differentiated lesions (fig. 183–185). The undifferentiated neoplasms (fig. 186, 187) are difficult to assess although the arrangement of the nuclei and the cytoplasmic border evident in figure 186 is suggestive of a glandular pattern.

Other Primary Malignant Neoplasms

Although the two types described account for the vast majority of esophageal neoplasms some of the less common malignancies listed in the classification at the beginning of this section do present with characteristic cytological manifestations. Examples of the following have been seen in our laboratory and are probably best illustrated by individual case histories.

Adenosquamous Carcinoma

In these carcinomas two histological patterns are apparent with sheets of both squamous carcinoma and adenocarcinoma tissue each of variable differentiation. They usually occur in the region of the cardio-esophageal junction and are analogous to those neoplasms that occur in other parts of the body where two different types of epithelium meet. These malignant neoplasms may at times represent the so-called ‘collision carcinomas’ representing a meeting and intermingling of a squamous carcinoma of the esophagus and an adenocarcinoma of the stomach. Alternatively, and more probably, they may be due to simultaneous stimulation of the adjacent squamous and glandular epithelium to produce a malignant tumour of mixed pattern. Very rarely a tumour may arise from the submucosal esophageal glands and differentiate to form both glandular and squamous components thus resembling the muco-epidermoid tumours of salivary glands. The specimen depicted in figure 188 was from a 61-year-old male who presented with a history of dysphagia of 12 months duration, odynophagia and considerable loss of weight. He had a long history of excessive alcohol ingestion. Endoscopic examination showed a stricture at 30 cm and biopsies from this site indicated a carcinoma.

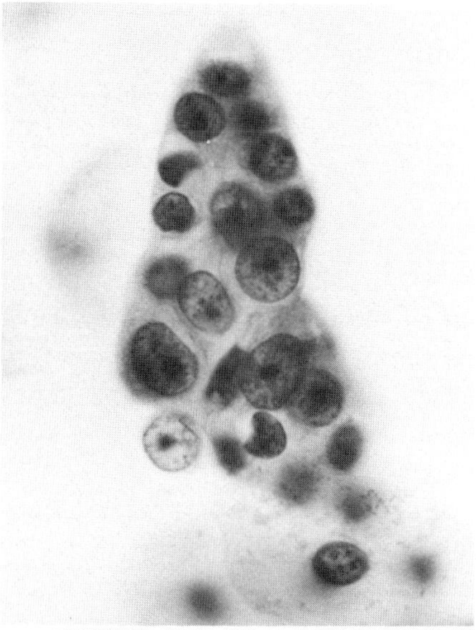

177a

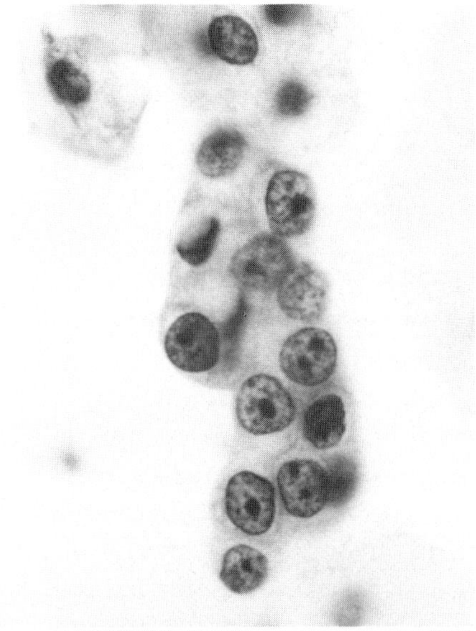

177b

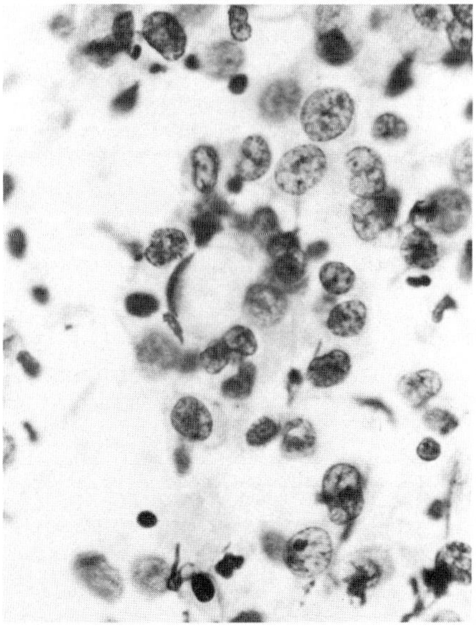

178

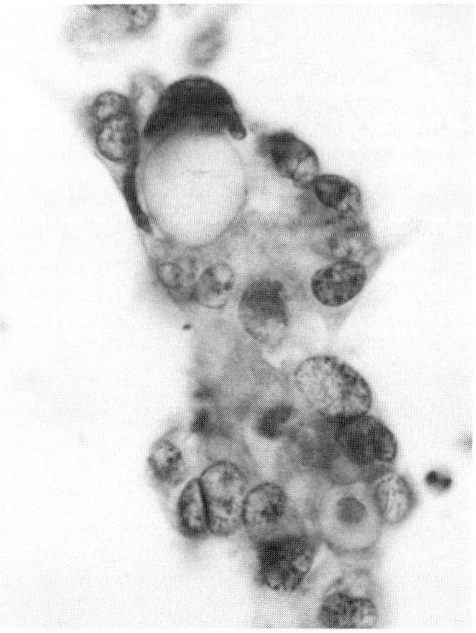

179

Fig. 177. a, b Papillary fragments in esophageal washing from patient with well differentiated papillary adenocarcinoma. × 545.

Fig. 178. Group of cells from tubular adenocarcinoma of esophagus showing acinar structure. × 545.

Fig. 179. Group of carcinoma cells from tubular adenocarcinoma of esophagus showing prominent acinar structure. × 545.

The operative specimen consisted of a segment of esophagus measuring 10 cm in length and including the cardio-esophageal junction. Situated 1.5 cm from the proximal line of resection was a polypoid mass which measured 2.0 × 1.0 × 1.0 cm in maximum dimensions. The surface of the mass had a rather shaggy appearance and the surrounding mucosa showed irregular fissures (fig. 189). A transverse section through the esophagus at the site of the mass confirmed its polypoid nature (fig. 190). The squamous epithelium adjacent to the mass showed extensive carcinoma in situ, of squamous type (fig. 191) with areas of superficial invasion. Histological examination of the mass itself revealed a mixture of both squamous (fig. 192) and adenocarcinoma (fig. 193). Two para-esophageal lymph nodes were infiltrated by carcinoma the metastatic deposits showing also a mixture of squamous and glandular elements.

The brushing specimen in a case of adenosquamous carcinoma contains, predictably, a variety of cell patterns. Squamous carcinoma cells, distinguished by nuclear hyperchromasia and cytoplasmic keratinization, may be seen alone (fig. 194) as may adenocarcinoma cells, cytoplasmic vacuolation being prominent (fig. 195). Not uncommonly the two cell types are present in the one microscopic field a close admixture being evident (fig. 196). Undifferentiated carcinoma cells are usually seen also (fig. 197).

Cases such as these, which appear to represent true composite carcinomas, should be separated from a predominantly glandular carcinoma in which foci of squamous metaplasia have occurred (fig. 198). Such a neoplasm is probably best referred to as an adenoacanthoma. In addition, it is important for the diagnostic cytologist to appreciate that an adenocarcinoma in the gastric cardia may be associated with, and indeed presumably stimulate, an atypical reaction in the contiguous esophageal squamous epithelium particularly if there is carcinomatous infiltration beneath that epithelium (fig. 199). In such cases interpretation of the obviously malignant glandular cells may be rendered more difficult by the presence also of atypical squamous cells. However, in the former example the abnormal squamous cells are few in number whilst in the latter the abnormal cells tend to be dyskaryotic rather than truly malignant.

Fig. 180. Moderately well differentiated adenocarcinoma of esophagus with prominent cytoplasmic vacuolation. × 545.

Fig. 181. Cells from moderately well differentiated adenocarcinoma of esophagus showing cytoplasmic vacuolation. Note prominent nuclear malignant criteria. × 545.

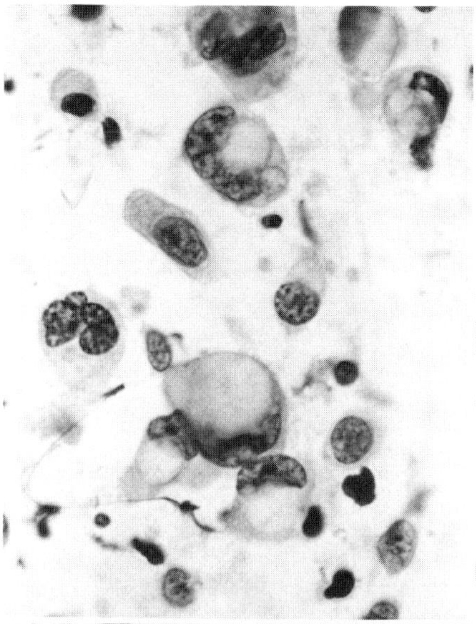

180

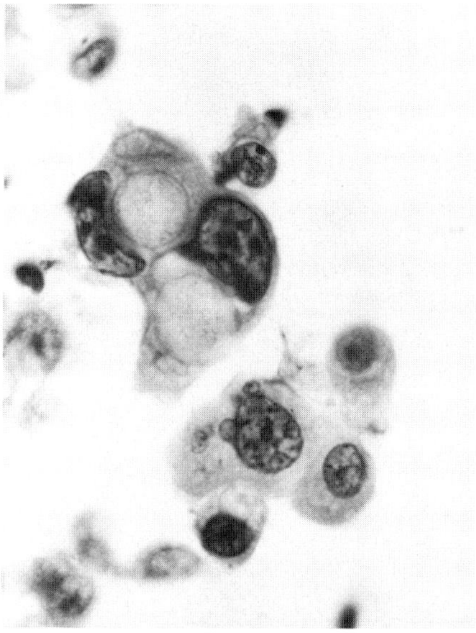

181

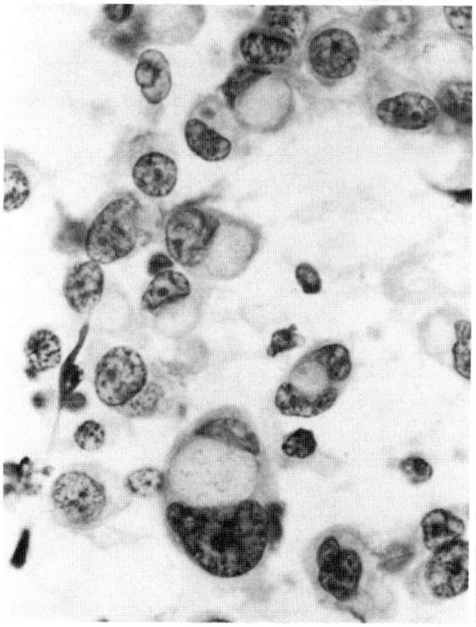

182

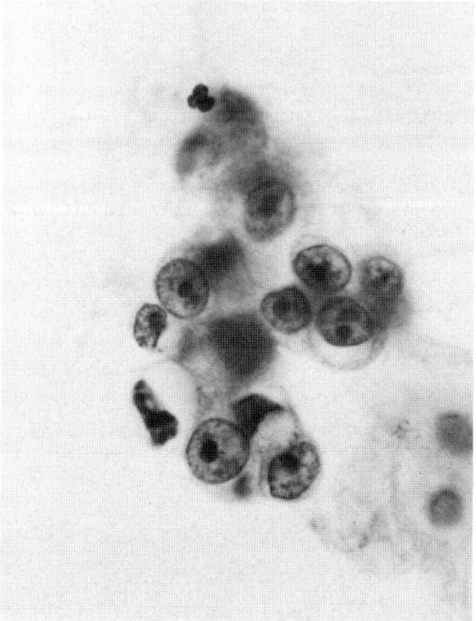

183

Fig. 182. Cells from moderately well to poorly differentiated esophageal adenocarcinoma. Cytoplasmic vacuoles are clearly evident as are nuclear malignant criteria. × 545.

Fig. 183. Poorly differentiated adenocarcinoma of esophagus. Cytoplasmic vacuolation is discernible although not a prominent feature. × 545.

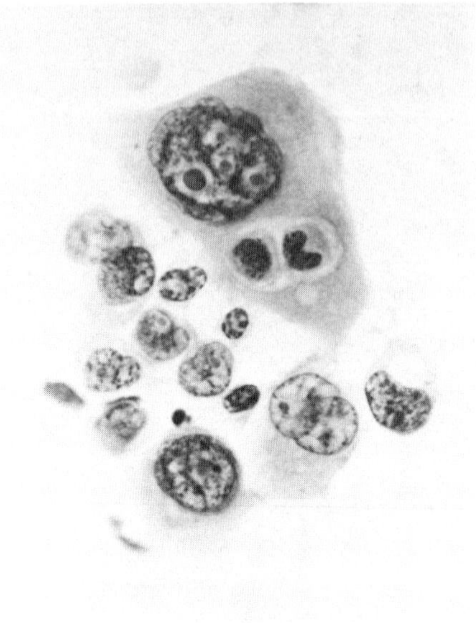

184

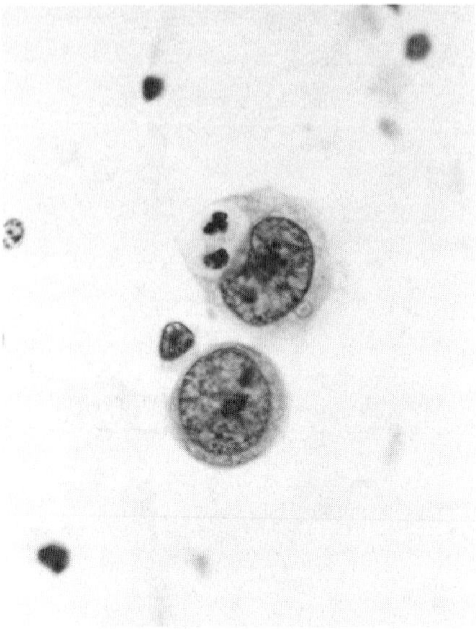

185

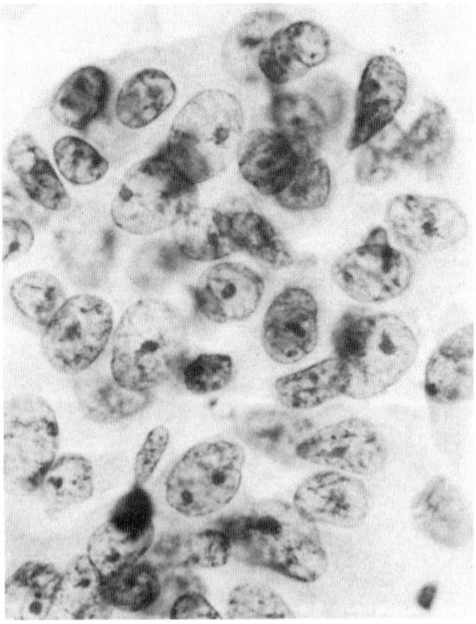

186

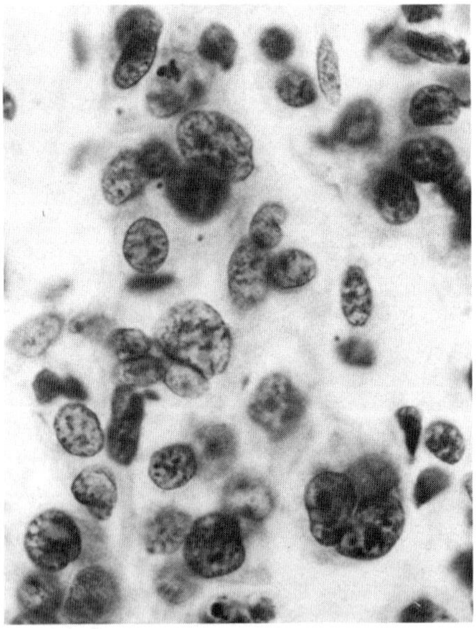

187

Fig. 184. Poorly differentiated adenocarcinoma of esophagus. Cytoplasmic vacuolation is not prominent but cells in upper picture have abundant, finely vacuolated or 'foamy' cytoplasm. × 545.

Fig. 185. Cells from poorly differentiated esophageal adenocarcinoma. Note cytoplasmic vacuole in upper cell. Polymorphs are present within the vacuole. × 545.

Fig. 186. Very poorly differentiated esophageal adenocarcinoma. There is no evidence of cytoplasmic specialization although the smooth curved border in the upper left hand corner of the photograph suggests glandular differentiation. × 545.

188

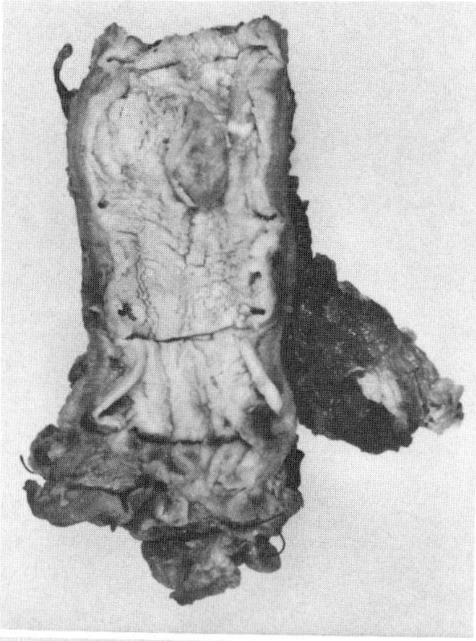

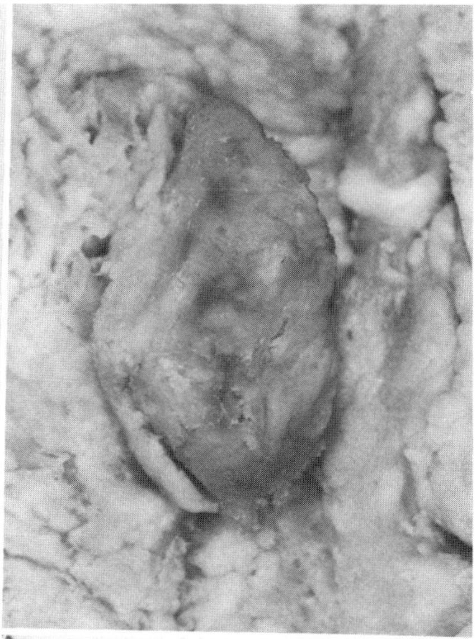

189

190

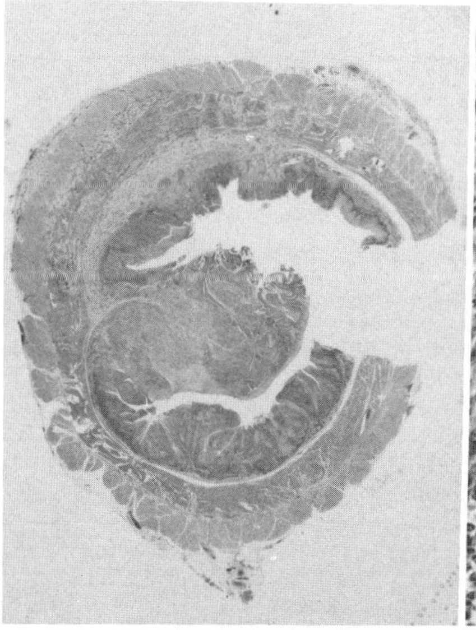

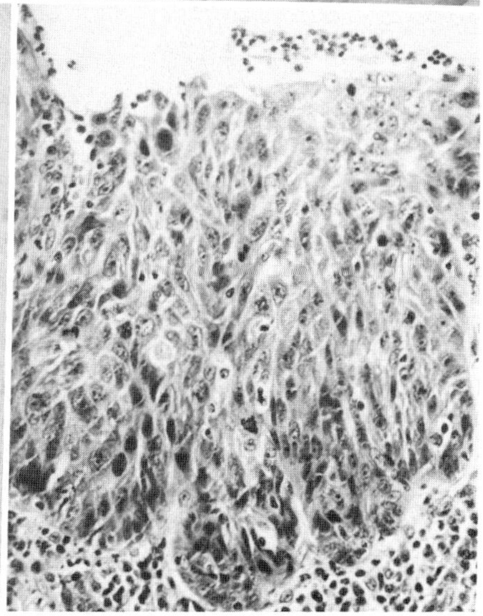

191

Fig. 187. Very poorly differentiated adenocarcinoma or undifferentiated carcinoma of esophagus. × 545.

Fig. 188. Operative specimen from patient with adenosquamous carcinoma of esophagus.

Fig. 189. Close-up view of esophageal adenosquamous carcinoma.

Fig. 190. Section through esophagus at site of origin of adenosquamous carcinoma showing prominent polypoid structure. × 2.

Fig. 191. Extensive carcinoma in situ of squamous type in esophageal mucosa adjacent to main tumour mass. × 173.

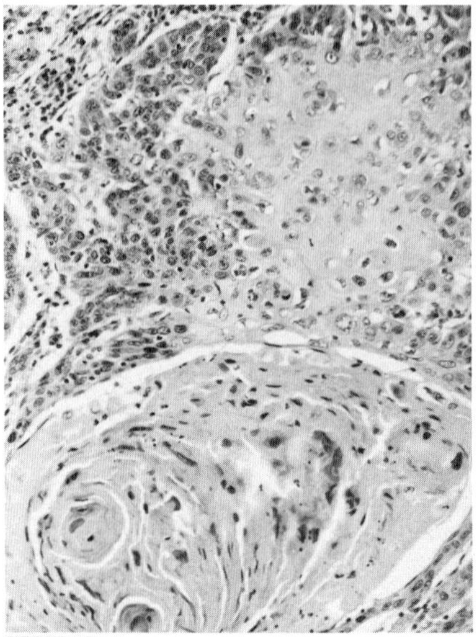

192

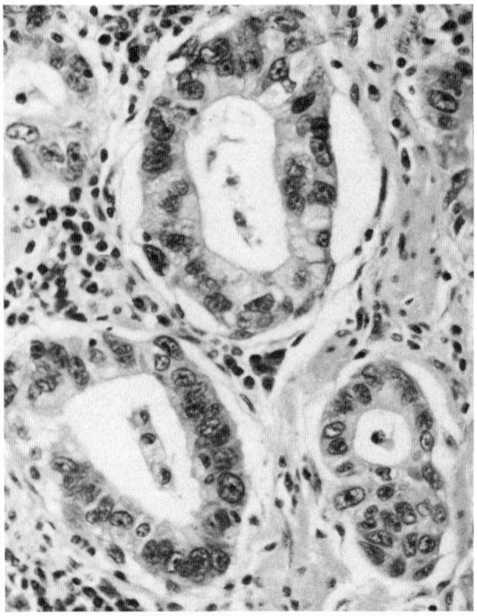

193

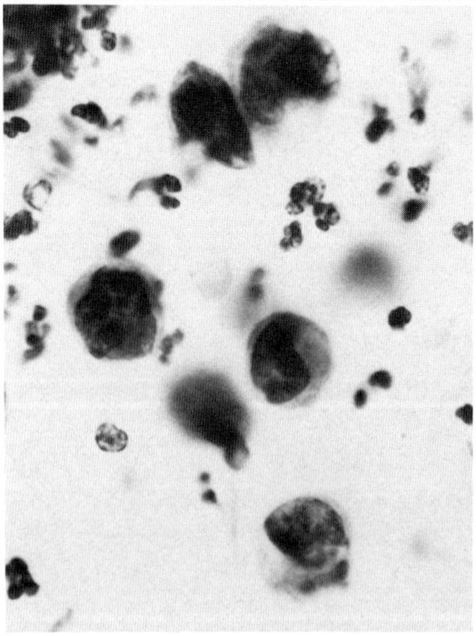

194

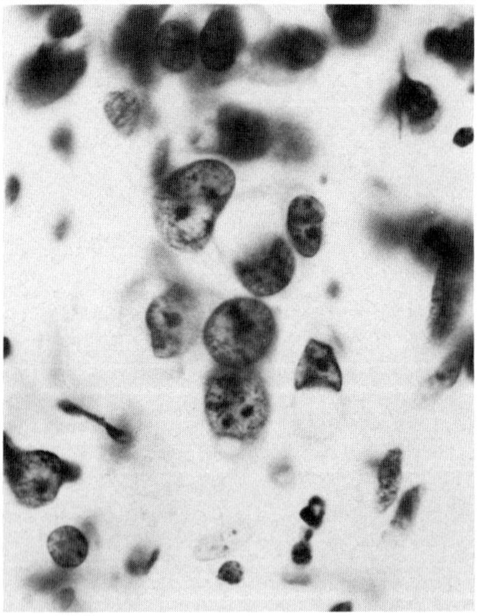

195

Fig. 192. Well-differentiated squamous cell carcinoma component of adenosquamous carcinoma. × 110.

Fig. 193. Well-differentiated glandular component of adenosquamous carcinoma of esophagus. × 205.

Fig. 194. Squamous carcinoma cells in esophageal brushing from patient with adenosquamous carcinoma of esophagus. × 545.

Fig. 195. Carcinoma cells in esophageal brushings from patient with adenosquamous carcinoma of esophagus. Cytoplasmic vacuolation is evident indicating glandular component of tumour. × 545.

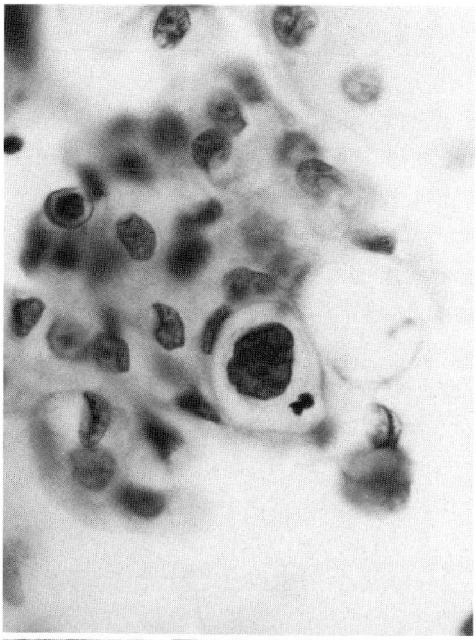

196

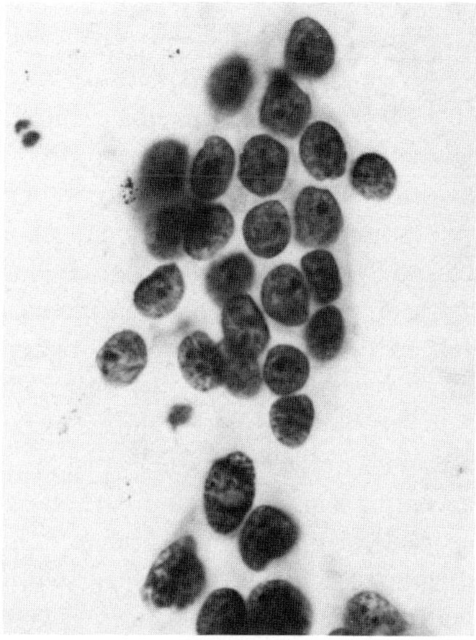

197

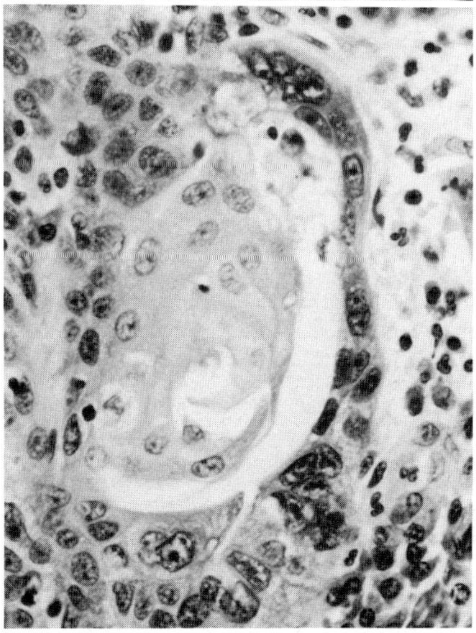

198

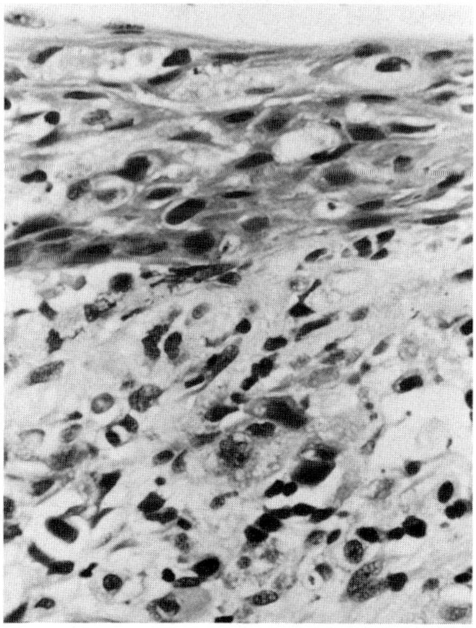

199

Fig. 196. Cells showing evidence of both glandular and squamous differentiation in esophageal brushings from patient with adenosquamous carcinoma of esophagus. × 545.

Fig. 197. Group of undifferentiated carcinoma cells from patient with adenosquamous carcinoma of esophagus. × 545.

Fig. 198. Small focus of squamous differentiation in adenocarcinoma of esophagus. × 345.

Fig. 199. Abnormal squamous cells in esophageal mucosa overlying poorly differentiated adenocarcinoma. × 275.

Small-Cell Undifferentiated Carcinoma

Most undifferentiated carcinomas are of the large-cell type and belong to the squamous cell series. Indeed many of these neoplasms, if sampled extensively, can be shown to be very poorly differentiated squamous cell carcinomas. The morphologic features of these have already been discussed. Very occasionally, however, a small-cell undifferentiated, or 'oat-cell' carcinoma may arise in the esophagus [37]. These resemble the 'oat-cell' carcinoma seen much more commonly in the bronchus and, like the bronchogenic type they are occasionally associated with hormone production. The possibility that the esophagus is involved by direct extension from a bronchogenic carcinoma must always be considered and excluded before accepting a small-cell undifferentiated or 'oat-cell' carcinoma as a primary esophageal neoplasm. However, that they do occur primarily and may be diagnosed cytologically is suggested by the following case of a male aged 77, who presented with a history of dysphagia present for 8 weeks. This was associated with epigastric pain and vomiting after ingestion of food. There had also been considerable weight loss. A barium meal showed a large mass at the lower end of the esophagus and in the gastric cardia.

Esophageal washings revealed numerous small cells with darkly stained nuclei and little or no cytoplasm (fig. 200). In places the cells were arranged in a linear fashion although, for the most part, they were seen in small groups. The nuclei were generally round or oval and nuclear moulding was prominent. The nuclei were generally small as can be seen by comparison with the squamous cell nucleus present in figure 201. Degenerative changes or pyknosis obscured the nuclear detail in most tumour cells but occasional groups appeared to be better preserved and in these nuclear malignant criteria were evident (fig. 202).

Endoscopic biopsy was performed and the histological examination showed strips of squamous epithelium with attached tumour tissue. The

---

Fig. 200. Large group of carcinoma cells with small, darkly staining nuclei and little or no discernible cytoplasm. Note the *Leptothrix* in association with the carcinoma cells. ×545.

Fig. 201. Group of cells from small-cell undifferentiated carcinoma of esophagus. Compare the tumour cell nuclei with the nucleus of the intermediate squamous cell visible centrally. ×545.

Fig. 202. Group of cells from small-cell undifferentiated carcinoma of esophagus showing better preservation of nuclear malignant criteria. Note also prominent nuclear moulding. ×545.

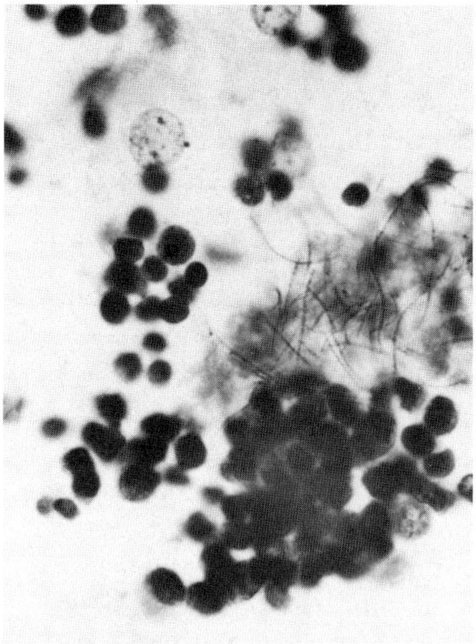

200

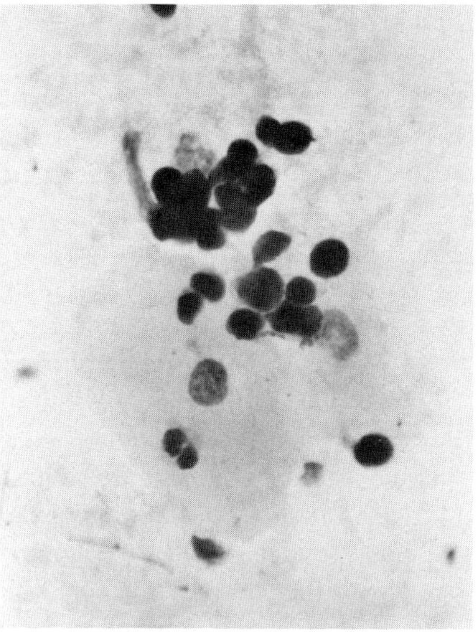

201

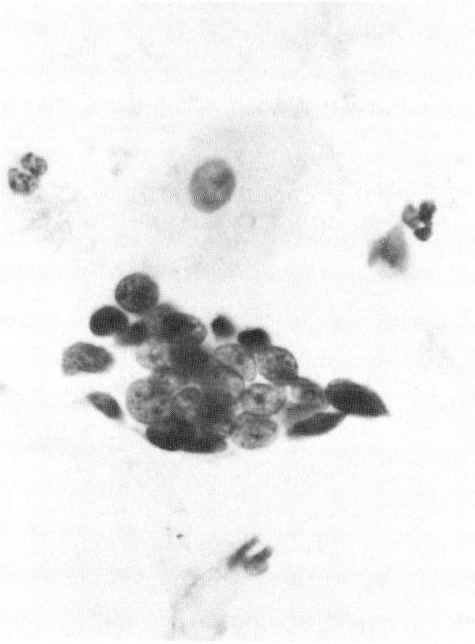

202

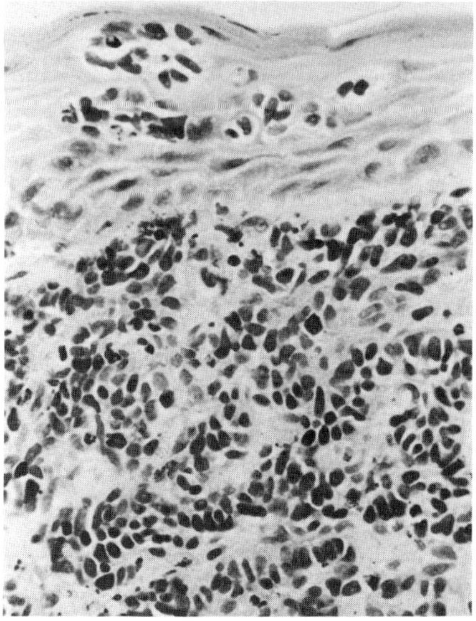

203

Fig. 203. Section of operative specimen from patient from whom cells depicted in previous three figures were derived. Note the infiltration of the esophageal mucosa and submucosa by cells characteristic of a small-cell undifferentiated or 'oat-cell' carcinoma. ×275.

tumour cells were small, had ovoid pyknotic nuclei, and were arranged in clumps or strands. Infiltration of the surface squamous epithelium was evident (fig. 203). The histological appearances were considered to be those of a small-cell undifferentiated or 'oat-cell' carcinoma. Extensive investigation at the time showed no evidence of a primary carcinoma elsewhere.

Carcinosarcoma and Pseudosarcoma

A true carcinosarcoma is a malignant tumour in which both epithelial and connective tissue elements are mixed. This type of tumour is rare but several examples of esophageal carcinosarcomas have been described [15, 27, 28, 32]. It is extremely difficult to differentiate this type of tumour from the so-called 'pseudosarcoma' – a lesion where the surface squamous epithelium has the appearances of an in situ carcinoma but the bulk of the tumour is made up of sarcoma or sarcoma-like tissue. Indeed *Matsusaka* et al. [36] suggest that the neoplasms are essentially the same pathologic entity because of their morphologic and biologic similarities.

The term pseudosarcoma was suggested by *Stout and Lattes* [61]. In 1957 they described two pseudosarcomas of the esophagus as independent pathologic entities thus distinguishing them from esophageal carcinomas. Since then a number of cases of pseudosarcoma of the esophagus have been described [16, 20, 56, 62, 63].

The histogenesis of these tumours is somewhat controversial and several possibilities have been suggested. Thus they may be true double tumours or they may represent a carcinoma which has become complicated by the development of a sarcoma either independently or as a result of an exaggerated stromal reaction to the carcinoma. A number of authors conclude that most such neoplasms are squamous cell carcinomas the bulk of the tumour having an unusual 'sarcoma-like' appearance or demonstrating a 'peculiar metaplasia'. Thus the spindle cells are regarded as morphologic variants of epithelial cells. However, it seems probable that some cases, at least, are truly double neoplasms being composed of both malignant epithelial and malignant mesenchymal elements.

The features of this unusual neoplasm are illustrated by a female patient aged 56 who presented with a history of dysphagia for 2 weeks. A bari-

---

Fig. 204. Single carcinoma cell from esophageal washing showing squamous differentiation. Note large hyperchromatic nucleus and abundant keratinized cytoplasm. × 545.

Fig. 205. Group of tumour cells with elongated or fusiform nuclei. × 345.

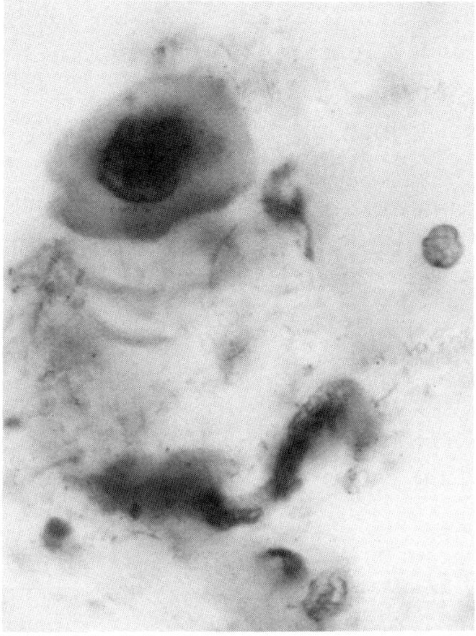
204

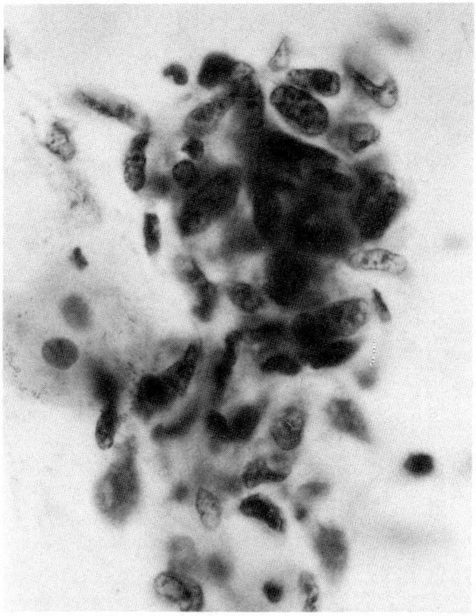
205

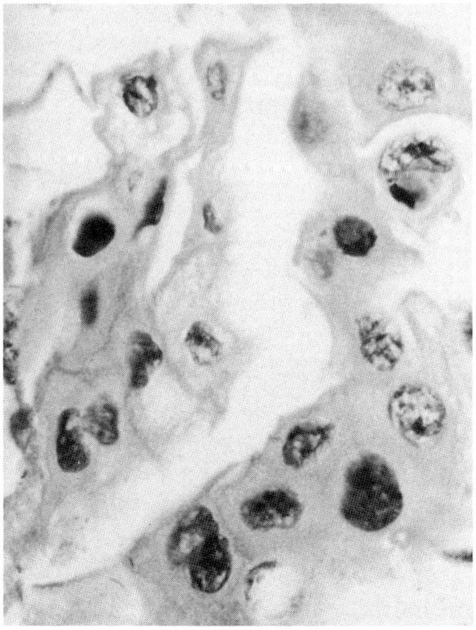
206

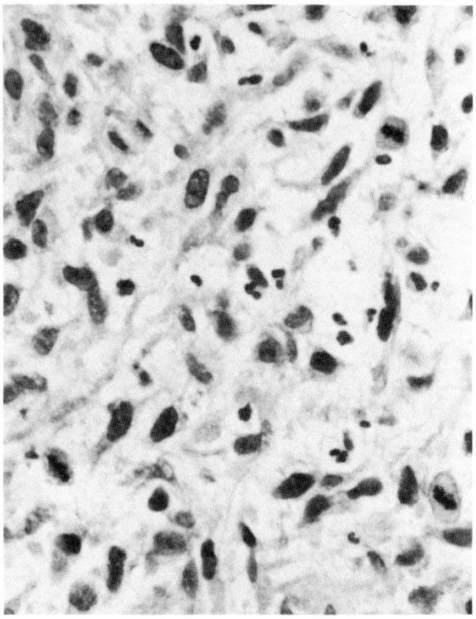
207

Fig. 206. Small island of well differentiated squamous cell carcinoma tissue. × 545.

Fig. 207. Sarcomatous tissue from same biopsy as that depicted in previous figure. Note similarity of cells to those present in the esophageal washings and depicted in figure 205. × 345.

um meal, prior to admission to hospital, showed a large irregular filling defect in the lower third of the esophagus. There was some dilatation of the esophagus but no evidence of stricture formation. The findings were considered to be consistent with the presence of a benign tumour projecting into the esophageal lumen with minimal involvement of the esophageal wall.

Esophageal washings revealed two distinct cell types although all cells were rather degenerate. There were a number of single cells present with large, irregular, hyperchromatic nuclei and relatively abundant keratinized cytoplasm (fig. 204). These were clearly squamous carcinoma cells. More numerous, and more difficult to interpret, were the cells with ovoid or, more commonly, elongated or fusiform nuclei, scanty cytoplasm, and poorly defined cell borders (fig. 205). Malignant criteria were evident in many of these nuclei and cytologically the cells were considered to be of sarcomatous origin.

Subsequent to this esophageal lavage an endoscopy was performed and this revealed a friable polypoid lesion at the 32-cm mark. Biopsies were taken from this lesion and the histological findings are depicted in figures 206 and 207. There was considerable necrosis present but scattered amongst the necrotic material were occasional small islands of recognizable well differentiated squamous carcinoma tissue (fig. 206). Larger masses of sarcomatous tissue were also recognizable (fig. 207).

Malignant Melanoma

Malignant melanoma of the esophagus is usually secondary from elsewhere. However, examples of melanotic tumours apparently confined to the esophagus and arising therein have been reported. Thus in a recent report of such a case the authors [63] state that 74 cases of primary esophageal melanoma are recorded in the world literature. The demonstration of junctional change within the esophageal mucosa is usually demanded as evidence that a melanoma is arising within that organ. Those that have been described have occurred in older persons of either sex. The neoplasm presents as a polypoid tumour in the lower esophagus and the histologic and cytologic features can again be best demonstrated by reference to a specific

---

Fig. 208. Operative specimen from patient with malignant melanoma of esophagus. Note large fleshy, darkly coloured, tumour encircling the lower esophagus.

Fig. 209. Section of tumour mass showing sheets of cells with large vesicular nuclei and prominent nucleoli. × 345.

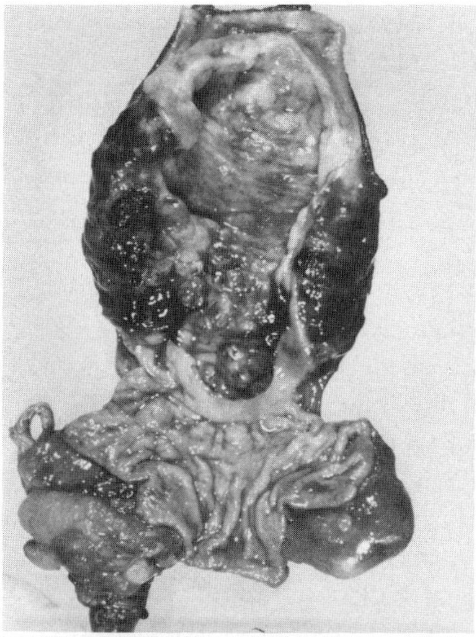

208

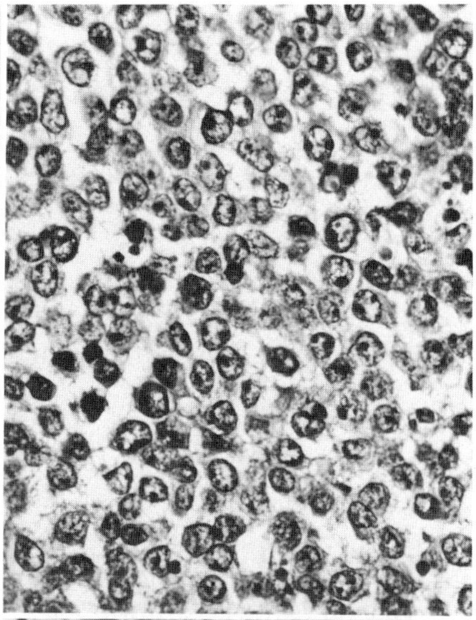

209

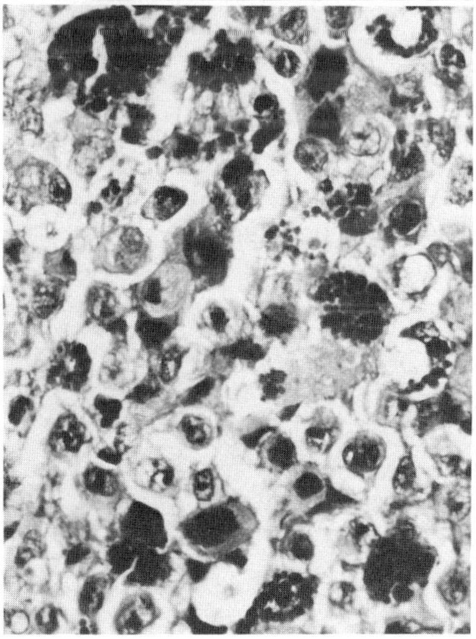

210

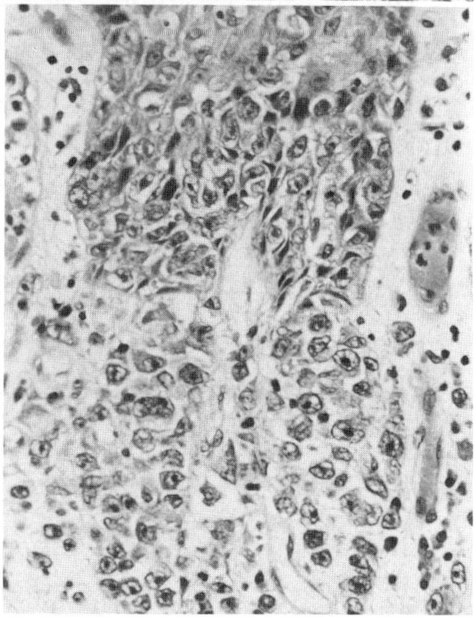

211

Fig. 210. Similar cells to those depicted in previous figure but abundant intra-cytoplasmic melanin present. × 345.

Fig. 211. Section of esophageal squamous mucosa showing prominent junctional activity indicating primary origin of tumour. × 220.

case study. A 47-year-old male with a long history of excessive alcohol ingestion was admitted to hospital following a small haematemesis. A barium swallow showed an irregular malignant stricture in the lower esophagus. This extended over approximately 4.0 cm and was situated immediately above a small hiatus hernia. Endoscopy confirmed the presence of an esophageal malignancy but a biopsy could not be carried out because of large esophageal varices.

Esophageal washings revealed many cells of the type depicted in figures 214–219. At the time the cytological diagnosis was that of poorly differentiated adenocarcinoma although, in retrospect, the correct diagnosis could have been made on the material available for examination. Subsequently the lower two thirds of the esophagus and the upper portion of the stomach was resected and the operative specimen is shown in figure 208. The distal portion of the esophagus was completely encircled by a large fleshy tumour which extended over a longitudinal distance of 11.0 cm. The tumour infiltrated the full thickness of the esophageal wall.

Histological sections of the tumour mass showed sheets of cells with large vesicular nuclei and prominent nucleoli (fig. 209). Much of the tumour was amelanotic but, in some areas, abundant melanin was present (fig. 210). A striking feature was the presence of prominent junctional activity in the esophageal squamous epithelium associated with the neoplasm (fig. 211). As already indicated, this feature is very strong evidence indeed that the melanoma originated in the esophagus as a primary lesion.

The high power views of the tumour tissue, as seen in figures 212 and 213, show a striking resemblance to the cells obtained by pre-operative esophageal lavage and illustrated in figures 214–219. The most prominent feature in all cells is the presence of a very large nucleolus. In many of the cells little or no cytoplasm was visible. However, in some of the cells with more abundant cytoplasm a slight 'dusky' appearance was evident. This feature is suggestive of melanin pigmentation although confirmation by special staining techniques is desirable.

The cytological features of malignant melanoma are probably best studied in a direct touch preparation or body fluid when the serosal surfaces

---

Fig. 212. Section of melanoma showing binucleate cell and vesicular nuclei with prominent nucleoli. × 345.

Fig. 213. Section of malignant melanoma showing characteristic vesicular nuclei and prominent nucleoli. × 545.

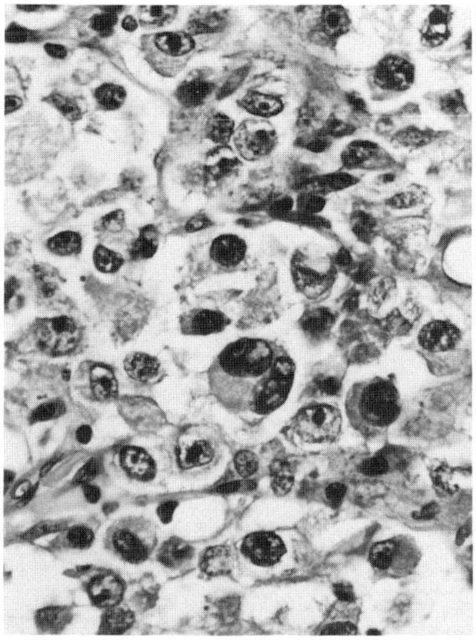

212

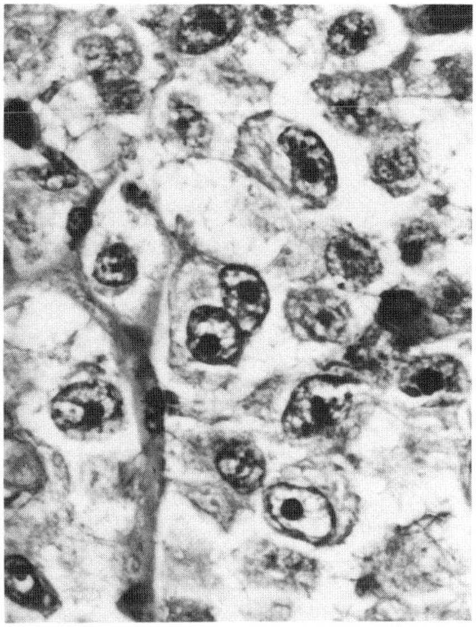

213

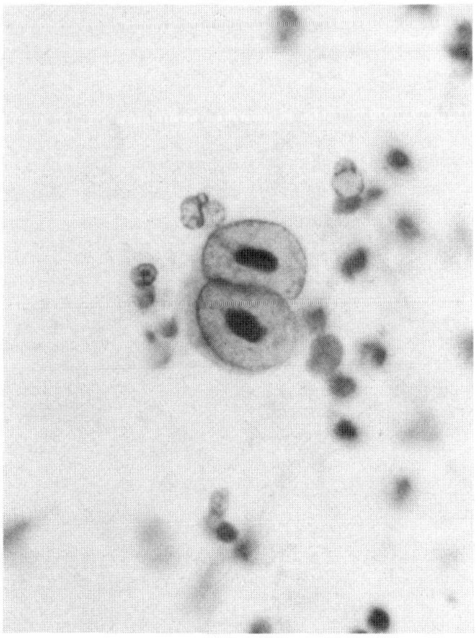

214

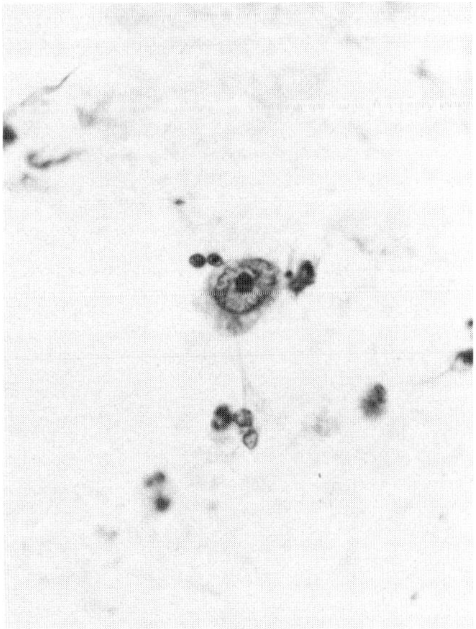

215

Fig. 214. Binucleate cell from esophageal washing with very prominent nucleoli. Cytoplasm, although scanty, has slight 'dusky' appearance. × 545.

Fig. 215. Single cell from esophageal washing. Note vesicular nucleus, prominent nucleolus and finely vacuolated, slightly 'dusky' cytoplasm. × 545.

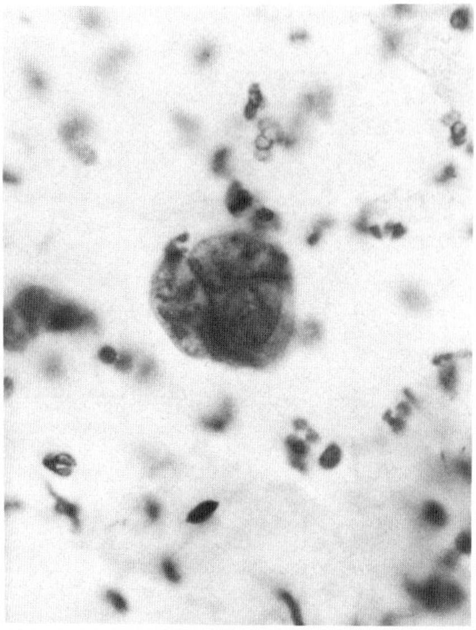

216

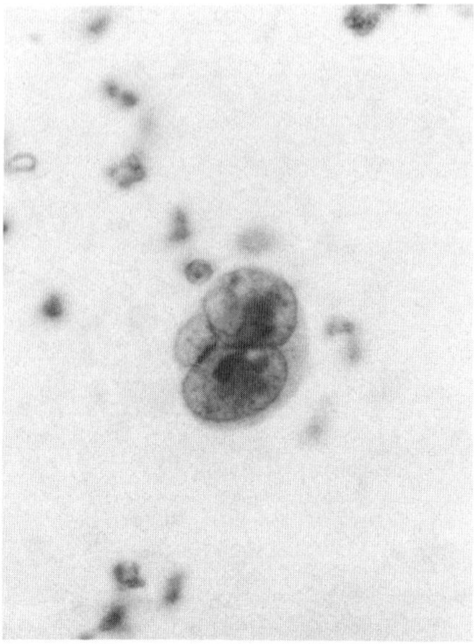

217

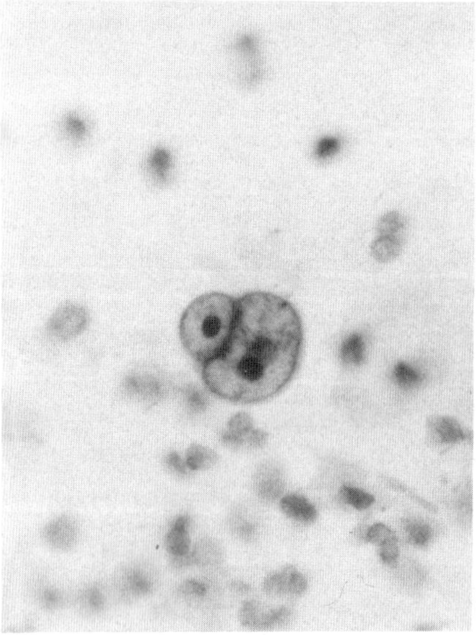

218

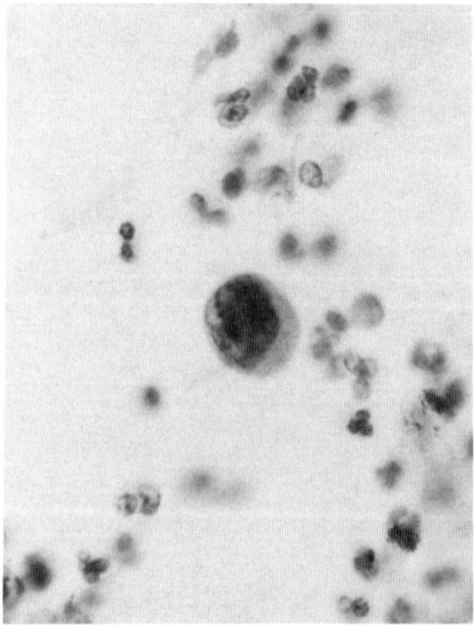

219

Fig. 216. Binucleate cell from esophageal washings. Note clearly defined pigment granules within cytoplasm. × 545.

Fig. 217. Binucleate cell from esophageal washings. Note prominent intranuclear vacuole in upper nucleus. × 545.

Fig. 218. Binucleate cell in esophageal washings. Note vesicular nuclei, prominent nucleoli, and nuclear moulding. × 545.

Fig. 219. Single malignant melanoma cell in esophageal washings. × 545.

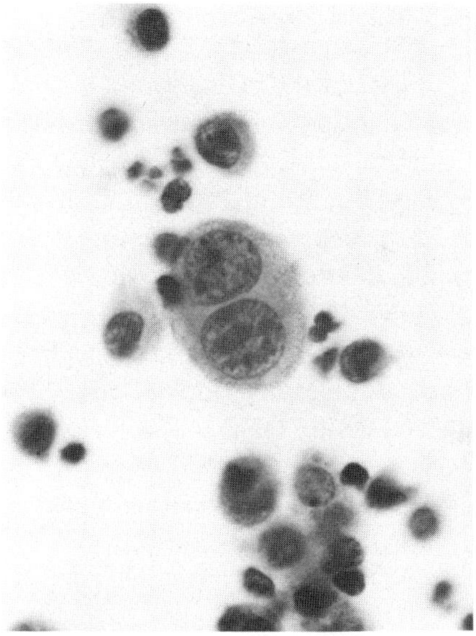

220

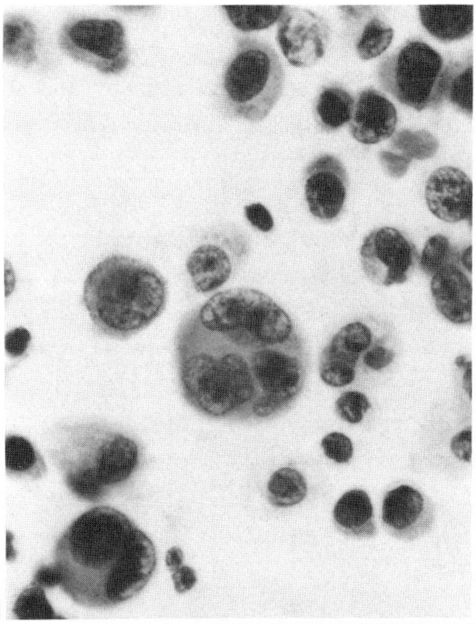

221

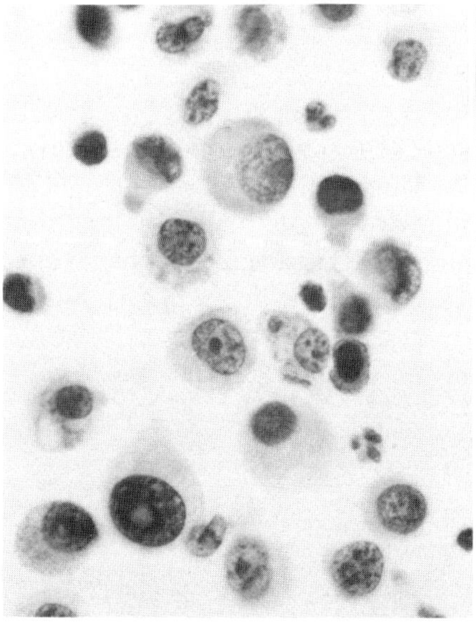

222

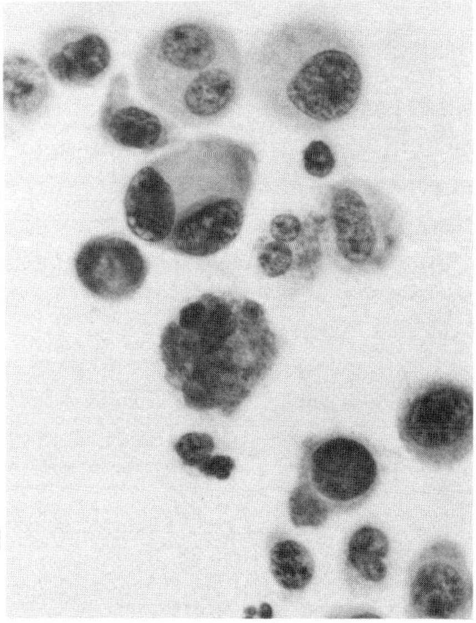

223

Fig. 220. Binucleate malignant melanoma cell from pleural effusion – metastatic from primary melanoma of skin. × 545.

Fig. 221. Bi- and multi-nucleate melanoma cells from pleural effusion. × 545.

Fig. 222. Malignant melanoma cells from pleural effusion. Note vesicular nuclei, prominent nucleoli, and, in lower part of photograph, clearly defined intra-nuclear vacuole. × 545.

Fig. 223. Malignant melanoma cells from pleural effusion. Note melanin pigment granules in cytoplasm of cell near centre of photograph. × 545.

are involved by metastatic disease. The nuclei of the cells usually display excellent malignant criteria. The cells vary considerably in size and shape and bizarre forms are often seen. They usually occur singly but may be seen in small clusters. They are often mononuclear but bi-nucleate and multinucleate forms are quite common (fig. 220, 221). A characteristic feature is the presence of nuclear vacuoles (fig. 222) which are due to invagination of the cytoplasm into the nucleus. Nucleoli are often prominent whilst the presence of melanin pigment in the cytoplasm, although not invariable, is diagnostic. This may be seen as clearly defined granules or 'droplets' (fig. 223) but sometimes the pigment is finely dispersed within the cytoplasm giving a characteristic 'dusky' appearance.

Although the cells in our case of esophageal melanoma were not always well preserved they displayed many of these diagnostic features.

Secondary Neoplasms

True metastatic tumours of the esophagus are rare. However, because of its intimate relationship to other organs, the esophagus may be involved in direct spread from an adjacent organ. Thus, as already discussed, a carcinoma of the gastric cardia may spread proximally to present in the lower esophagus whilst a bronchogenic carcinoma may invade the esophagus and, on rare occasions, diagnostic cells may be seen in the esophageal specimen.

References

1 Ashley, D.J.B.: Evans' histological appearances of tumours; 3rd ed. (Churchill Livingstone, Edinburgh 1978).

2 Barrett, N.R.: The lower esophagus lined by columnar epithelium. Surgery *41:* 881–894 (1957).

3 Barrett, N.R.: Achalasia of the cardia. Reflections upon a clinical study of over 100 cases. Br. med. J. *i:* 1135 (1964).

4 Behar, J.; Sheahan, D.G.: Histological abnormalities in reflux esophagitis. Archs Path. *99:* 387–391 (1975).

5 Bender, M.D.; Allison, J.; Cuartas, F.; Montgomery, C.: Glycogenic acanthosis of the esophagus: a form of benign epithelial hyperplasia. Gastroenterology *65:* 373 (1973).

6 Berenson, M.; Riddell, R.H.; Skinner, D.B.; Freston, J.W.: Malignant transformation of esophageal columnar epithelium. Cancer *41:* 554–562 (1978).

7 Berg, J.W.: Esophageal herpes: a complication of cancer therapy. Cancer *8:* 731–740 (1955).
8 Bosch, A.; Frias, Z.; Caldwell, W.L.: Adenocarcinoma of the esophagus. Cancer *43:* 1557–1561 (1979).
9 Brown, J.W.; McKee, W.M.: Acute monilial oesophagitis occurring without underlying disease in a young male. Am. J. dig. Dis. *17:* 85 (1972).
10 Buss, D.H.; Scharyj, M.: Herpes virus infection of the esophagus and other visceral organs in adults: incidence and clinical significance. Am. J. Med. *66:* 457–462 (1979).
11 Cohen, S.; Soloway, R.D.: Diseases of the esophagus. Contemporary issues in gastroenterology (Churchill Livingstone, New York 1982).
12 Deprew, W.T.; Prentice, R.S.A.; Beck, I.T.; et al.: Herpes simplex ulcerative esophagitis in a healthy subject. Am. J. Gastroent. *68:* 381–385 (1977).
13 Dodge, O.G.: Intraesophageal adenocarcinoma. Gut *1:* 351–356 (1960).
14 Dow, C.J.: Oesophageal tuberculosis: four cases. Gut *22:* 234–236 (1981).
15 Elton, S.E.; Joannides, M., Jr.: Carcinosarcoma of the esophagus. A review and case report. Dis. Chest *41:* 111 (1962).
16 Enrile, F.T.; Dejusus, P.O.; Baskt, A.A.; Baluyot, R.: Pseudosarcoma of the esophagus. Cancer *31:* 1197–1202 (1973).
17 Eras, P.; Goldstein, M.J.; Sherlortz, P.: Candida infection of the gastro-intestinal tract. Medicine *51:* 367–379 (1972).
18 Fishbein, P.G.; Tuthill, R.; Kressel, H.; Friedman, H.; Snape, W.J.: Herpes simplex esophagitis. Dig. Dis. Sci. *24:* 540 (1979).
19 Foroozan, P.; Enta, T.; Winship, D.; Trier, J.S.: Loss and regeneration of the esophageal mucosa in pemphigoid. Gastroenterology *52.* 548 (1967).
20 Fraser, C.M.; Kinley, C.E.: Pseudosarcoma with carcinoma of the esophagus. Archs Path. *85:* 325–330 (1968).
21 Haggitt, R.C.; Tryzelaar, J.; Ellis, F.H.; Colcher, H.: Adenocarcinoma complicating columnar epithelium-lined (Barrett's) esophagus. Am. J. clin. Path. *70:* 1–5 (1978).
22 Herschman, B.R.; Uppaputhangkule, V.; Maas, L.; Gelzayd, E.: Esophageal leukoplakia: a rare entity. J. Am. med. Ass. *239:* 2021 (1978).
23 Holt, J.M.: Candida infection of the esophagus. Gut *9:* 227–231 (1968).
24 Ismail-Beigi, F.; Horton, P.F.; Pope, C.E., II: Histological consequences of gastroesophageal reflux in man. Gastroenterology *58:* 163–174 (1970).
25 Ito, Y.; Kobayashi, S.; Kasugai, T.: Tuberculosis of the esophagus. Am. J. Gastroent. *65:* 454 (1976).
26 Jones, J.M.: Necrotizing Candida esophagitis. J. Am. med. Ass. *244:* 2190–2191 (1980).
27 Kenneweg, D.J.; Cimmino, C.V.: Carcinosarcoma of the esophagus. Am. J. Roentg. *101:* 482–484 (1967).
28 King, H.A.; Koerner, J.A.: Carcinosarcoma of the esophagus. Surgery *42:* 389–393 (1957).
29 Kodsi, B.E.; Wickremesinghe, P.C.; Kozinn, P.J.: Candida esophagitis. Gastroenterology *71:* 715–719 (1976).
30 Lasser, A.: Herpes simplex virus esophagitis. Acta cytol. *21:* 301–302 (1977).
31 Lightdale, C.J.; Wolf, D.J.; Marcucci, R.A.; Slayer, W.R.: Herpetic esophagitis in

patients with cancer: ante mortem diagnosis by brush cytology. Cancer *39:* 223–226 (1977).

32 Lin, M.H.; Luna-Munoz, M.I.; Kraft, J.R.; Marks, L.M.: Carcinosarcoma of the esophagus. Am. J. Gastroent. *55:* 249–256 (1972).

33 Livolsi, V.A.; Jaretski, A.: Granulomatous esophagitis. Gastroenterology *64:* 313–319 (1973).

34 Livstone, E.; Sheahan, D.G.; Behar, J.: Studies of esophageal epithelial proliferation in patients with reflux esophagitis. Gastroenterology *73:* 1315–1319 (1977).

35 Marsden, R.A.; Sambrook Gower, F.J.; MacDonald, A.F.; Main, R.A.: Epidermolysis bullosa of the oesophagus with oesophageal web formation. Thorax *29:* 287 (1974).

36 Matsusaka, T.; Watanabe, M.; Enjoji, M.: Pseudosarcoma and carcinosarcoma of the oesophagus. Cancer *37:* 1546 (1976).

37 Matsusaka, T.; Watanabe, M.; Enjoji, M.: Anaplastic carcinoma of the esophagus. Report of three cases and their histogenetic consideration. Cancer *37:* 1352–1358 (1976).

38 Medak, H.; Burkalow, P.; McGrew, E.A.; Tiecke, R.: The cytology of vesicular conditions affecting the oral mucosa: pemphigus vulgaris. Acta cytol. *14:* 11–21 (1970).

39 Ming, Si-Chun: Tumors of the esophagus and stomach. Atlas of tumor pathology, 2nd series, fascicle 7 (Armed Forces Institute of Pathology, Bethesda 1971).

40 Mirra, S.S.; Bryan, J.A.; Butz, W.C.; Miles, M.L.: Concomitant herpes – monilial esophagitis: case report with ultrastructural study. Human Pathol. *8:* 760–763 (1982).

41 Morson, B.C.; Dawson, I.M.P.: Gastrointestinal pathology; 2nd ed. (Blackwell, Oxford 1979).

42 O'Morchoe, P.J.; Lee, D.C.; Kozak, C.A.: Esophageal cytology in patients receiving cytotoxic drug therapy. Acta cytol. *27:* 630–634 (1983).

43 Oota, K.; Sobin, L.H.: Histological typing of gastric and oesophageal tumors. International histological classification of tumours, No. 18 (World Health Organization, Geneva 1977).

44 Owensby, L.C.; Stammer, J.L.: Esophagitis associated with herpes simplex infection in an immunocompetent host. Gastroenterology *74:* 1305 (1978).

45 Paull, A.; Trier, J.S.; Dalton, M.D.; et al.: The histologic spectrum of Barrett's esophagus. New Engl. J. Med. *295:* 476–480 (1976).

46 Pattison, J.N.; Osborn, G.; Morson, B.C.: Hiatus hernia with adenocarcinoma arising in the region of the cardia. J. Fac. Radiol. *7:* 90–101 (1955/1956).

47 Pazin, G.J.: Herpes simplex esophagitis after trigeminal nerve surgery. Gastroenterology *74:* 741 (1978).

48 Pearce, J.; Dagradi, A.: Acute ulceration of the esophagus with associated intranuclear inclusion bodies. Report of four cases. Archs Path. *35:* 889–897 (1943).

49 Poleynard, G.D.; Marty, A.T.; Birnbaum, W.B.; et al.: Adenocarcinoma in the columnar-lined (Barrett) esophagus. Archs Surg. *112:* 997–1000 (1977).

50 Prolla, J.C.; Kirsner, J.B.: The gastrointestinal lesions and complications of leukemias. Ann. intern. Med. *61:* 1084–1103 (1964).

51 Raque, C.J.; Stein, K.M.; Samitz, M.H.: Pemphigus vulgaris involving the esophagus. Archs Derm. *102:* 371–373 (1970).

52 Rosen, P.; Hajdu, S.I.: Visceral herpes virus infections in patients with cancer. Am. J. clin. Path. *56:* 459–465 (1971).
53 Ruzzuk, M.A.; Urshel, H.C.; Race, G.J.; Nathan, M.J.; Paulson, D.L.: Pseudosarcoma of the esophagus. J. thorac. cardiovasc. Surg. *61:* 650–653 (1971).
54 Shah, S.M.; Schaefer, R.J.; Araoz, E.: Cytologic diagnosis of herpetic esophagitis. A case report. Acta cytol. *21:* 109–111 (1977).
55 Sharp, G.S.: Leukoplakia of the esophagus. Am. J. Cancer *15:* 2029 (1931).
56 Shields, T.W.; Eilert, J.B.; Battifora, H.: Pseudosarcoma of the esophagus. Thorax *27:* 472–479 (1972).
57 Smith, J.L.: Pathology of adenocarcinoma of the esophagus and gastroesophageal region and Barrett's esophagus as a predisposing condition; in Stroehlein, Romsdahl, Gastro-intestinal cancer (M.D. Anderson Clinical Conferences on Cancer, Series No. 25), pp. 125–135 (Raven Press, New York 1981).
58 Smithers, D.W.: Adenocarcinoma of the esophagus. Thorax *11:* 257 (1956).
59 Springer, D.J.; Da Costa, L.R.; Beck, I.T.: A syndrome of acute self-limiting ulcerative esophagitis in young adults probably due to herpes simplex virus. Dig. Dis. Sci. *24:* 535 (1979).
60 Stone, J.; Friedberg, S.A.: Obstructive syphilitic esophagitis. J. Am. med. Ass. *177:* 711 (1961).
61 Stout, A.P.; Lattes, R.: Tumors of the esophagus. Atlas of tumor pathology, fasc. 20 (Armed Forces Institute of Pathology, Washington 1957).
62 Takubo, K.; Tsuchiya, S.; Nakagawa, H.; Futatsuki, K.; Ishibashi, I.; Hirata, F.: Pseudosarcoma of the esophagus. Human Pathol. *5:* 504–505 (1982).
63 Takubo, K.; Kando, Y.; Ishii, M.; Nonose, N.; Saida, T.; Fujita, K.; Nakagawa, H.; Fujiwara, M.: Primary malignant melanoma of the esophagus. Human Pathol. *14:* 727–730 (1983).
64 Wychulis, A.R.; Woolam, G.L.; Andersen, H.A.; Ellis, F.H.: Achalasia and carcinoma of the esophagus. J. Am. med. Ass. *215:* 1638–1641 (1971).

5. Gastric Cytology – Normal and Abnormal

Normal Cytology

The only type of mucosal cell seen commonly is the surface epithelial cell. These are rarely seen as single cells but almost invariably appear in small sheets. When viewed from above a sheet of cells has a characteristic honeycomb appearance (fig. 224). This honeycomb structure can be accentuated by adjusting the plane of focus of the microscope to highlight the cell borders (fig. 225). In profile the cells are columnar and sometimes have a tapering tail (fig. 226). Not uncommonly small clusters of cells are seen arranged in a circular fashion to produce a 'cartwheel' or rosette appearance. These cells may be joined by their tails or by their lumenal borders (fig. 227). The nuclei of the normal gastric mucosal cells are rounded or oval with uniform granular chromatin, a distinct nuclear membrane, and often a small nucleolus. The cytoplasm is relatively abundant, rather delicate in appearance, and granular or finely vacuolated.

Very commonly gastric mucosal cells may show degenerative changes and it is important to be familiar with the full range of such changes. The earliest change, and one that is seen frequently, is a shrinkage of the nucleus with irregularity of the nuclear outline and variable hyperchromasia (fig. 228). More advanced degeneration may lead to fragmentation of the cytoplasm, the 'bare' nuclei being enlarged and pale staining. Sometimes, prior to the complete disappearance of the cytoplasm, the nuclei may become eccentric within the cell and undergo karyorrhexis. With margination of the chromatin material these degenerate mucosal cells may be mistaken for plasma cells (fig. 229). On other occasions the enlarged pale nucleus, in association with abundant foamy cytoplasm, gives the appear-

Fig. 224. Sheet of normal gastric mucosal cells viewed 'from above'. Note characteristic 'honeycomb' appearance, round or oval nuclei, well defined nuclear membrane, and small nucleoli. × 370.

Fig. 225. Normal gastric mucosal cells viewed 'from above'. Plane of focus adjusted to highlight cell borders. × 545.

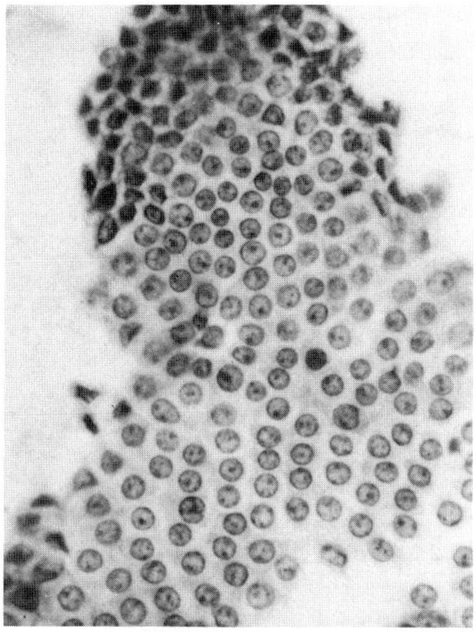

224

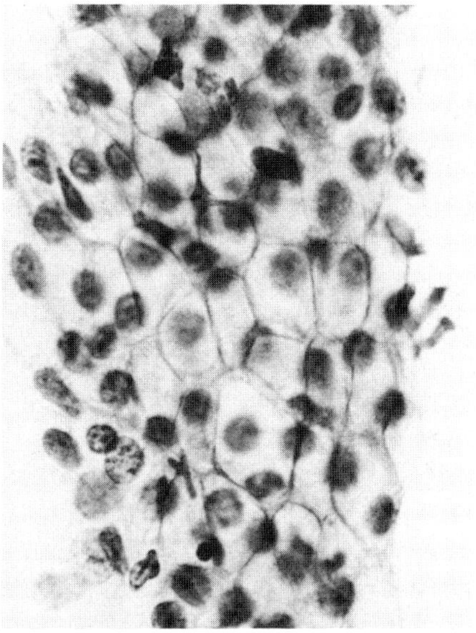

225

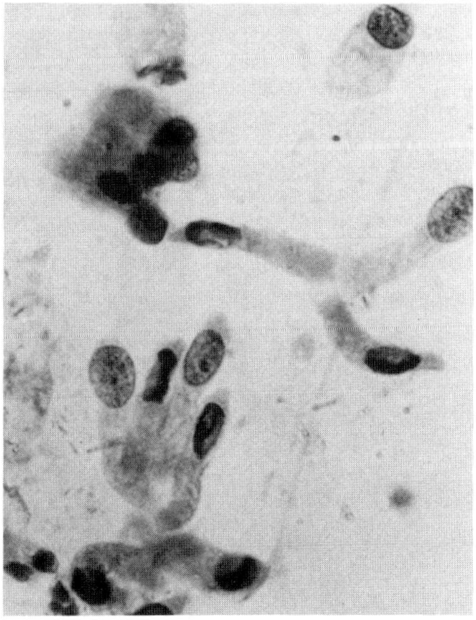

226

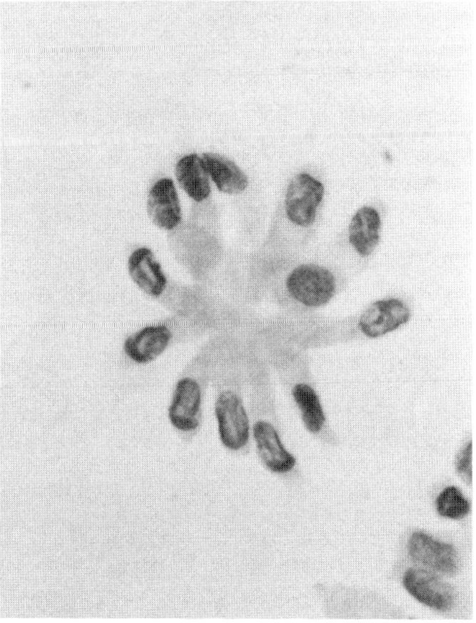

227

Fig. 226. Normal gastric mucosal cells in profile. Note columnar shape and tapering tail. × 550.

Fig. 227. Normal gastric mucosal cells joined by lumenal border to produce 'cart-wheel' or rosette formation. × 545.

ance of macrophages or histiocytes (fig. 230). Finally karyorrhexis associated with pyknosis of the nuclear fragments leads to an unusual appearance originally described by *Gibbs* [16] as 'mercury drop' nuclei (fig. 231). Subsequently these fragments are extruded leaving anucleate remnants or 'ghost' cells. These unusual degenerative changes, which also occur in columnar cells from other body sites, may lead to diagnostic difficulties.

The parietal and chief cells are occasionally seen in gastric cytology specimens, particularly those collected by the brushing technique. *Henning and Witte* [24] described chief cells in air-dried gastric smears as plump cells with numerous coarse basophilic granules whilst the same authors described the parietal cells as small cylindrical cells with marked vacuolation of the cytoplasm. Conversely *Nieburgs and Glass* [46] identified parietal cells in gastric brushings as large, pale, round or triangular cells. In our laboratory parietal cells have been demonstrated in a number of specimens. They are characterized by their round or oval nucleus, and their abundant granular or finely vacuolated cytoplasm (fig. 232). With the routine Papanicolaou stain this abundant cytoplasm stains brightly eosinophilic (plate I). The chief cells are less commonly seen and rarely if ever in isolation. Occasionally, however, they are seen in association with parietal cells, the chief cells being smaller with granular cytoplasm and indistinct cell borders (fig. 233). The nuclei are also round or ovoid. The chief cells may be identified more readily by the use of Romanowsky staining techniques such as the Giemsa stain, the cytoplasmic granules staining deeply basophilic. In the routine Papanicolaou stain the basophilic cytoplasm of the chief cells contrasts with the eosinophilic staining of the parietal cells (plate I).

Gastric washings are contaminated frequently by material from the oral cavity and esophagus and also by swallowed sputum. The oral cavity and esophagus yield squamous cells whilst swallowed sputum may contain ciliated columnar cells from the respiratory epithelium. These cells are usually well preserved, sometimes remarkably so (fig. 234) – possibly a reflection of the protective properties of the mucus swallowed with them. These well preserved cells are readily recognized particularly by means of their cilia. When swallowed respiratory epithelial cells are degenerate and

---

Fig. 228. Gastric mucosal cells showing early degenerative changes characterized by nuclear shrinkage and hyperchromasia. × 545.

Fig. 229. Degenerate gastric mucosal cells with nuclear fragmentation or karyorrhexis. × 545.

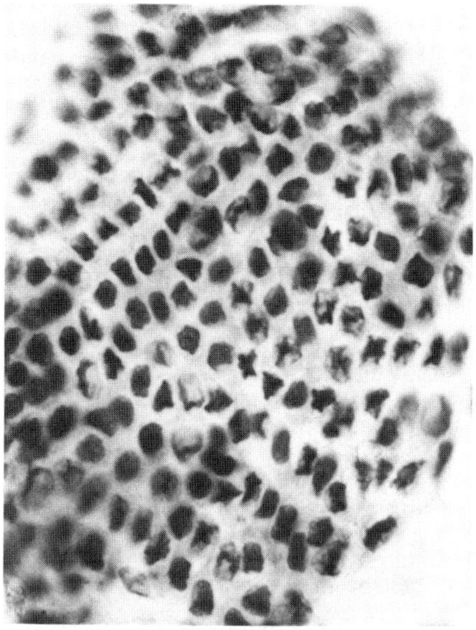

228

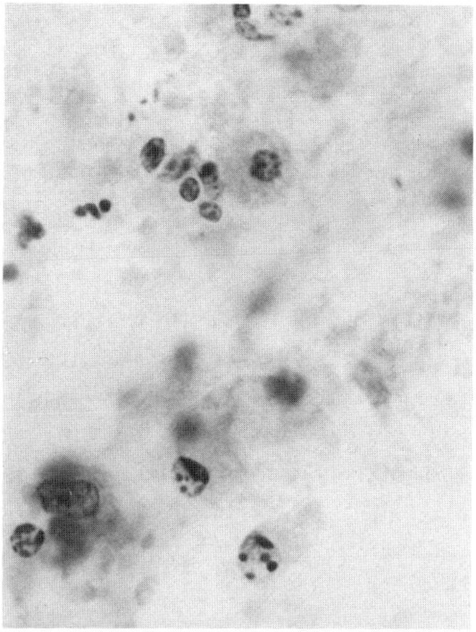

229

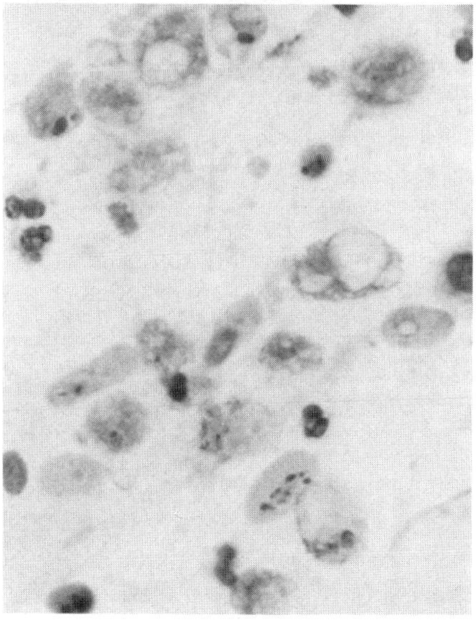

230

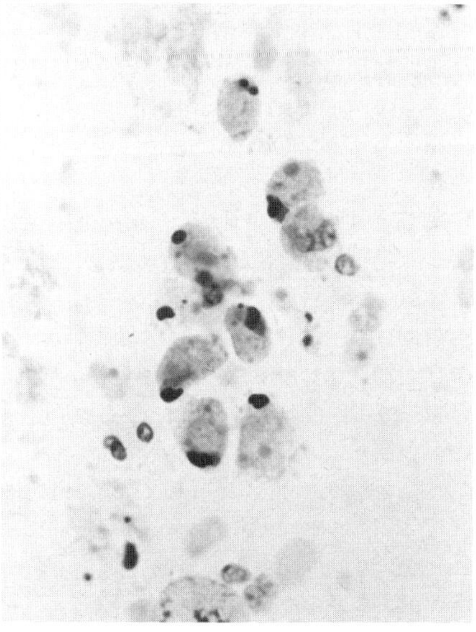

231

Fig. 230. Degenerate gastric mucosal cells showing cytoplasmic vacuolation and fragmentation and some evidence of karyorrhexis. × 545.

Fig. 231. Degenerate gastric mucosal cells showing cytoplasmic vacuolation and nuclear fragmentation producing 'mercury drop' appearance. × 545.

clumped they may cause some anxiety but careful examination usually reveals the presence of some recognizable cilia, these structures aiding cell recognition (fig. 235). Swallowed sputum may also contain Curschmann's spirals (fig. 236) and histiocytes or macrophages. The latter are usually recognized by their characteristic eccentric nucleus and abundant, often finely vacuolated, cytoplasm, particularly if this cytoplasm contains ingested dust (fig. 237).

Food particles are commonly seen in gastric washings and are usually identifiable. Thus well preserved meat fibres show prominent cross-striation and well defined sarcolemmal nuclei (fig. 238) whilst vegetable cells are usually recognizable by their characteristic structure and staining properties (fig. 239). Nevertheless vegetable cells, in particular, may display a very wide range of appearances on occasions resembling gastric mucosal cells and causing some diagnostic difficulties. In addition food particles may be a problem if there is any impediment to gastric emptying as they may obscure cytological detail.

Red blood cells may simply indicate trauma during collection of the specimen. Degenerate blood, however, may be more significant since it indicates bleeding prior to the procedure and hence suggests the presence of a gastric lesion. Polymorphonuclear leucocytes, lymphocytes and plasma cells may be present in small numbers but, if plentiful, may indicate the presence of inflammation.

Finally, a variety of micro-organisms of dubious significance may be seen. Some, such as *Alternaria* (fig. 240), are always contaminants and their presence may represent contamination of fluids or stains used during the preparation of the specimen. *Actinomyces* is another relatively common contaminant and is readily recognized as is the parasite *Giardia lamblia* with its characteristic double nuclei (fig. 241). This organism is more commonly seen in duodenal specimens but does occur in the stomach also. *Candida* species may also be demonstrated but are more commonly associated with mucosal lesions such as ulceration. Indeed there have been occasional case reports of death due to erosion of the wall of a major artery in the ulcer floor by fungal invasion [49]. Finally the organism *Leptothrix* is

---

Fig. 232. Parietal or oxyntic cells in gastric brushing. Note round or oval nuclei and abundant granular or, in places, finely vacuolated cytoplasm. × 545.

Fig. 233. Sheet of parietal and chief cells in gastric brushings. The chief cells are smaller than the parietal cells and have indistinct cell borders. × 545.

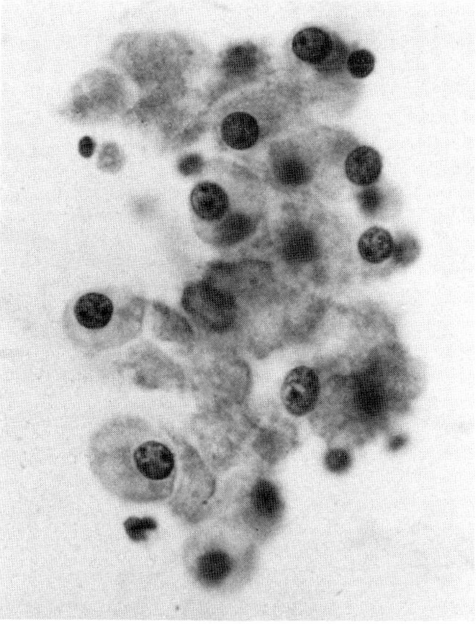

232

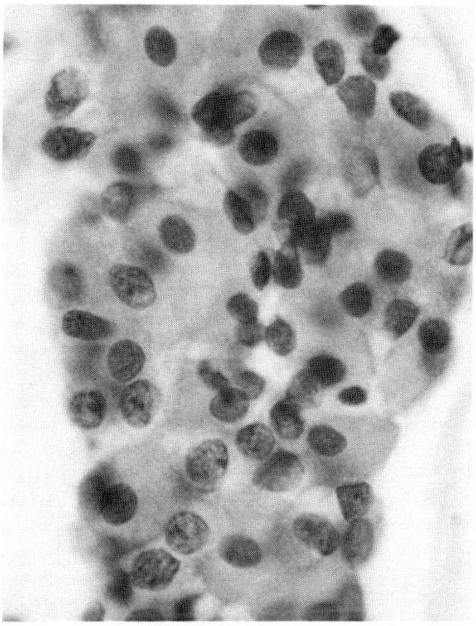

233

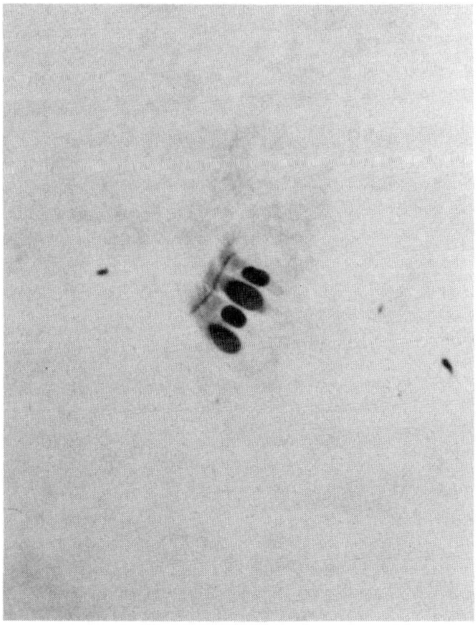

234

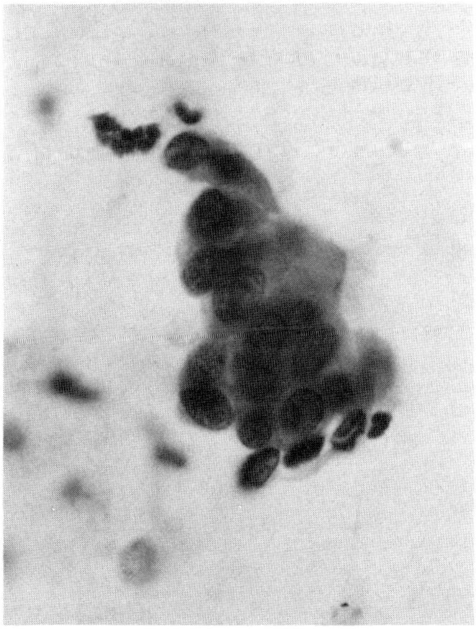

235

Fig. 234. Swallowed respiratory epithelial cells in gastric washings showing extremely good preservation. × 375.

Fig. 235. Group of swallowed respiratory epithelial cells in gastric washings. Although less well preserved than those in previous figure cilia are still discernible along the right-hand margin of the cell group. × 545.

236

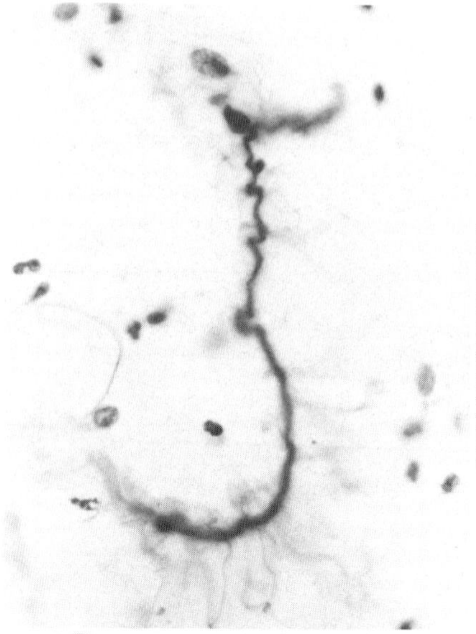

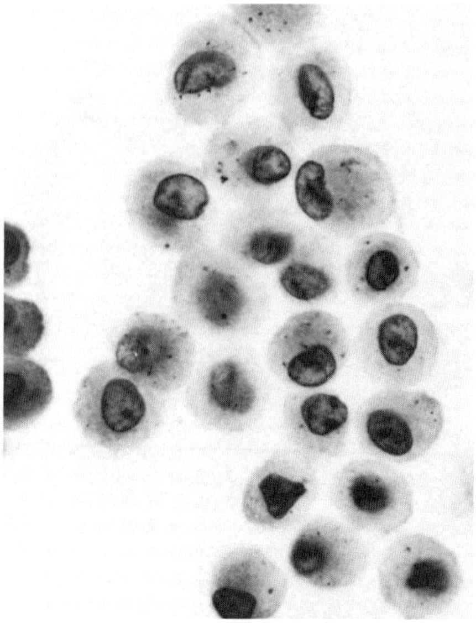

237

238

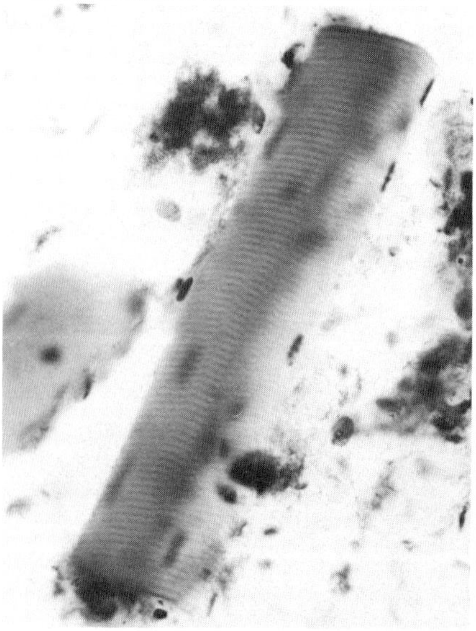

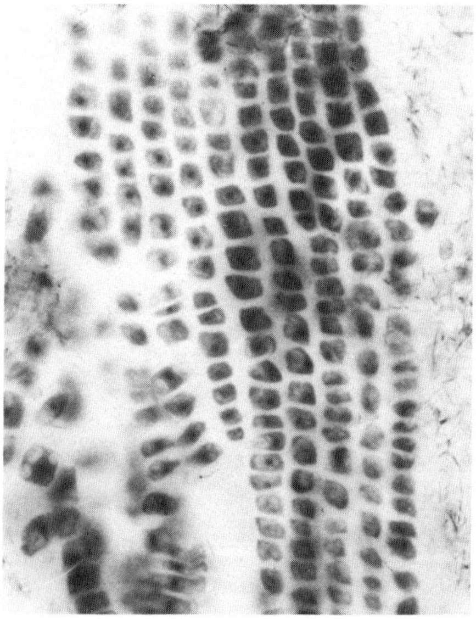

239

Fig. 236. Curschmann's spiral in gastric washings – presumably from swallowed sputum. × 345.

Fig. 237. Histiocytes or macrophages in gastric washings – presumably from swallowed sputum. Note 'dust' particles in cytoplasm. × 545.

Fig. 238. Meat fibre in gastric washings showing prominent cross striation and sarcolemmal nuclei. × 345.

Fig. 239. Vegetable cells from gastric washings. Note superficial resemblance to gastric mucosal cells. × 345.

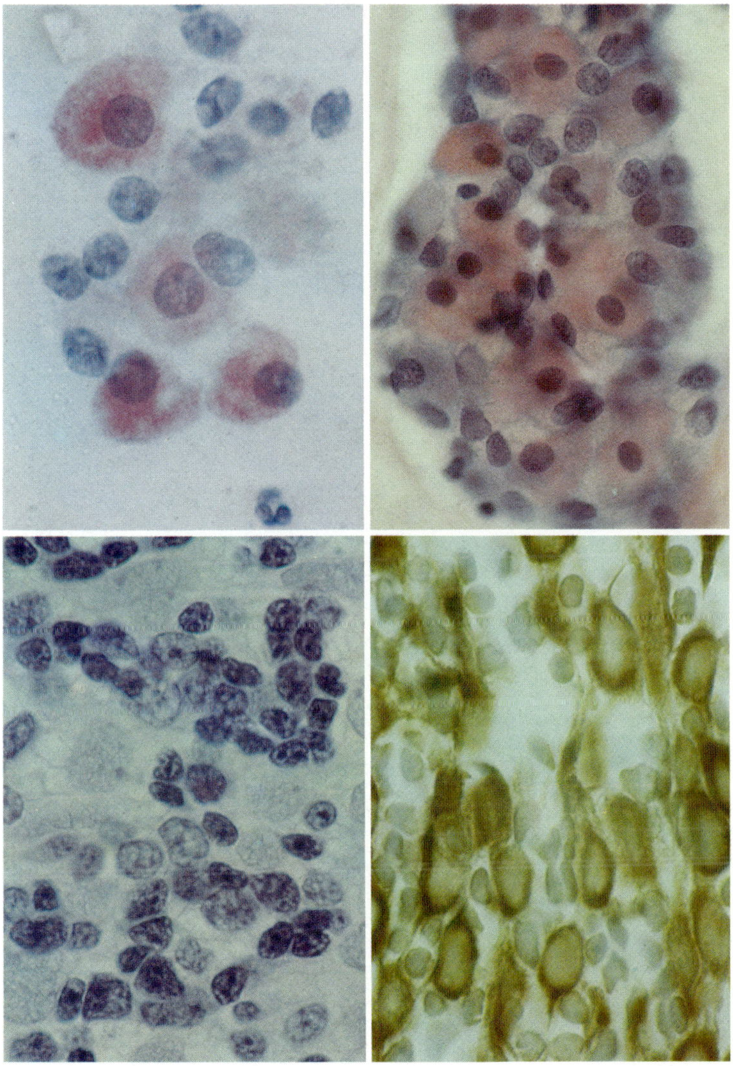

a b

c d

For legends, see reverse side.

Plate I.

a Parietal or oxyntic cells stained by the routine Papanicolaou method. Gastric brushings. × 1,000.

b Group of parietal and chief cells stained by Papanicolaou method. Parietal cells are distinguished by their eosinophilic cytoplasm whilst cytoplasm of chief cells is basophilic. Gastric brushings. × 400.

c Section of mediastinal mass – differential diagnosis malignant lymphoma vs. undifferentiated carcinoma. HE. × 1,580.

d Section of mediastinal mass stained by immunoperoxidase technique using antibody to prekeratin – an epithelial marker. Brown staining indicates epithelial nature of neoplasm. × 1,580. Specimen c and d by courtesy of Dr. *Wayne Hancock,* Prince Henry's Hospital, Melbourne.

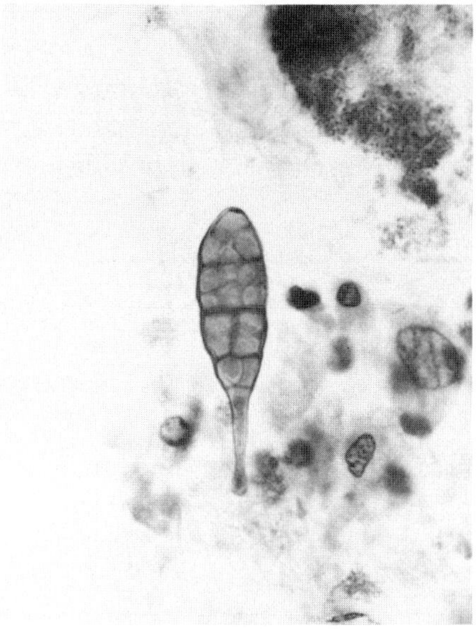

240

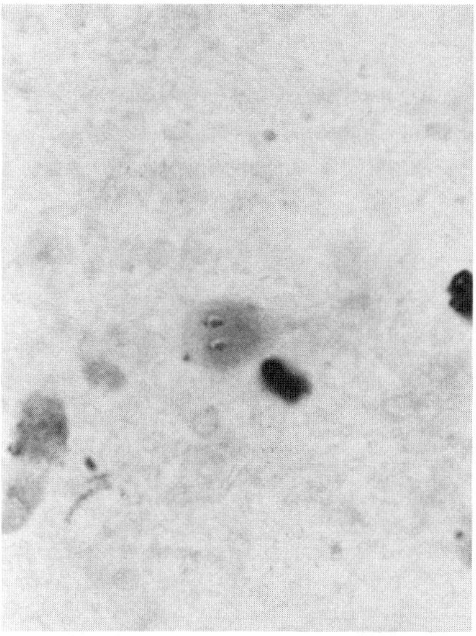

241

242

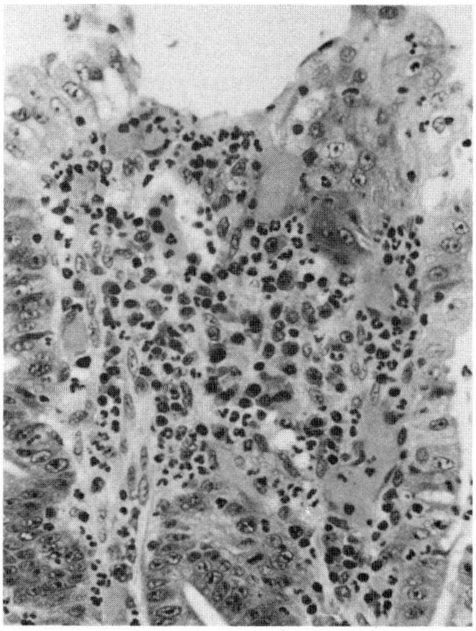

243

Fig. 240. Alternaria as contaminant of gastric washings. × 545.

Fig. 241. Giardia lamblia in gastric washings. Note characteristic double nuclei. × 860.

Fig. 242. Partial gastrectomy specimen showing multiple mucosal erosions characteristic of acute gastritis.

Fig. 243. Acute gastritis manifested by dense infiltration of lamina propria by inflammatory cells and superficial erosion. × 173.

seen quite frequently in normal gastric specimens and very commonly in association with peptic ulcers and malignancies, both carcinomas and lymphomas. This organism will be seen frequently in the photographs illustrating these conditions.

Non-Neoplastic Abnormalities

Gastritis and Peptic Ulceration

The term gastritis is a general one used to denote any inflammatory disorder of the stomach. The clinician may apply it in a very loose manner to explain a variety of upper gastro-intestinal symptoms where clinical signs are absent and radiological investigations negative. However, it is preferable to apply the term gastritis only to those cases where pathological changes are demonstrable. On this basis we recognize two broad categories of gastritis – acute and chronic – the former, by definition, being a relatively transient event whilst the latter is a condition of long standing with a rather complex pattern of histological changes.

Acute Gastritis

Since this condition is one of short duration histologic and cytologic studies are few in number. The primary disease may kill the patient rapidly or, much more commonly, recovery occurs within hours or days. Often the cause of acute gastritis is obscure although conversely it may be a feature of a severe bacterial or viral infection, either localized or generalized. Acute gastritis may be associated with a variety of non-infective conditions such as hiatus hernia, uremia, duodenal ulcer or cirrhosis of the liver. Finally, and perhaps most frequently, acute gastritis may be the consequence of the ingestion of irritant substances including alcohol and various therapeutic substances such as ferrous sulphate and salicylates. The latter, in the form of insoluble aspirin, can cause a particularly florid gastritis, small fragments of the substance becoming embedded in the gastric mucosa and producing an intense local destructive reaction.

In those cases of acute gastritis that have been examined the gastric mucosa appears swollen and oedematous and is usually covered with mucus. Cleaning the mucus away reveals a reddened mucosa often with numerous petechial haemorrhages and small erosions or well defined but shallow ulcers (fig. 242). Histological examination usually shows involvement of the superficial mucosa only often with a patchy or focal distribution of lesions. There is an intense infiltration by polymorphonuclear leucocytes and mucosal cell necrosis with associated haemorrhage and superficial ulcer formation (fig. 243).

As the condition is of short duration it is seldom subject to cytological examination. However, the inflamed gastric mucosa may be brushed during the acute phase and the cytological features reflect the histological changes. Polymorphs are numerous as are gastric mucosal cells (fig. 244). The epithelial cells may be seen singly but often small tissue fragments are shed. All cells are degenerate the degenerative changes being particularly marked in the gastric epithelial cells. Often the latter are fragmented and frequently appear as 'bare' nuclei or as nuclei with attached cytoplasmic remnants (fig. 245).

Aspirin (acetylsalicylic acid) undoubtedly has harmful effects on the gastric mucosa as can be documented by means of the gastrocamera [37]. In most people small areas of mucosal damage with slight haemorrhage are visible within 30 min of taking two ordinary aspirin tablets. In unusually susceptible people the mucosal effects may be considerable.

Descriptions of the cytological features of acute gastritis following aspirin ingestion are few but *Koss* [31] refers to 'conspicuous cell and nuclear abnormalities' and indicates that distinction from gastric carcinoma may not be possible. However, he further states that recovery 'may be remarkably rapid with return of normal cytology within a few days'. Within our own laboratory we have seen several cases of aspirin-induced gastritis and the cytological manifestations of this condition are depicted in figures 246–248. As indicated by *Koss* the nuclear abnormalities are marked and distinction from malignancy may be difficult. Other analgesic/anti-inflammatory drugs may also produce considerable cytological abnormalities. The cells depicted in figure 249 were derived from a patient who was taking a combination of 'Naprosyn' (naproxen) and 'Indocid' (indomethacin). The cells are virtually indistinguishable from those indicative of an undifferentiated or anaplastic carcinoma. 'Indocid' alone may also produce considerable abnormalities. This drug is composed of indomethacin and is a non-steroidal anti-inflammatory preparation used predominantly in the treatment of various forms of arthritis. It is a recognized gastro-intestinal irritant although the effects of the drug on the gastric mucosa are somewhat controversial [9]. Regardless of the mechanism we have seen marked cytological abnormalities in patients on high dosages of indomethacin and these are depicted in figures 250 and 251.

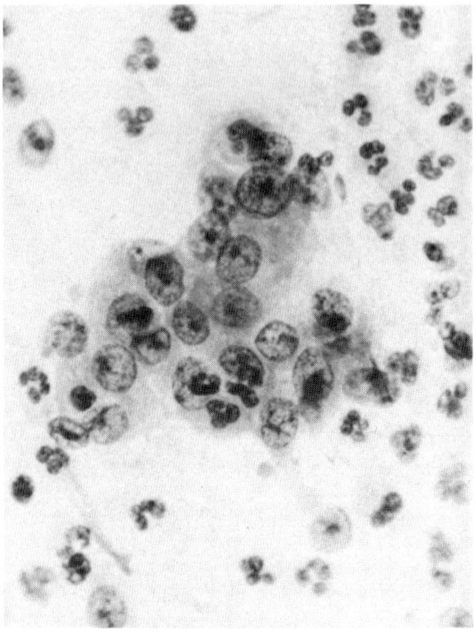

244

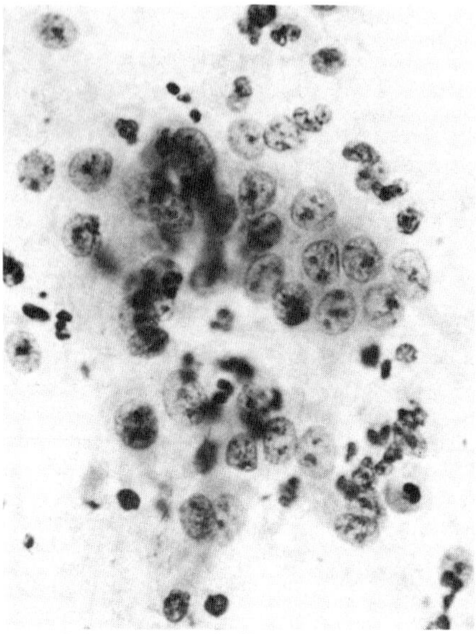

245

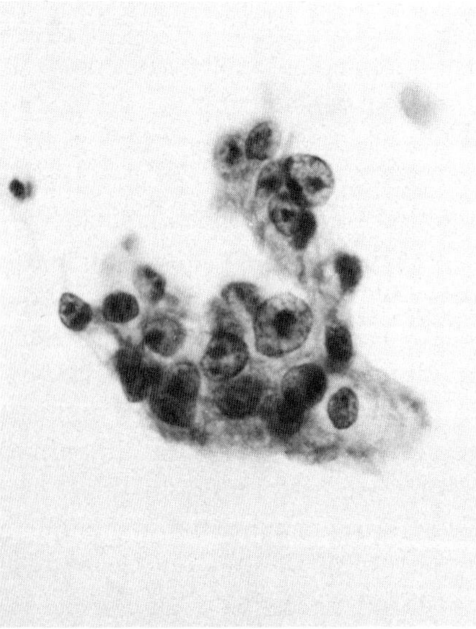

246

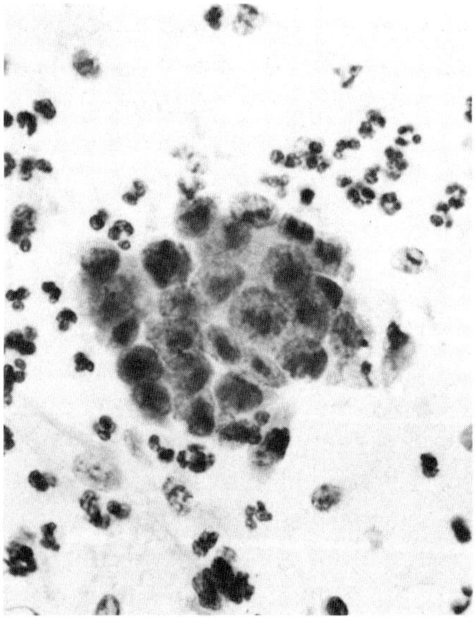

247

Fig. 244. Gastric brushings from patient with non-specific acute gastritis. Note numerous inflammatory cells and degenerate gastric mucosal cells. × 545.

Fig. 245. Gastric brushings from case of acute non-specific gastritis. Note inflammatory cells and degenerate gastric mucosal cells with, in particular, bare nuclei. × 545.

Fig. 246, 247. Groups of cells from patient presenting with acute gastritis. History of high aspirin intake for headaches over past 2 months. Symptoms resolved with aspirin withdrawal. × 545.

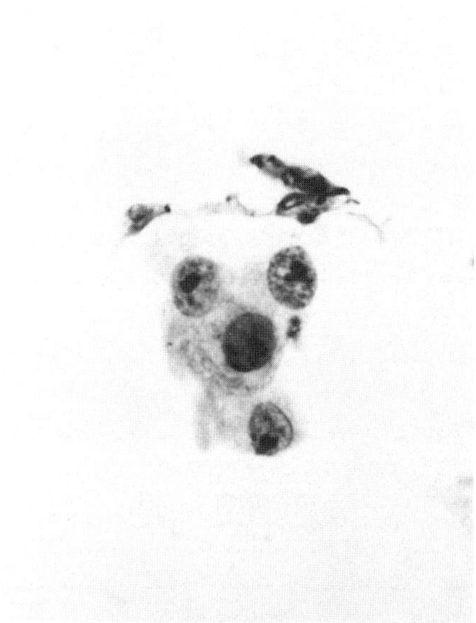

248

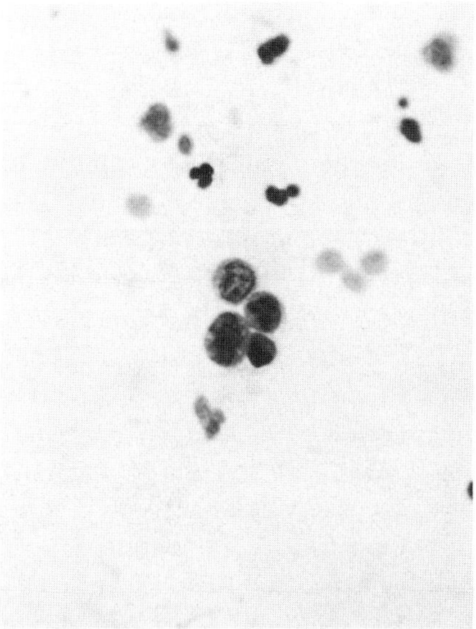

249

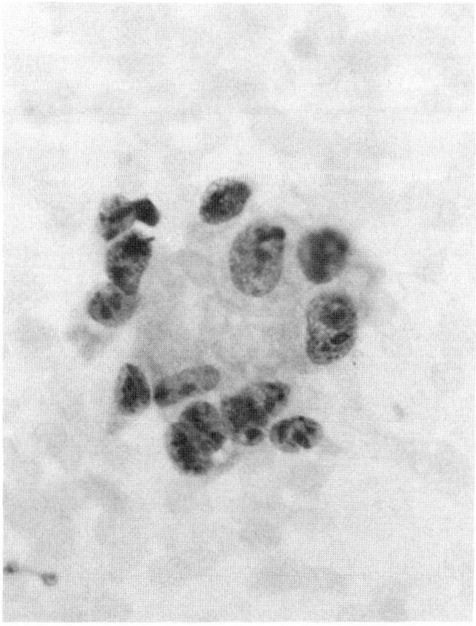

250

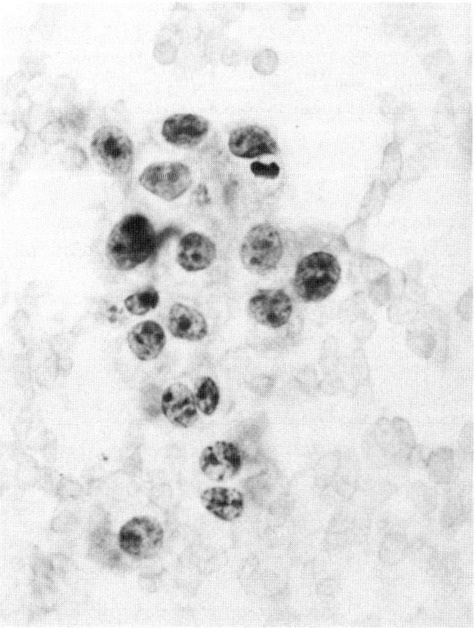

251

Fig. 248. Cells from patient who had been on soluble aspirin 200 mg twice daily for three months. Patient complained of epigastric discomfort after aspirin ingestion. × 545. Specimen courtesy of Dr. *Elaine Waters,* Royal Perth Hospital, Western Australia.

Fig. 249. Cells from patient taking a combination of naproxen and indomethacin for osteoarthritis. × 545.

Fig. 250, 251. Groups of cells from patient on high dosage of indomethacin for arthritis. × 545.

Chronic Infective and Granulomatous Gastritis

The stomach has been the site of a wide variety of chronic inflammatory disorders including a number of granulomatous lesions such as tuberculosis and syphilis [51], whilst the cytological features of one case of sarcoidosis have also been described [1]. In these few case reports multinucleate and epithelioid cells were described as being characteristic of the disease process.

Multinucleate cells (fig. 252) and epithelioid cells (fig. 253) are seldom seen in gastric washings or brushings and, when present, usually do not indicate a specific lesion of the gastric mucosa. The specimen illustrated in figure 254 is portion of a stomach from a female aged 58 with a long history of dyspepsia which had been investigated previously with negative results. She was admitted to hospital following several episodes of haematemesis and melaena. Endoscopy revealed a bleeding ulcer and subsequently a partial gastrectomy was carried out. On macroscopic examination the gastric mucosa was noted to be irregularly fissured with an unusual 'cobblestone' pattern. In addition there were two small areas of ulceration each measuring approximately 0.6 cm in diameter. Microscopic examination showed numerous non-caseating granulomata. These were scattered throughout all layers of the stomach wall, a number being situated within the gastric mucosa (fig. 255). The granulomata were composed of epithelioid cells and Langhans giant cells (fig. 256) the latter being particularly prominent (fig. 257). The diagnosis of Crohn's disease was made and, on the basis of further investigations and follow-up, it would appear that the disease was limited to the stomach.

Pre-operative cytological examination was not carried out. However, vigorous brushing of the operative specimen yielded occasional multinucleate giant cells (fig. 258) and epithelioid cells (fig. 259). The paucity of diagnostic material and the necessity for more vigorous brushing than would be possible in the clinical situation presumably reflects the relative inaccessibility of the granulomata.

Fig. 252. Multinucleate giant cell in gastric brushings from patient with carcinoma of esophagus. No gastric lesion demonstrated. × 545.

Fig. 253. Epithelioid cells from gastric brushings of patient with non-specific chronic superficial gastritis. × 545. Specimen courtesy of Dr. *S.K. Tang,* Alfred Hospital, Melbourne.

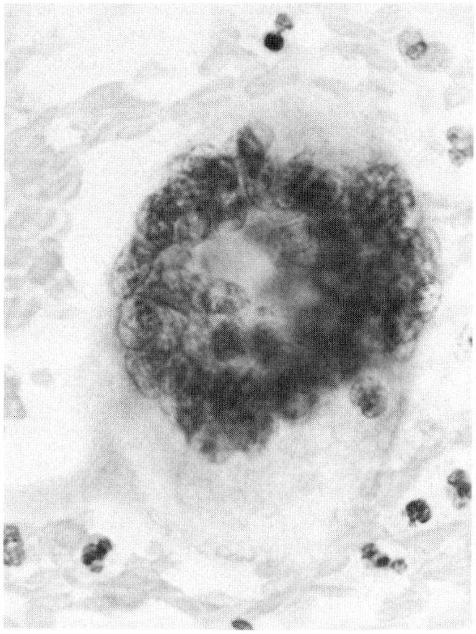

252

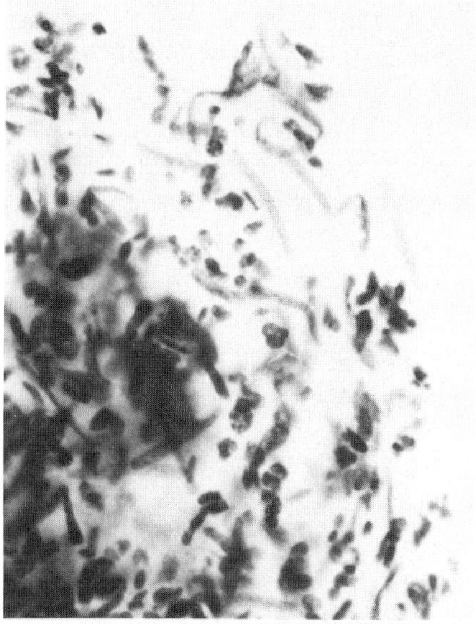

253

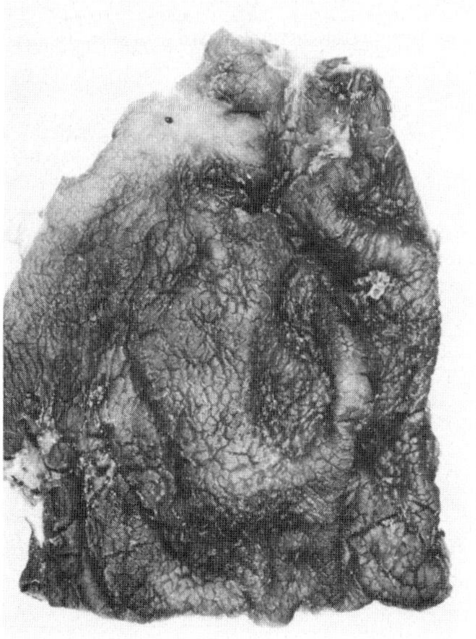

254

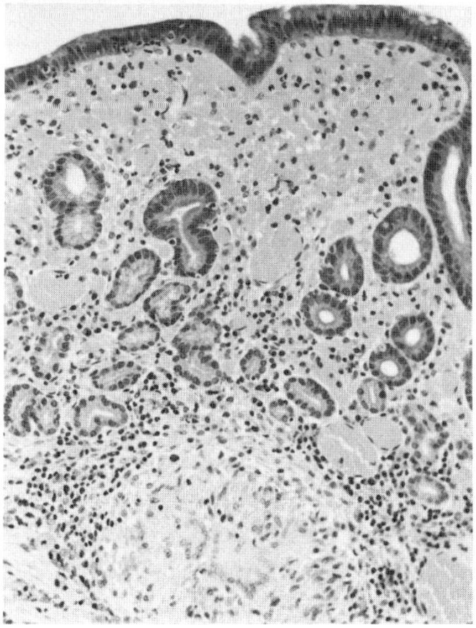

255

Fig. 254. Partial gastrectomy specimen from patient with Crohn's disease of stomach.

Fig. 255. Section of Crohn's disease of stomach showing intra-mucosal granuloma. × 110.

In one series of 300 patients with Crohn's disease the incidence of gastro-duodenal involvement was 4% [42]. All but one of the patients comprising this 4% had evidence of Crohn's disease elsewhere at some time during the period of observation. The remaining patient had a gastric lesion alone. Hence primary Crohn's disease of the stomach is an unusual lesion and unlikely to pose a cytological diagnostic problem.

Chronic Non-Specific Gastritis

Although it is necessary to be aware of the occurrence of the disease processes described above, for practical purposes the interest of the cytopathologist lies in the non-specific type of chronic gastritis, the causes of which are poorly understood and the classification and descriptions of which are correspondingly complex. Nevertheless there are a number of papers available which describe the various histological patterns of chronic gastritis in considerable detail and several useful classifications are available. In this descriptive section the morphological classification adopted by *Morson and Dawson* [42] is the most useful although other classifications will be considered at a later stage when gastric cancer precursor lesions are being discussed.

Accordingly chronic gastritis may be divided on histological grounds as follows:

Chronic superficial gastritis
- Active
- Quiescent

Chronic atrophic gastritis
- Active
- Quiescent

Gastric atrophy

Within this histological spectrum several subtypes or patterns such as follicular gastritis and chronic cystic gastritis, are sometimes described. However, these entities have no cytological significance. Similarly the so-

---

Fig. 256. Intra-mucosal granuloma in patient with Crohn's disease of stomach. Multinucleate giant cells and epithelioid cells are prominent. × 275.

Fig. 257. Characteristic Langhans' giant cell of intra-mucosal granuloma in patient with Crohn's disease of stomach. × 545.

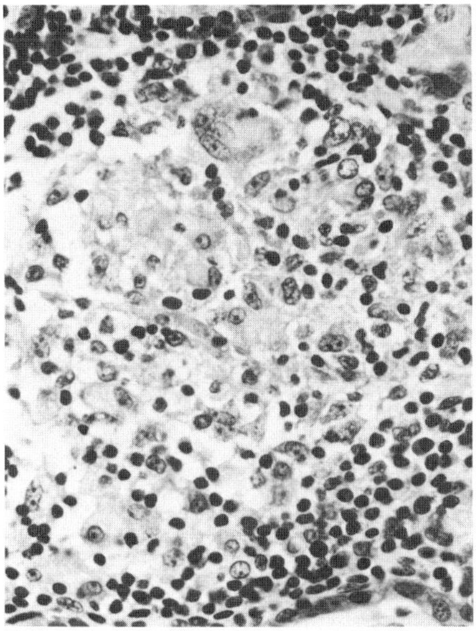

256

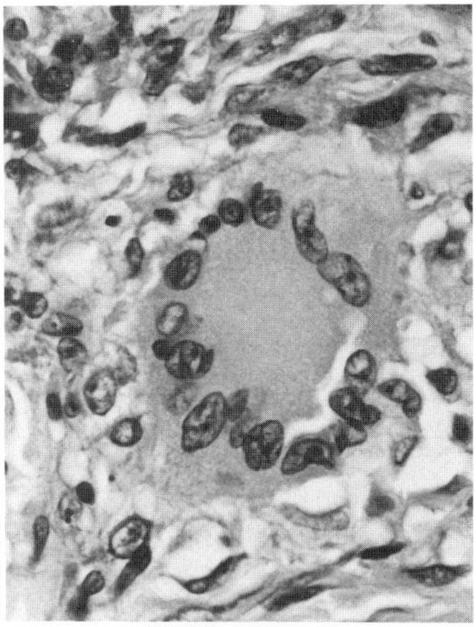

257

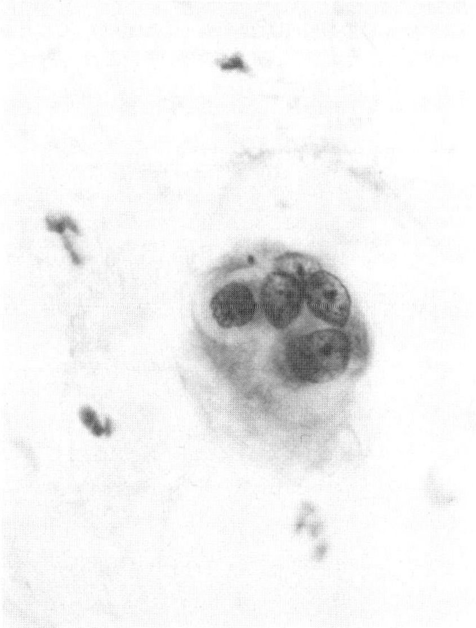

258

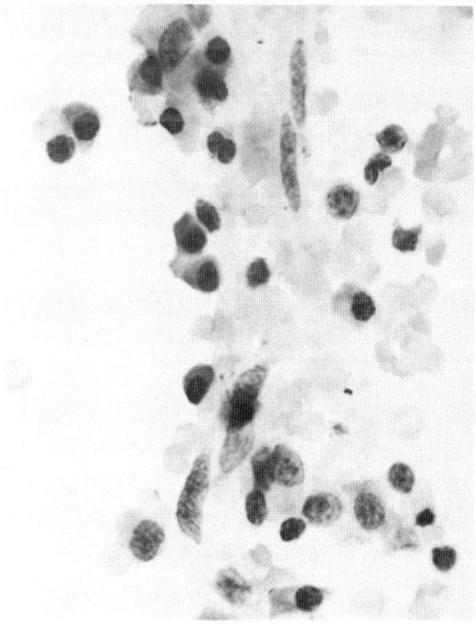

259

Fig. 258. Primary Crohn's disease of stomach – multinucleate giant cell in brushings of operative specimen. × 545.

Fig. 259. Primary Crohn's disease of stomach – epithelioid cells in brushings of operative specimen. × 545.

called giant hypertrophic gastritis or Ménétrier's disease is of little cytological importance and hence will not be discussed further. It is an uncommon lesion and, although of obscure nature, is almost certainly not a true gastritis.

Chronic Superficial Gastritis

The earliest histologically recognizable lesion is chronic superficial gastritis. In this condition the mucosal thickness may be normal or slightly increased and the changes are confined to the superficial zone of the mucosa (fig. 260). The lamina propria is congested and oedematous with dilatation of the small blood vessels (fig. 261). There is a marked increase of inflammatory cells within the lamina propria, including polymorphonuclear leucocytes and lymphocytes (fig. 262). Polymorphs also accumulate within the gastric pits the appearances being reminiscent of the crypt abscesses seen in inflammatory bowel disease (fig. 263). The surface and glandular epithelial cells show marked abnormalities frequently being cuboidal in shape with diminished cytoplasm, and enlarged, irregular, hyperchromatic nuclei (fig. 264). These abnormal cells may be heaped up to form a multilayered epithelium (fig. 265). The abnormal cells also desquamate into the surface exudate (fig. 266) thus producing the characteristic cytological manifestations of the disease process. Extensive desquamation of surface epithelial cells will result in small superficial erosions (fig. 267).

When the disease process is quiescent mucosal oedema and congestion is less apparent, the inflammatory cell infiltrate is much less dense and comprises almost exclusively lymphocytes and plasma cells. Similarly the epithelial cell changes are absent or of minimal degree.

Chronic Atrophic Gastritis

This entity appears to follow chronic superficial gastritis and indeed the two conditions are not clearly separable. Again the lamina propria of the mucosa shows vascular congestion and inflammatory cell infiltrates (fig. 268) but the degree of such infiltration is extremely variable. As the name implies the mucosa is usually reduced in thickness but this is not an

---

Fig. 260. Chronic superficial gastritis – histological changes confined to superficial zone of mucosa. ×140.

Fig. 261. Chronic superficial gastritis – congestion and oedema of lamina propria of mucosa. ×435.

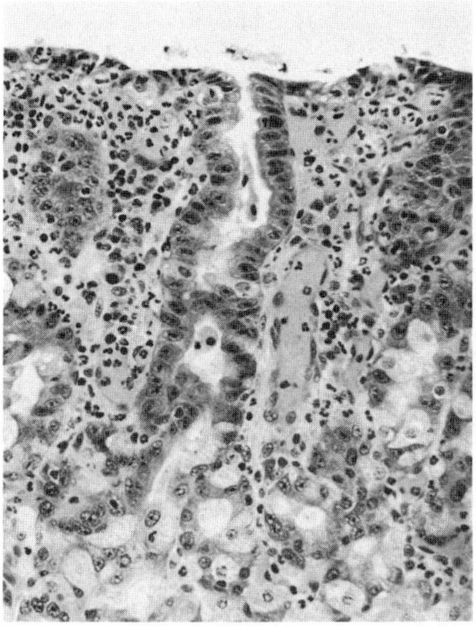

260

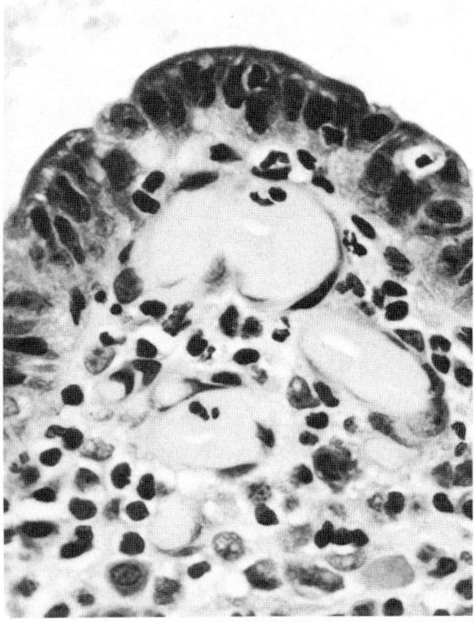

261

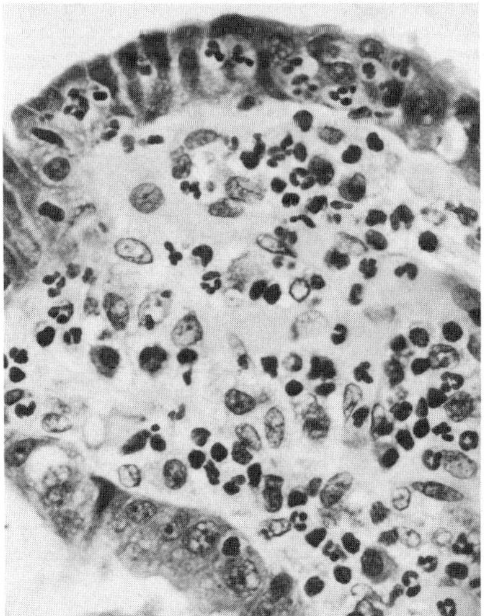

262

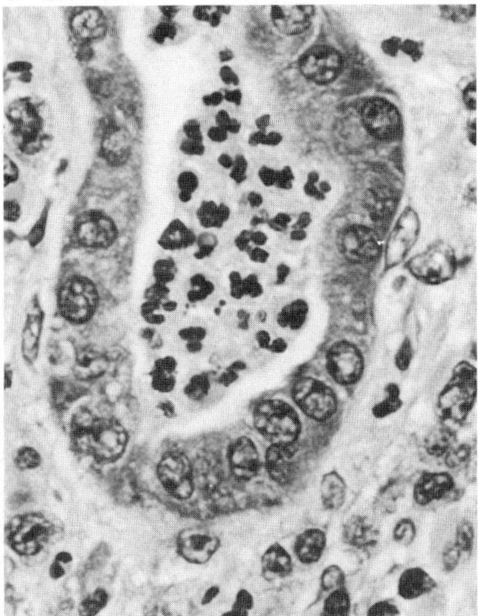

263

Fig. 262. Chronic superficial gastritis – congestion and oedema of lamina propria with a dense infiltration of inflammatory cells. Note polymorphonuclear leucocytes within surface epithelium. × 345.

Fig. 263. Chronic superficial gastritis – polymorphs within gastric pit resembling 'crypt abscess'. × 545.

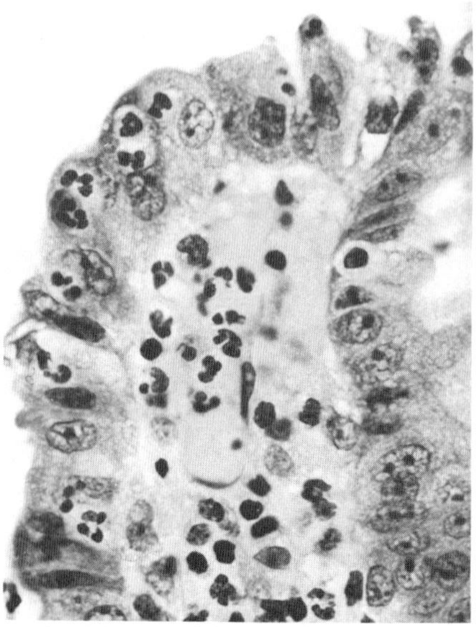
264

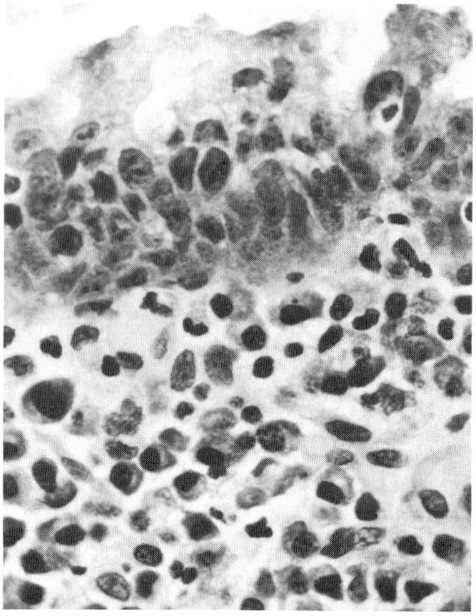
265

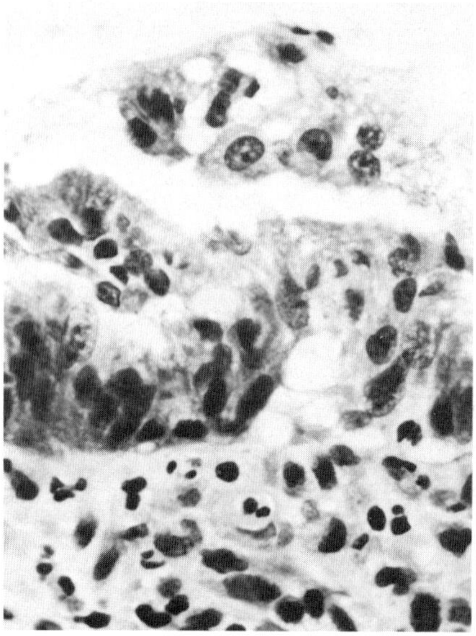
266

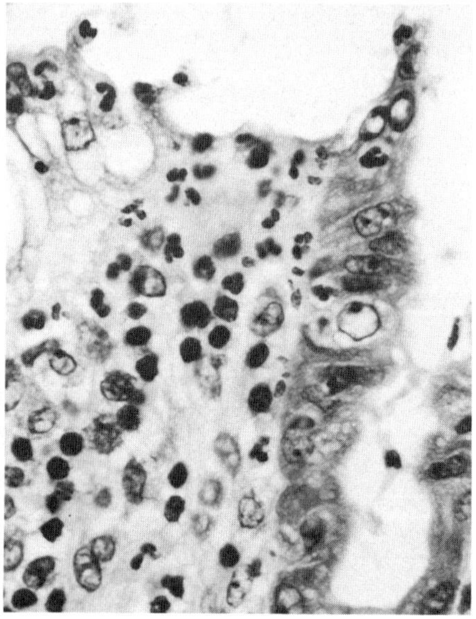
267

Fig. 264. Chronic superficial gastritis – infiltration by inflammatory cells and marked abnormalities of surface epithelial cells. ×435.

Fig. 265. Chronic superficial gastritis – markedly abnormal surface epithelial cells with multiple layering. ×435.

Fig. 266. Chronic superficial gastritis – desquamation of abnormal epithelial cells into surface exudate. ×435.

Fig. 267. Chronic superficial gastritis – superficial erosion of mucosa. ×435.

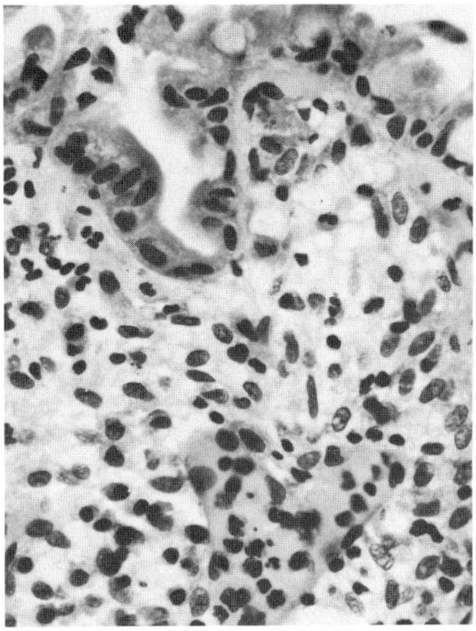

268

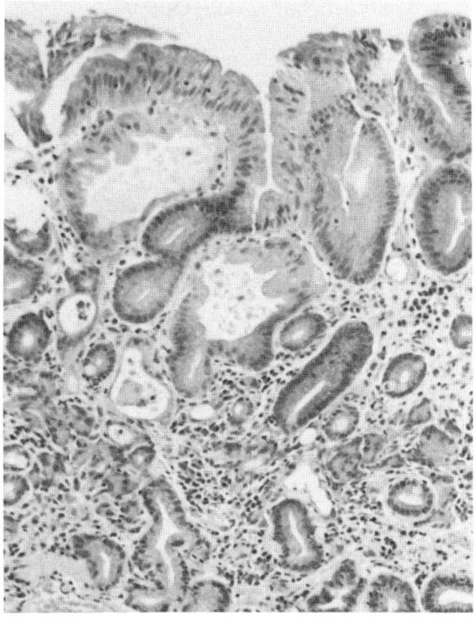

269

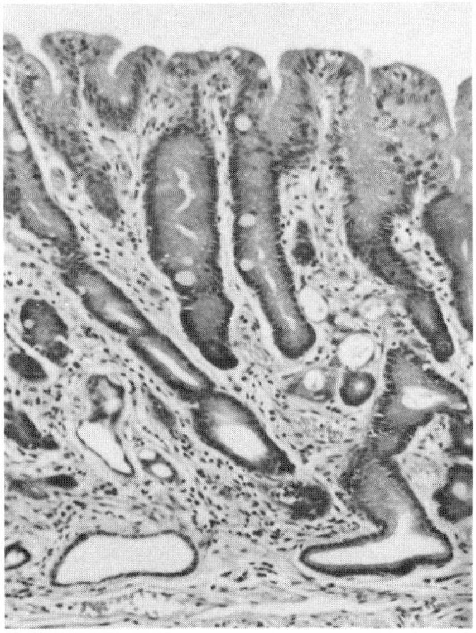

270

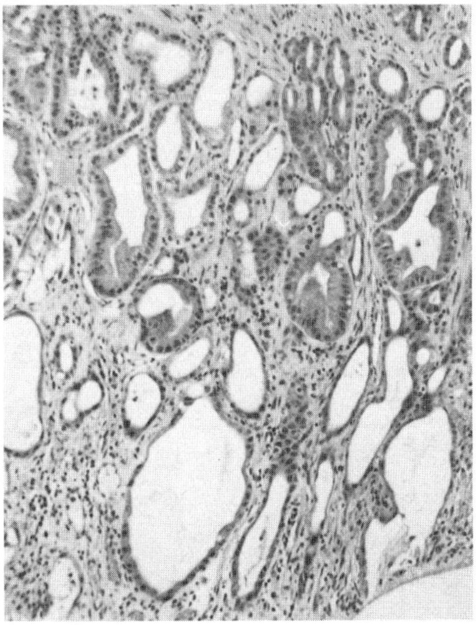

271

Fig. 268. Chronic atrophic gastritis – infiltration of lamina propria by inflammatory cells and abnormal surface epithelial cells. × 435.

Fig. 269. Chronic atrophic gastritis – diffuse infiltration by inflammatory cells and considerable mucosal atrophy. × 86.

Fig. 270. Gastric atrophy with marked decrease in thickness of mucosa and loss of specialized glandular cells. Note evidence of intestinal metaplasia. × 86.

Fig. 271. Gastric atrophy showing cystic dilatation of mucosal glands. × 69.

invariable feature. There is, however, an atrophy of glands in the deep zone of the mucosa and a considerable reduction in the numbers of parietal and chief cells (fig. 269). During the active phase of the disease epithelial cell degeneration and regeneration is again a feature. This is not seen in the quiescent phase nor are inflammatory cells necessarily increased in number. However, usually there is a modest increase in the number of lymphocytes and plasma cells.

Gastric Atrophy

This is usually regarded as the end result of chronic gastritis. However, it may occur as an ageing phenomenon in the elderly and it is the earliest demonstrable lesion in pernicious anaemia. The most striking feature is the reduction in the width, or atrophy of the mucosa. There is an absence of chief and parietal cells, the gastric tubules being lined by simple columnar epithelium with scattered intestinal goblet cells (fig. 270). Cystic dilatation of the glands is common and may be extreme (fig. 271). Lymphoid aggregates are commonly seen in the lamina propria but inflammatory cell infiltrates are absent.

Peptic Ulceration

Peptic ulcers may also be acute or chronic. The acute ulcers are usually multiple and may involve any part of the gastric mucosa (fig. 272). They are frequently small and shallow (fig. 273) but the depth of penetration varies and indeed an acute ulcer may perforate the gastric wall.

However, it is the chronic variant that is of greatest interest to the diagnostic cytologist. Within the stomach they are usually seen within the antral and prepyloric regions and along the lesser curvature of the body of the stomach (fig. 274). They can be clearly demonstrated radiologically (fig. 275).

Although multiple chronic ulcers may occur they are usually single, round or oval in shape, and usually less than 2 cm in diameter (fig. 276). Occasionally they may be extremely large measuring 10 cm or more in main dimension (fig. 277). The edges are clear-cut, in contrast to the raised

---

Fig. 272. Multiple acute peptic ulcers of gastric mucosa. Photograph courtesy of Professor *P. Bhathal,* Royal Melbourne Hospital.

Fig. 273. Acute peptic ulcer of stomach. Note continuity of muscle layer and absence of dense fibrosis characteristic of chronic ulceration.

272

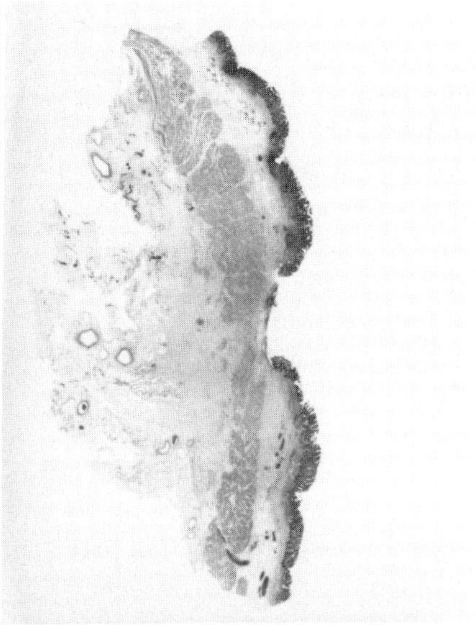

273

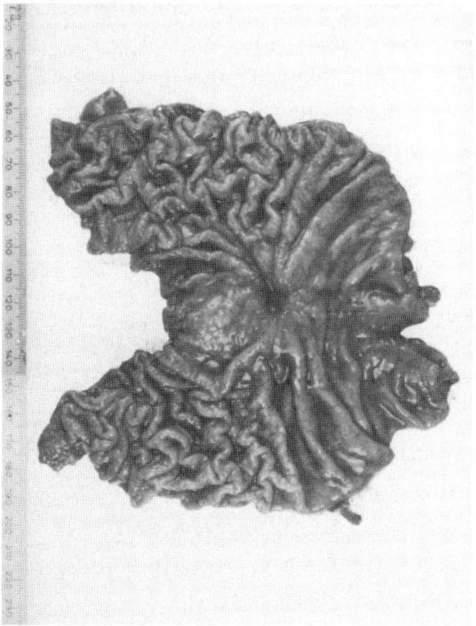

274

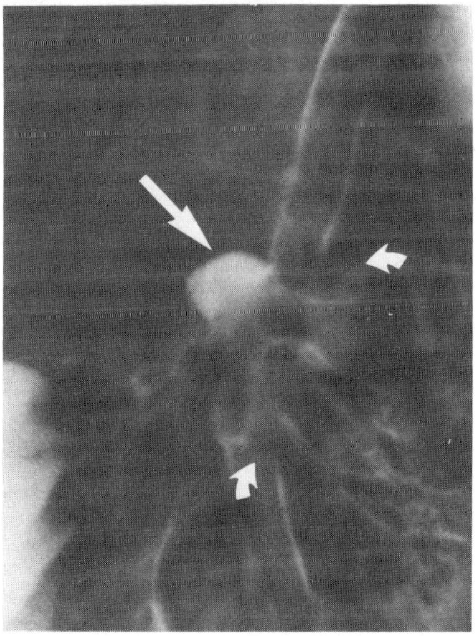

275

Fig. 274. Chronic peptic ulcer on lesser curvature of stomach.

Fig. 275. Radiological demonstration of chronic peptic ulcer of lesser curvature of stomach. The straight arrow indicates the ulcer crater filled with contrast medium. Note the relatively smooth outline of the crater and also the mucosal folds radiating outwardly from the ulcer as indicated by the two small curved arrows.

everted edges of a malignant ulcer, and they frequently overhang the ulcer crater to produce a flask-shaped lesion (fig. 278). The surrounding gastric mucosa may appear flat and atrophic but frequently there is a zone of congested reddened mucosa around the ulcer.

The microscopic structure of the chronic peptic ulcer is also characteristic. Four zones or layers are usually recognizable in the base – an outer layer of fibrous tissue, a layer of granulation tissue, a layer of necrotic tissue and, resting on this and forming the floor of the ulcer, a variable thickness of fibrino-purulent exudate (fig. 279). The latter is of particular interest to the diagnostic cytologist as necrotic debris and purulent exudate is frequently seen in the cytological specimen whether it be collected by direct brushing of the ulcer floor or by blind lavage.

Of much greater interest, however, are the epithelial cell changes that are seen at the periphery of the ulcer. Here there is an ongoing process of epithelial cell degeneration and necrosis accompanied by intense regenerative activity (fig. 280). Regenerating cells may also extend as a single layer across the floor of the ulcer (fig. 281). As chronic peptic ulcers are subject to exacerbations and remissions of activity so too does the process of necrosis and regeneration wax and wane. However, as the patient harbouring the ulcer is usually investigated during a period of activity the cells derived from the edges and floor of the ulcer, which are usually shed in strips (fig. 282) and fragments (fig. 283), can cause considerable diagnostic difficulty.

Cytological Manifestations of Chronic Gastritis and Peptic Ulceration

Chronic superficial gastritis and peptic ulceration are considered together because they have a similar cytological presentation. Cell debris is usually present as are micro-organisms and a variety of inflammatory cells. More important, however, are the epithelial cell changes that result from co-existent tissue destruction and regeneration. The histological features of these processes have already been described and illustrated and it is not surprising that the resultant cell abnormalities, when seen in cytological preparations, may cause considerable diagnostic difficulties.

Fig. 276. Typical chronic peptic ulcer of gastric mucosa. Note the relatively small size and the clear-cut edges.

Fig. 277. Unusually large chronic peptic ulcer of gastric mucosa. Note the large blood vessel visible in the floor of the ulcer.

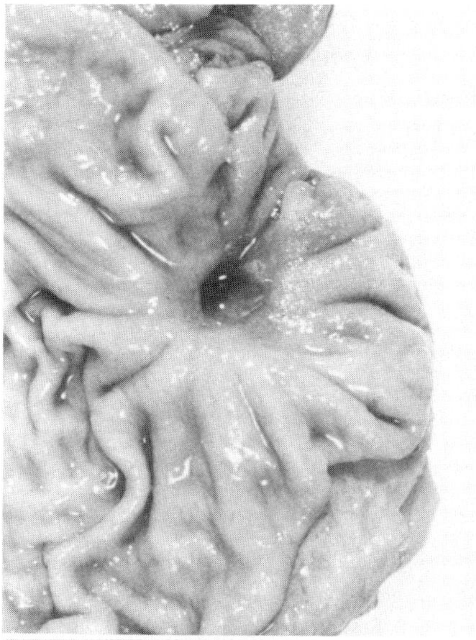

276

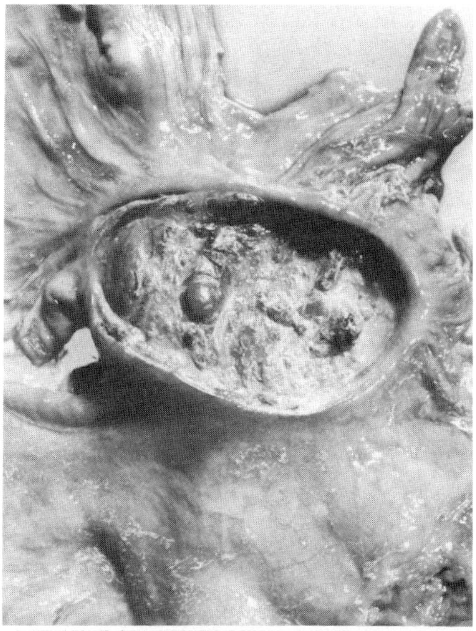

277

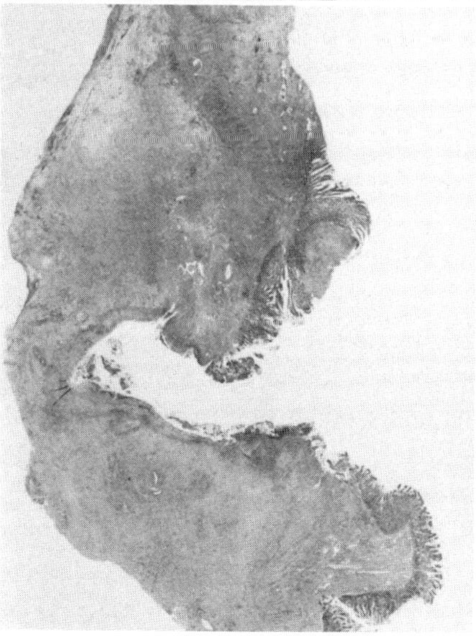

278

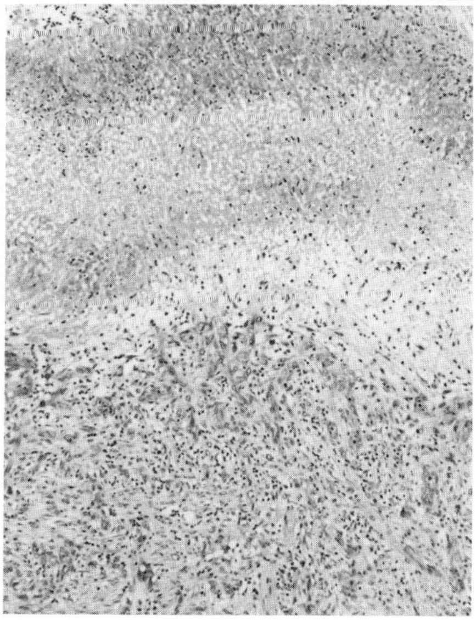

279

Fig. 278. Cross section of chronic peptic ulcer of stomach showing overhanging edges producing a flask-shaped lesion. × 2. Specimen courtesy of Dr. *Brian Essex,* Alfred Hospital, Melbourne.

Fig. 279. Base of chronic peptic ulcer of stomach. Note layer of granulation tissue inferiorly, necrotic tissue, and, forming the floor of the ulcer, fibrino purulent exudate. × 55.

The nuclei of the abnormal cells may or may not be enlarged although, contrary to what has been stated [31] it is our experience that enlargement is usual. The enlarged nuclei vary considerably in staining capacity being most commonly hyperchromatic (fig. 284) but sometimes they may be hypochromatic. The extreme nuclear hypochromasia depicted in figure 285 is quite commonly seen, the changes being largely degenerative in nature. Disturbance of the nuclear chromatin pattern is common and nucleoli are usually prominent. These prominent nucleoli are frequently irregular in shape (fig. 286) thus suggesting the possibility of malignancy. To compound the problem the nucleoli of truly malignant cells, although usually irregular in outline, may be deceptively regular (fig. 287).

Attempts have been made to categorize these cytological abnormalities into grades or degrees. For example *Prolla and Kirsner* [52] suggest that there are four degrees of atypia whilst other workers have spoken of mild, moderate and marked atypia. Such precision is somewhat unrealistic. As illustrated in figures 288–295 one must recognize a spectrum of abnormalities varying from slight to very marked, the latter being difficult to distinguish from malignancy. However, despite the wide variation in cytological presentation, there is usually a preservation of tissue fragments and a general uniformity of cell size and shape within these fragments. Figures 296 and 297 depict respectively a strip and fragment of cellular material shed from the edge of a peptic ulcer. The general uniformity of cell size and shape and the preservation of polarity of cells is evident. This contrasts with the tendency of carcinoma cells to separate from each other due to poor cell adhesion as depicted in figure 298. In addition, within these groups of carcinoma cells there is a loss of cell polarity and also evidence of considerable pleomorphism. Difficulties may arise with the interpretation of single cells shed from the edge of a peptic ulcer, as depicted in figure 299, but fortunately this is a relatively rare phenomenon.

Several authors [5, 6] have described cytological changes that they regard as indicative of chronic atrophic gastritis. In addition to inflammatory cells, predominantly lymphocytes, and evidence of intestinal metaplasia, there are abnormal gastric mucosal cells referred to as 'pale cells'. The nuclei are described as being enlarged but with a concomitant enlargement

---

Fig. 280. Abnormal gastric mucosal cells at edge of chronic peptic ulcer. × 545.

Fig. 281. Single layer of abnormal epithelial cells covering floor of chronic peptic ulcer of stomach. × 545.

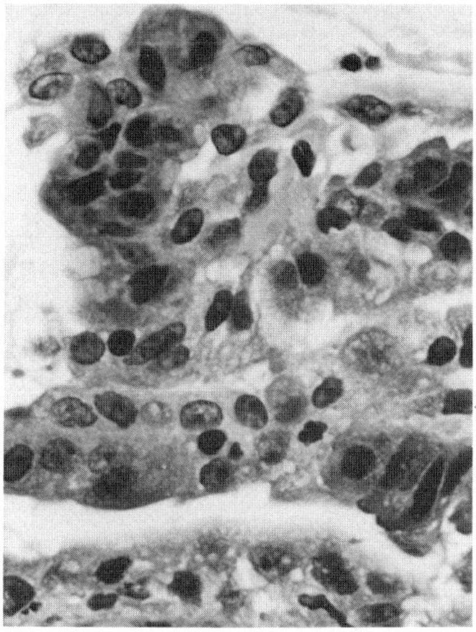

280

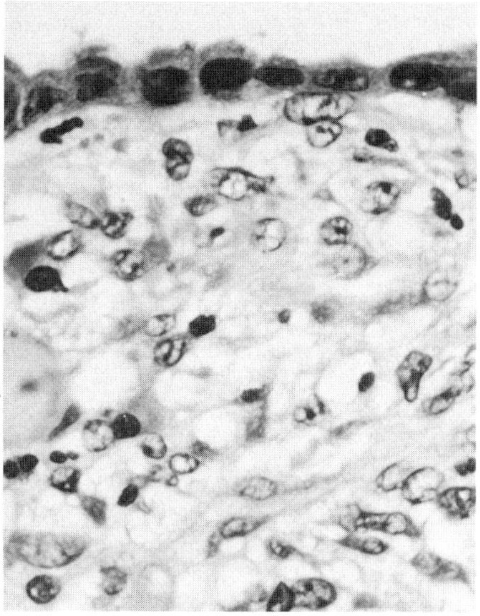

281

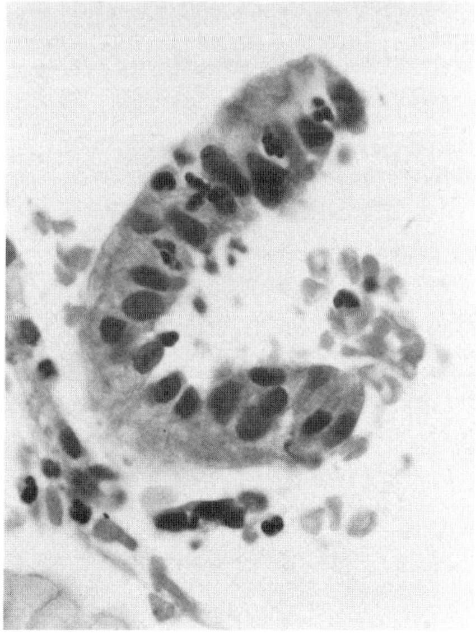

282

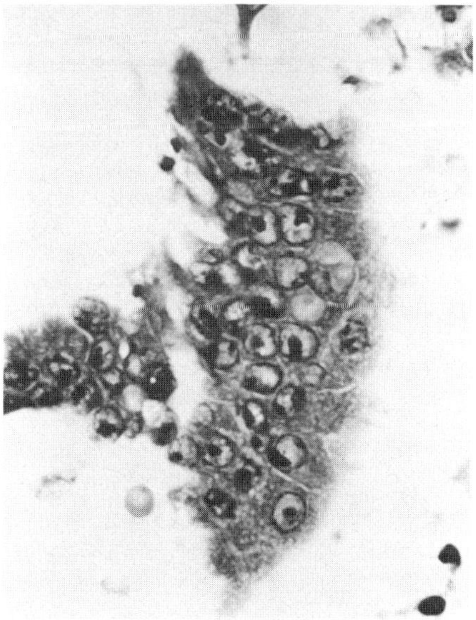

283

Fig. 282. Section of chronic peptic ulcer of stomach showing desquamated strip of abnormal gastric epithelium. × 590.

Fig. 283. Section of chronic peptic ulcer of stomach showing desquamated fragment of abnormal mucosal tissue. × 545.

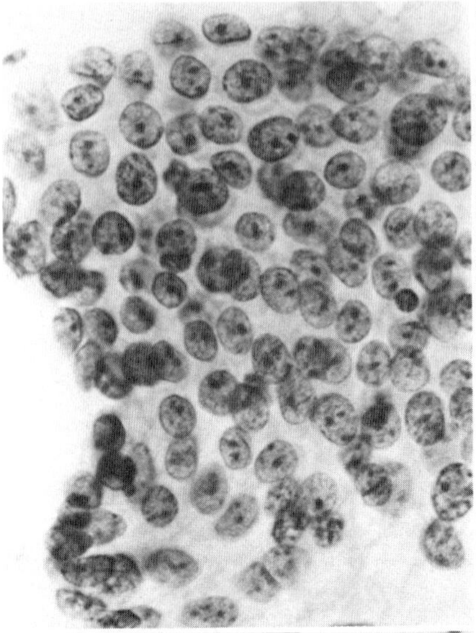
284

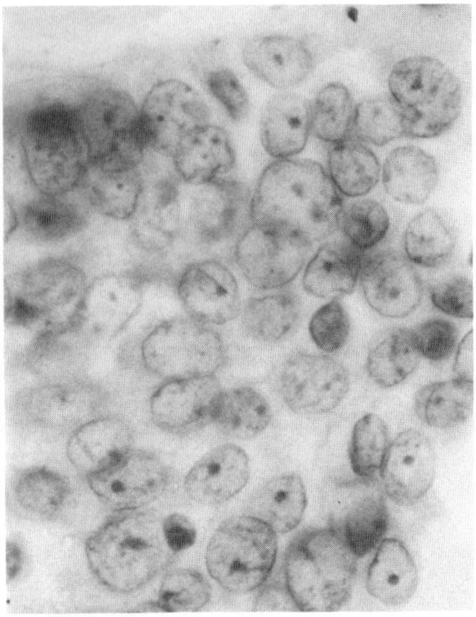
285

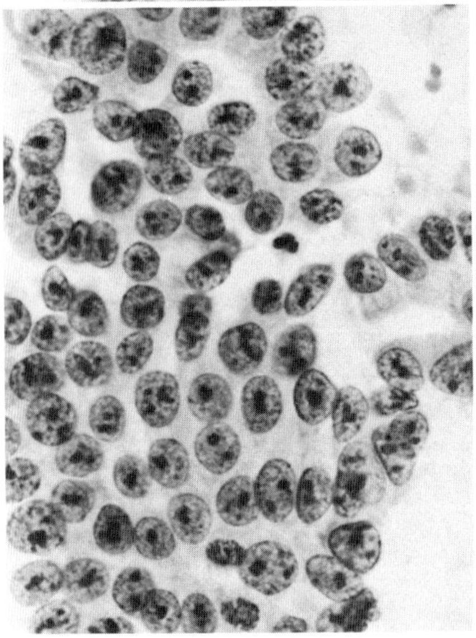
286

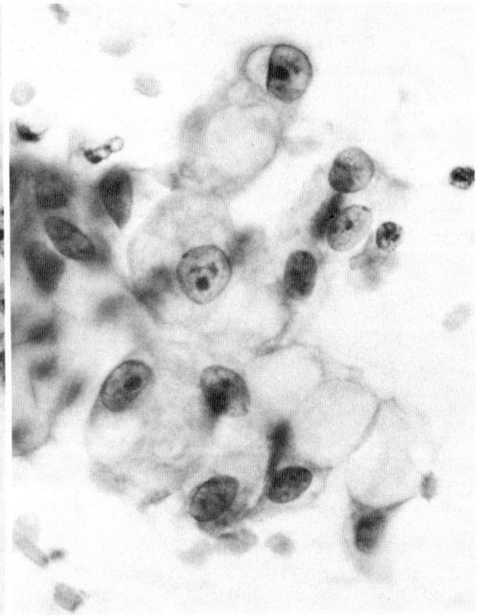
287

Fig. 284. Cells in gastric brushing from patient with chronic peptic ulcer of stomach. Note slight nuclear enlargement and abnormalities of chromatin distribution. × 545.

Fig. 285. Abnormal cells in gastric washings from patient with chronic peptic ulcer. Nuclear hypochromasia is prominent – largely a degenerative phenomenon. × 545.

of the cell as a whole resulting in a normal nuclear-cytoplasmic ratio. The chromatin material, which is frequently arranged in bands with clear spaces between the bands, is unusually pale thus imparting a general appearance of nuclear hypochromasia. Nucleoli are commonly present and unusually prominent. These changes have been interpreted as a specific marker of atrophy and indicative of maturation disorders [46]. Although these cytological changes undoubtedly are seen in association with chronic atrophic gastritis they are probably largely degenerative in nature.

There is no doubt that the epithelial cell abnormalities associated with chronic gastritis and peptic ulceration, in a phase of activation or exacerbation, do produce the single most difficult diagnostic problem in gastric cytology. The problem is due to those terrible cytological twins, degeneration and regeneration which, as is the habit of twins, frequently go hand in hand. This problem, of course, is not unique to gastric cytology as witness, for example, the diagnostic dilemma associated with some cases of acute or chronic inflammation of the uterine cervix.

Fig. 286. Abnormal cells from patient with chronic superficial gastritis. Nuclear enlargement is minimal but abnormalities of chromatin pattern and nucleoli are prominent. × 545.

Fig. 287. Group of cells from adenocarcinoma of stomach. Note smooth rounded nucleoli which contrast with those in previous figure. × 545.

Fig. 288. Cells in gastric brushings from patient with chronic peptic ulcer. Note slight nuclear enlargement and abnormal chromatin pattern. × 545.

Fig. 289. Cells in gastric brushings from patient with chronic superficial gastritis. Nuclear enlargement is moderate and irregular nucleoli are prominent. × 545.

Fig. 290. Cells in gastric brushings from patient with chronic superficial gastritis. There is some enlargement and irregularity in outline of nuclei and moderate disturbances of chromatin pattern. × 545.

Fig. 291. Cells from gastric brushings of patient with chronic peptic ulcer. Nucleoli are prominent and nuclei are enlarged although hypochromasia indicates degenerative component. × 545.

Fig. 292. Cells in gastric brushings from patient with chronic superficial gastritis. Nuclear abnormalities are moderate. × 545.

Fig. 293. Cells in gastric brushings from patient with chronic peptic ulcer. Nuclei are normal in size or slightly enlarged but there is considerable disturbance of chromatin pattern. × 545.

Fig. 294. Cells in gastric brushings from patient with chronic peptic ulcer. Nuclear chromatin abnormalities are considerable. × 545.

Fig. 295. Cells in gastric brushings from patient with chronic superficial gastritis. Note nuclear enlargement and marked chromatin abnormalities. × 545.

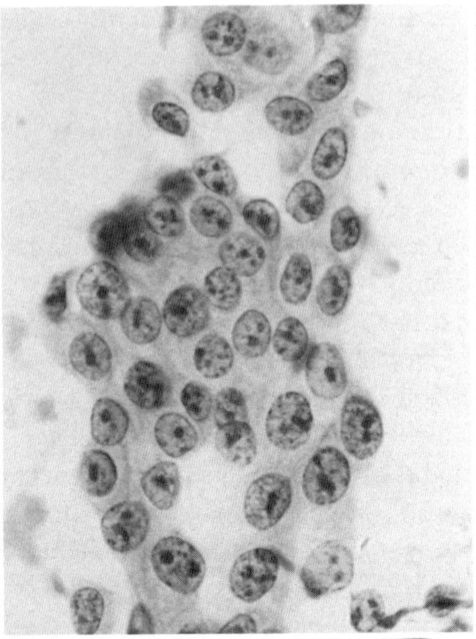

288

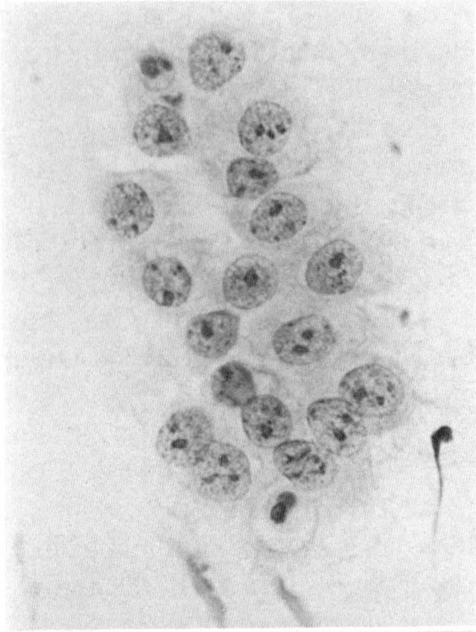

289

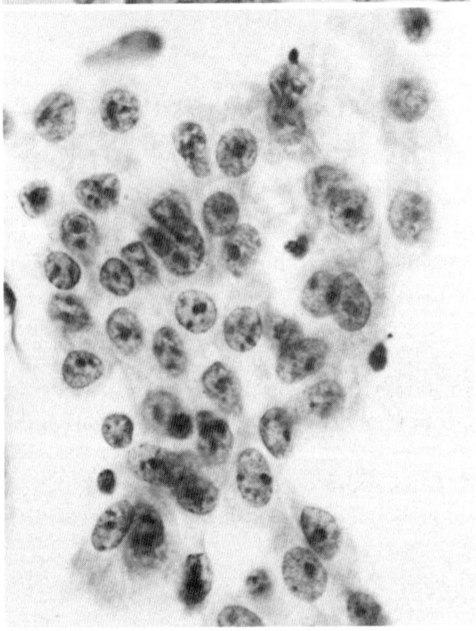

290

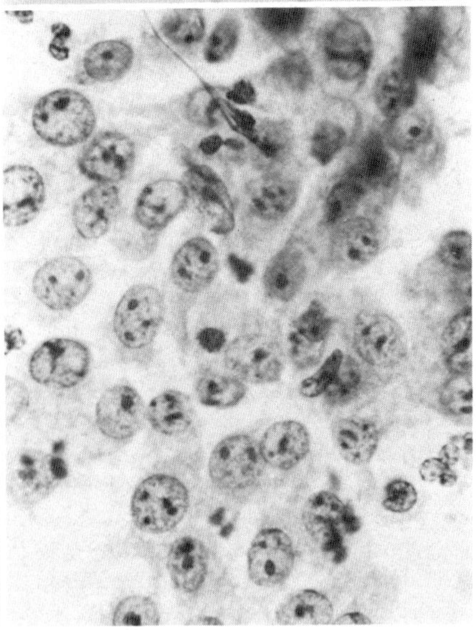

291

For legends, see p. 147.

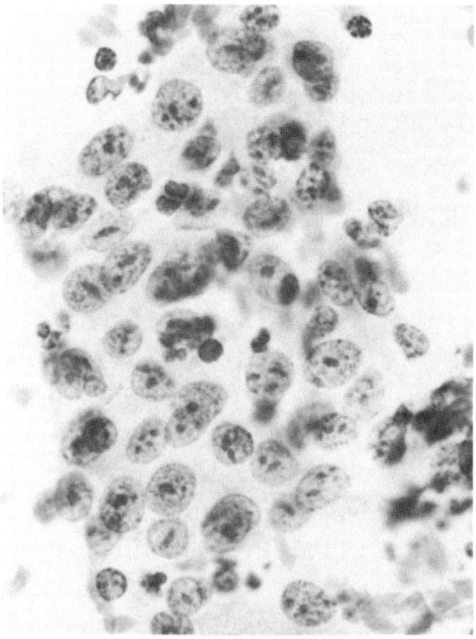

292

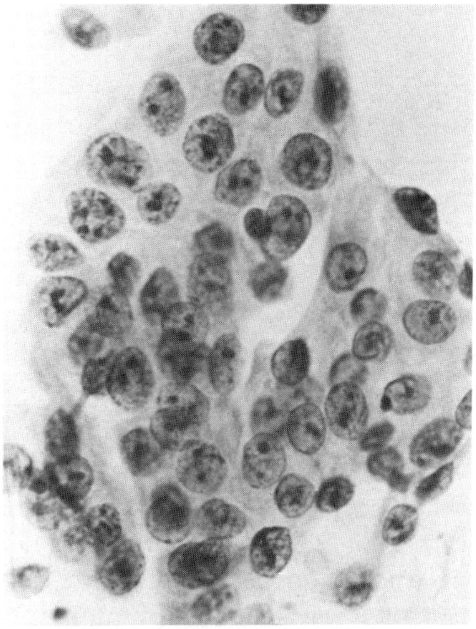

293

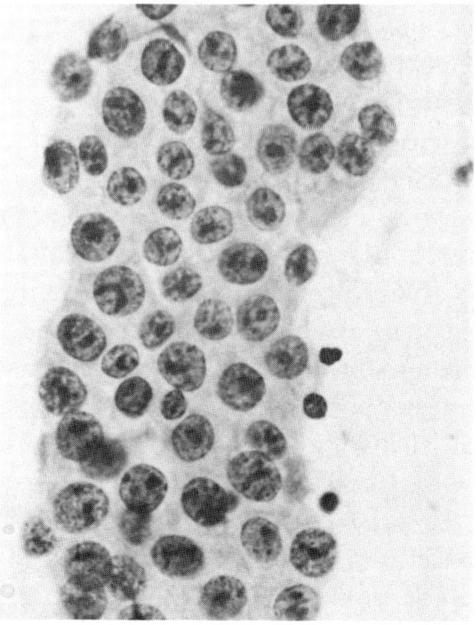

294

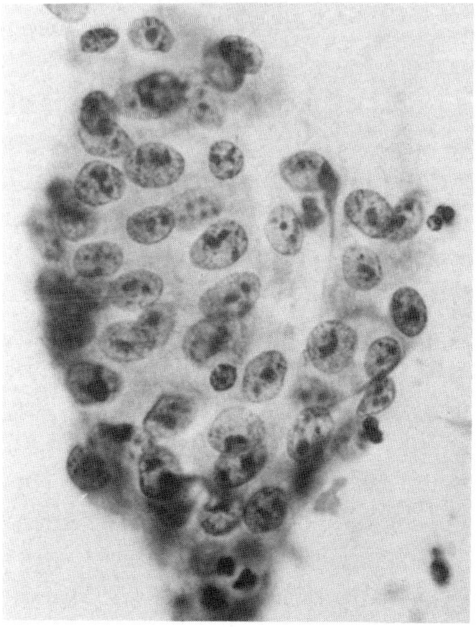

295

For legends, see p. 147.

Intestinal Metaplasia

Atrophic gastritis and gastric atrophy are almost invariably associated with the condition known as intestinal metaplasia (fig. 300). In this condition the normal gastric mucosa is replaced by cells of intestinal type-absorptive cells, goblet cells, Paneth cells and argentaffin cells (fig. 301). The goblet cells are particularly striking (fig. 302) as are the non-secretory columnar cells with the prominent microvilli or brush border so characteristic of the small bowel (fig. 303). That the metaplastic epithelium has the enzyme pattern and mucin content of normal intestinal cells can be shown by histochemical methods and the contrast between the largely neutral gastric mucus of the 'native' cells with the acid mucus of the metaplastic cells is particularly striking (plate II). Usually all morphologic and enzymatic characteristics of the metaplastic epithelium are identical with those of normal intestine and the metaplasia is then regarded as mature. Sometimes, however, the metaplastic epithelium may be less mature with, in particular, a lack of the complete set of intestinal enzymes. Such incomplete metaplasia may be of considerable significance in the genesis of gastric cancer and will be referred to again in the appropriate chapter.

The cytological recognition is of considerable importance since, as will be discussed in a subsequent chapter, it is commonly seen in association with gastric cancer and is regarded by many as a 'precancerous' condition [18, 22, 25, 26, 28, 29, 43, 48]. Its cytological manifestations are somewhat variable. Cells with a characteristic brush border may be seen in association with intestinal metaplasia, particularly if the specimen is collected by the brushing technique (fig. 304). However, there is no doubt that the mucus-secreting or goblet cells are the most common manifestation of intestinal metaplasia in both washings and brushings. The cells may be seen singly, in small groups or, occasionally, in small strips, the latter presumably torn intact from the mucosal surface (fig. 305). Occasionally cells with a brush border and goblet cells may be seen in the same microscopic field (fig. 306). A very characteristic feature is the presence of polymorphonuclear leucocytes within the cytoplasmic vacuole of the goblet cell (fig. 307).

Fig. 296. Strip of abnormal gastric mucosal cells in brushings from patient with chronic peptic ulcer. Note adhesion of cells and general retention of cell polarity. × 545.

Fig. 297. Fragment of abnormal gastric epithelium in brushings from patient with chronic peptic ulcer. Note general uniformity of nuclear size and shape and good adhesion of cells. × 545.

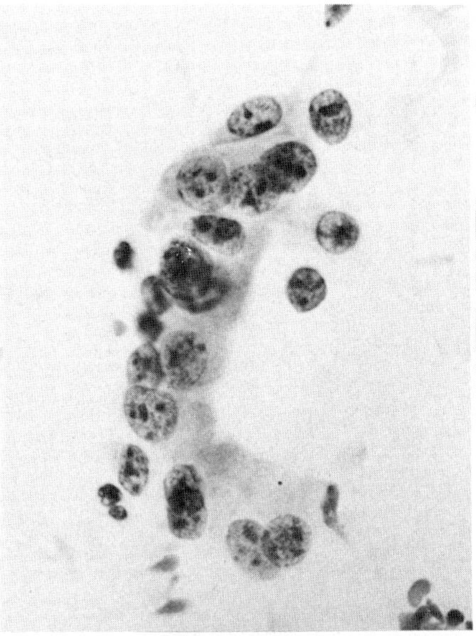

296

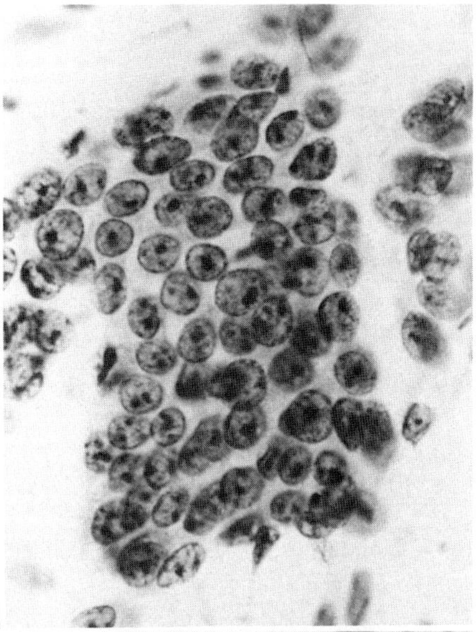

297

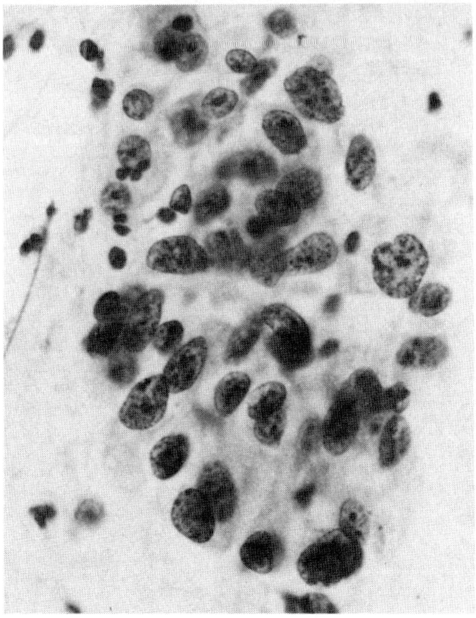

298

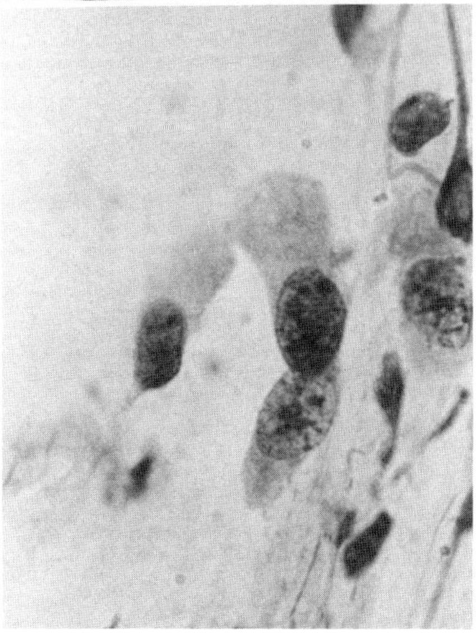

299

Fig. 298. Carcinoma cells in gastric brushings showing poor adhesion with separation of cells. Note also pleomorphism and cellular disarray. × 545.

Fig. 299. Single abnormal cells in gastric brushings from patient with chronic peptic ulcer. × 860.

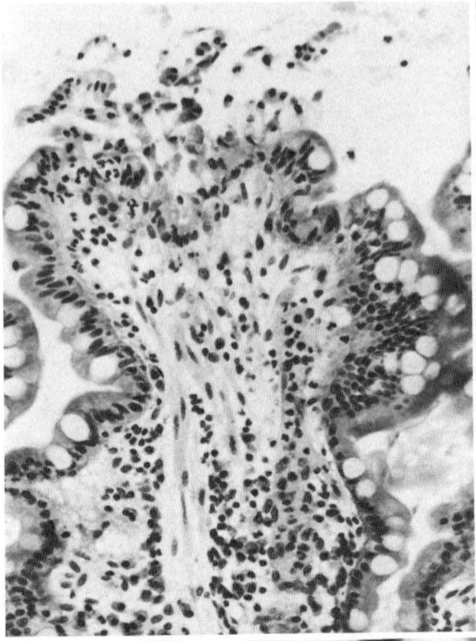

300

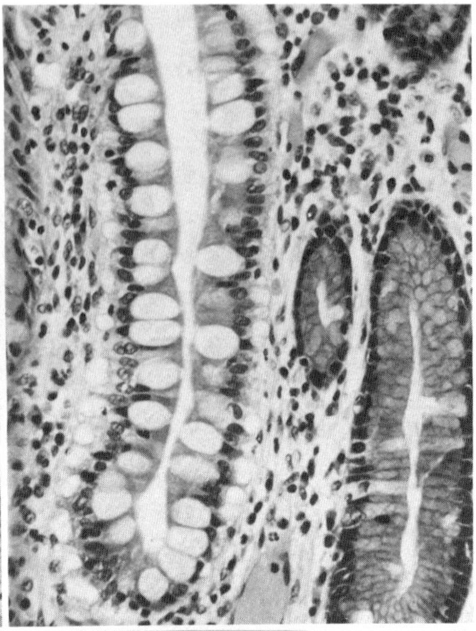

301

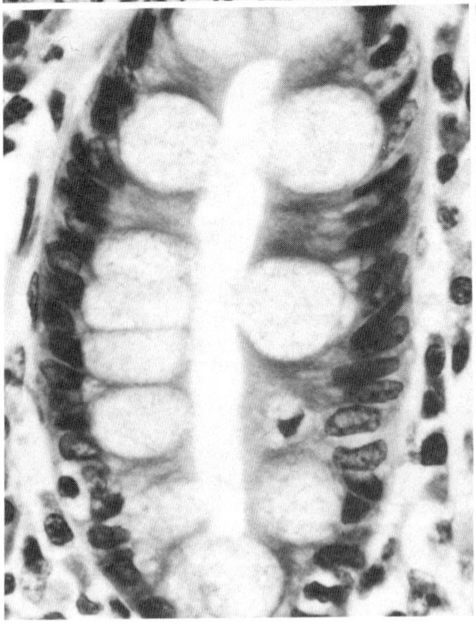

302

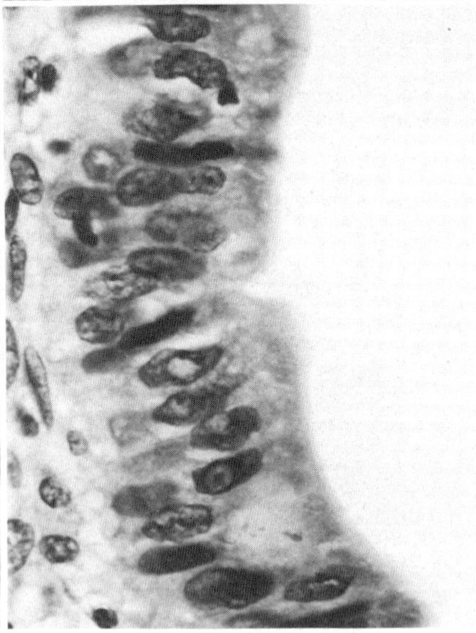

303

Fig. 300. Intestinal metaplasia associated with chronic atrophic gastritis. × 140.

Fig. 301. Intestinal metaplasia of gastric mucosa. Note characteristic goblet cells in gland in centre of photograph and compare with normal mucosal cells in other glands. × 173.

Fig. 302. Goblet cells in gastric gland characteristic of intestinal metaplasia. × 545.

Fig. 303. Area of intestinal metaplasia of gastric gland showing prominent microvilli or brush border. × 545.

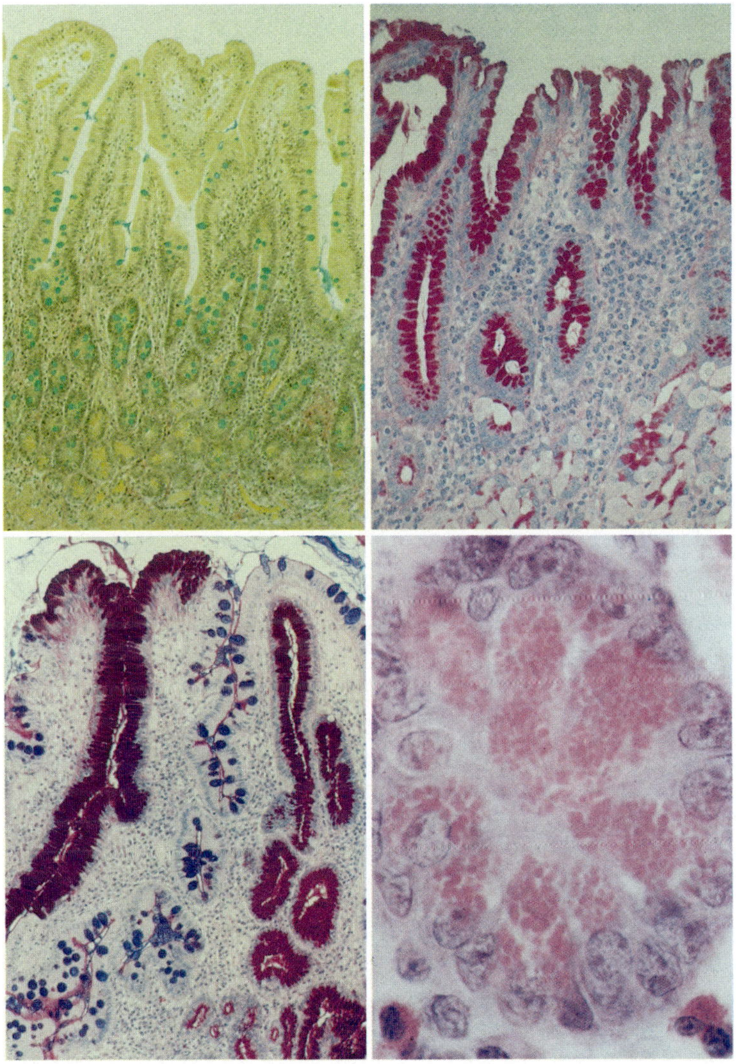

a *b* *c* *d*

For legends, see reverse side.

Plate II.

a Section of normal small intestine stained with Alcian blue. The goblet cells contain acid mucopolysaccharides and hence stain blue with this technique. × 200.

b Section of normal stomach stained by periodic acid-Schiff (PAS) reaction. The normal gastric mucins contain neutral mucopolysaccharides which do not stain with Alcian blue but react positively with PAS to appear brilliant red or magenta. × 400.

c Section of stomach with areas of intestinal metaplasia stained by combined Alcian blue/PAS technique. Note 'native' cells staining red and metaplastic cells staining blue. × 200.

d Gastric mucosal gland with evidence of intestinal metaplasia. Coarse eosinophilic granules indicate presence of Paneth cells – a normal component of intestinal glands. Pap. × 1,580.

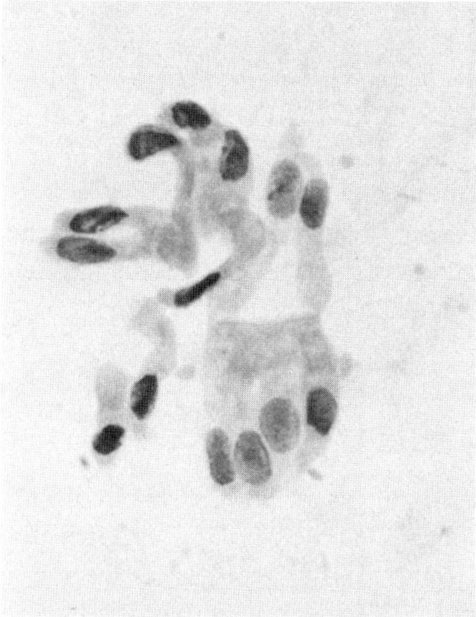

304

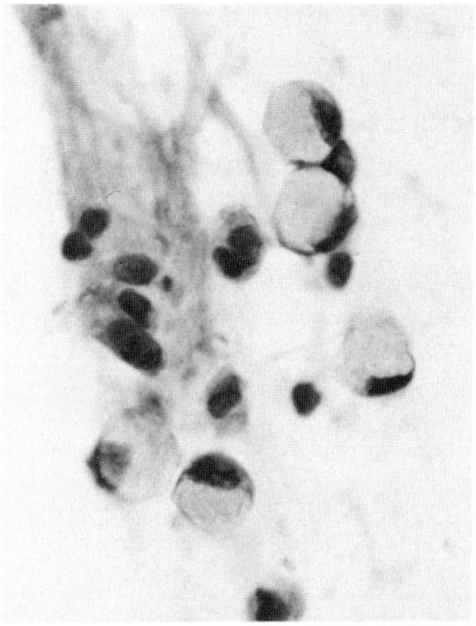

305

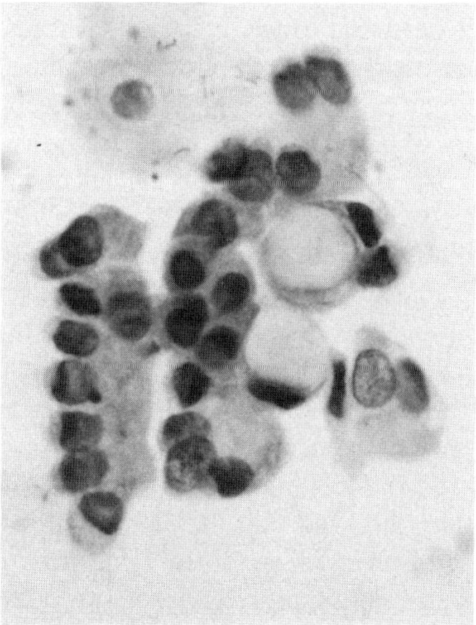

306

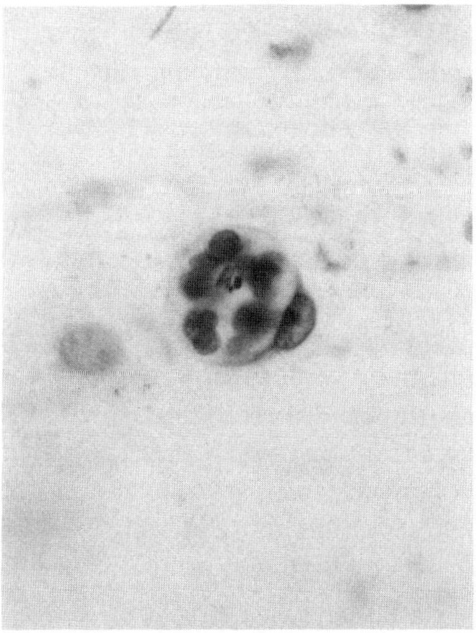

307

Fig. 304. Cells in gastric washings of patient with intestinal metaplasia. Note the prominent brush border particularly in the lower group of cells. × 545.

Fig. 305. Goblet cells in gastric washings from patient with intestinal metaplasia. × 545.

Fig. 306. Cells with brush border and goblet cells in gastric washings from patient with intestinal metaplasia. × 545.

Fig. 307. Goblet cell in gastric washings from patient with intestinal metaplasia. Note polymorphonuclear leucocytes within the cytoplasmic vacuole. × 545.

In evaluating the goblet cells shed from an area of intestinal metaplasia, particularly when they are seen as single cells, it is extremely important to distinguish them from the so-called signet ring cells characteristic of one form of adenocarcinoma. This distinction relies upon an accurate assessment of nuclear morphology and the relationship of the nucleus to the cytoplasmic vacuole. This differential diagnosis will be discussed and illustrated when signet-ring cell carcinoma is dealt with below.

Pernicious Anaemia

Pernicious anaemia is discussed separately as its cytological manifestations in gastric specimens are controversial. There is no doubt that it is a condition of considerable importance to the gastric cytologist as there is a well recognized association between pernicious anaemia and gastric cancer [33, 44, 58, 59, 66]. Thus the frequency of gastric cancer is higher in adults suffering from pernicious anaemia than in the general population and a higher incidence of stomach cancer is found among the relatives of subjects with pernicious anaemia than in families of patients not suffering from this haematological disorder. *Kaplan and Rigler* found that 12% of their patients with pernicious anaemia developed gastric cancer [27].

Changes in blood cells, both of the erythroid and myeloid series, in megaloblastic anaemia due to a deficiency of vitamin B_{12} have been recognized for a long time. In 1954, *Graham and Rheault* [19] described abnormalities of both squamous and columnar cells in patients with pernicious anaemia and these authors attributed the changes directly to the haematological disorder. The changes they described consisted of enlargement of both nuclei and cell body, multinucleation and an abnormal distribution of nuclear chromatin material. They further observed that the abnormalities disappeared after treatment of the anaemia. However, *Massey and Rubin* [38] claimed that some changes in the gastric columnar cells in patients with pernicious anaemia were permanent, these changes persisting after treatment of the condition. These authors coined the terms 'bland' and 'active' to describe the changes that they saw in the gastric washings from patients with pernicious anaemia.

Subsequently *Brandborg* et al. [3] described the 'bland' and 'active' cells of pernicious anaemia in some detail. Thus according to these authors the large bland cells have a delicate nuclear membrane which may be folded or creased. The nuclear chromatin material is distributed in fine aggregates in a relatively simple background. Nucleoli are small or absent. The nucleus and cytoplasm enlarge proportionately maintaining a normal nuclear-cytoplasmic ratio.

In contrast the large active cells have clumped irregularly distributed nuclear chromatin and the nuclear membranes are thick and irregular. These 'active' cells may closely resemble malignant cells although a normal nuclear-cytoplasmic ratio is usually maintained.

Takeda [63] illustrates cells from three cases of pernicious anaemia. In all cases the author refers to the abnormal cells as the 'bland cells' of pernicious anaemia. These cells all have enlarged nuclei with a 'washed-out' appearance and prominent but tiny red nucleoli. *Takeda* stresses the importance of the 'very fine ridges of the nuclear membrane' that are present. Examination of the photographs suggests that the 'fine ridges' are partial folds in the nuclear membrane and hence in accordance with Brandborg's original description of the 'bland' cell of pernicious anaemia.

Within our laboratory cells resembling the so-called 'bland' cells are readily identifiable in cases of pernicious anaemia both treated and untreated. They are characterized by large nuclei, with a distinct nuclear membrane but very little stainable chromatin. This latter feature gives the cells a characteristic 'empty' look. Frequently they have small nucleoli (fig. 308). Folds or 'very fine ridges' in the nuclear membrane are visible in many of the cells (fig. 309).

In some cases 'active' cells are also prominent. They have enlarged, somewhat hyperchromatic nuclei, one or more large nucleoli and abnormalities of the chromatin pattern which may be quite marked (fig. 310).

Whilst there is no doubt that cells of both types do occur in cases of pernicious anaemia their significance is uncertain. Several authors attribute the abnormalities specifically to the haematological disorder and indeed *Nieburgs and Glass* [46] attribute the changes to an interference in gastric cell maturation due to a disturbance of DNA synthesis. Most authors, however, are inclined to regard the changes as not specific to pernicious anaemia but as a reflection of the associated chronic atrophic gastritis or gastric atrophy and intestinal metaplasia [16, 57].

In addition to abnormalities of glandular cells gastric washings and brushings, particularly the former, from patients with pernicious anaemia frequently contain abnormal squamous cells presumably derived from the esophagus and oral cavity. The abnormalities consist of nuclear enlargement and hyperchromasia and, quite frequently, binucleate and multinucleate cells (fig. 311). Again there is some speculation as to the genesis of these cells. *Koss* [31] suggests that they may result from intubation of the esophagus or they may be accounted for by a number of dietary deficiencies such as vitamin B12 or folic acid deficiency. However, again it seems reasonable to suggest that as vitamin B12 is essential to the processes of DNA synthesis, a deficiency of this substance may result in disturbances of squamous cell maturation and consequent morphologic abnormalities.

It would appear to the author that there can be no final resolution of these problems. The evidence would suggest that some of the abnormalities are due to the pernicious anaemia per se, with the concomitant vitamin B12 deficiency, but that most of the changes are the result of the associated chronic atrophic gastritis or gastric atrophy and intestinal metaplasia. Since, as already discussed, carcinoma develops more frequently in stomachs of pernicious anaemia patients, with their gastric atrophy and intestinal metaplasia, and since the development of established cancer may well be preceded by a dysplastic stage the so-called 'active' cells may well be a reflection of this dysplasia. The most important thing, however, is not so much the speculation as to causal relationships but rather the practical approach to the patient who is known to be suffering from pernicious anaemia. Thus great care is required in the interpretation of cells in the gastric specimens from these patients particularly if they have not been treated for their haematological disorder. The care is necessary to avoid the risk of making a 'false-positive' diagnosis or, perhaps more importantly, the risk of attributing the appearances of carcinoma cells to those due to pernicious anaemia. Prolonged follow-up of such patients is always desirable and indeed, as discussed elsewhere, they may be subject to regular cytological screening.

Benign Neoplasms

Benign neoplasms of the stomach are relatively uncommon. However, the reported incidence depends, to a large extent, on the thoroughness of the examination as many such tumours are small, asymptomatic and easily overlooked. In one analysis of figures derived from several large series of

---

Fig. 308. 'Bland' cells in gastric washings from patient with pernicious anaemia. 'Washed-out' nuclei and nuclear moulding are prominent features. × 545.

Fig. 309. 'Bland' cells in gastric washings from patient with pernicious anaemia. Folds in nuclear membrane or 'fine ridges' are prominent in some cells. × 545.

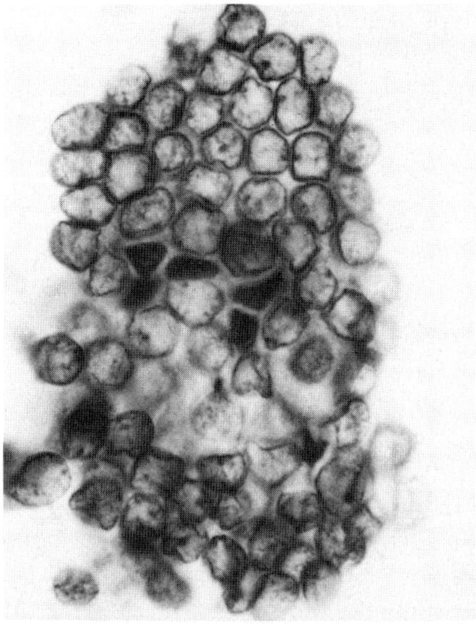

308

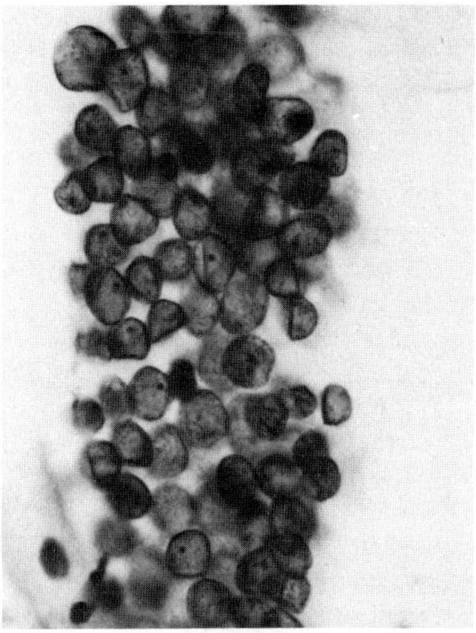

309

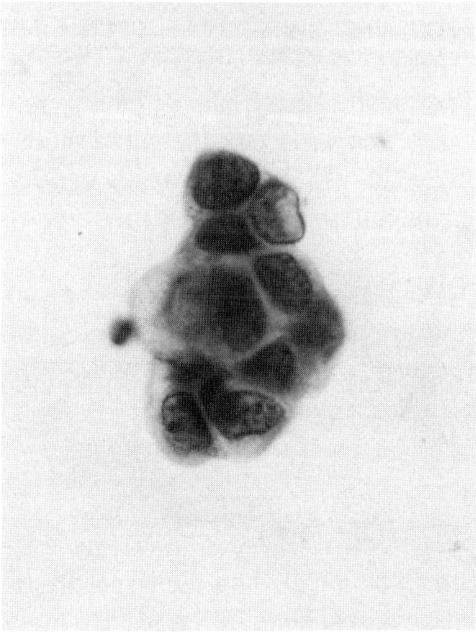

310

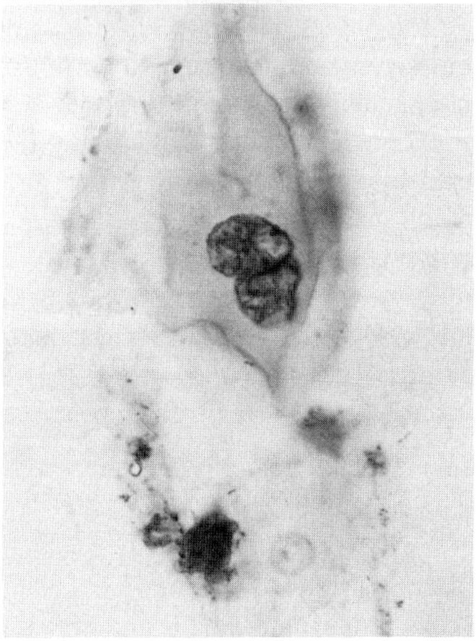

311

Fig. 310. 'Active' cells in gastric washings from patient with pernicious anaemia. Nuclear enlargement, moulding, and abnormal chromatin pattern suggests possibility of dysplasia or carcinoma. × 545.

Fig. 311. Abnormal squamous cells, presumably from oral cavity, in gastric washings from patient with pernicious anaemia. × 545.

case reports [41] benign epithelial polyps were most common, accounting for 41% of cases whilst a further 37% were leiomyomas. The remaining tumours were so rare as to be of little importance and indeed none has cytological implications.

Benign Epithelial Polyps

Gastric polyps, although representing the commonest benign tumour of the stomach, are not common. In a study of 7,000 autopsies *Lawrence* [35] found only 50 cases whilst, in an even larger series of 11,000 autopsies, *Stewart* [61] was able to demonstrate only 47 examples. Two main categories are recognized:

Hyperplastic (Regenerative) Polyps

The hyperplastic polyp is the commoner accounting for 80–90% of all gastric polyps. It is usually small, measuring approximately 1–2 cm in main diameter, and is frequently multiple (fig. 312). The lesions are usually pedunculated and are composed of hyperplastic gastric glands which may show prominent cystic dilatation (fig. 313). The surface of the polyp is covered by a single layer of gastric epithelial cells which usually show considerable mucus secretion (fig. 314). The malignant potential of this type of polyp is minimal.

Cytologically there are no characteristic diagnostic features. The surface cells shed readily and fragments may be seen in which there is some evidence of cell crowding (fig. 315). However, the arrangement of the cells is an orderly one and the appearances are unmistakably benign.

Adenomatous Polyps

The adenomatous polyp, which is generally regarded as a true neoplasm, is much less common. They tend to be larger than the hyperplastic polyp, are usually single, and are almost always sessile. In figure 316, how-

---

Fig. 312. Multiple hyperplastic or regenerative polyps of gastric mucosa.

Fig. 313. Section of hyperplastic polyp of gastric mucosa. Note clearly defined pedicle and prominent cystic dilatation of gastric glands. × 2.

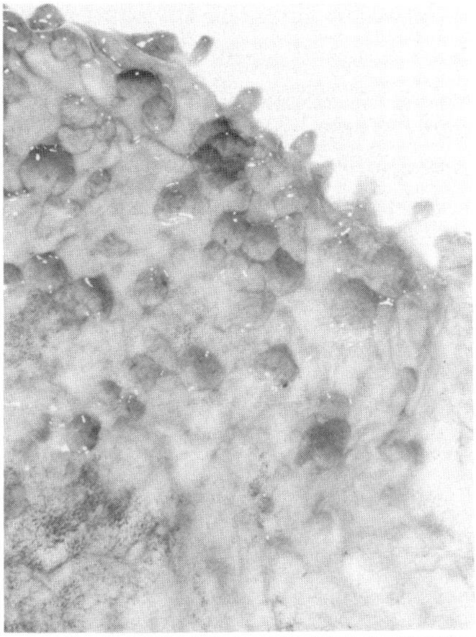

312

313

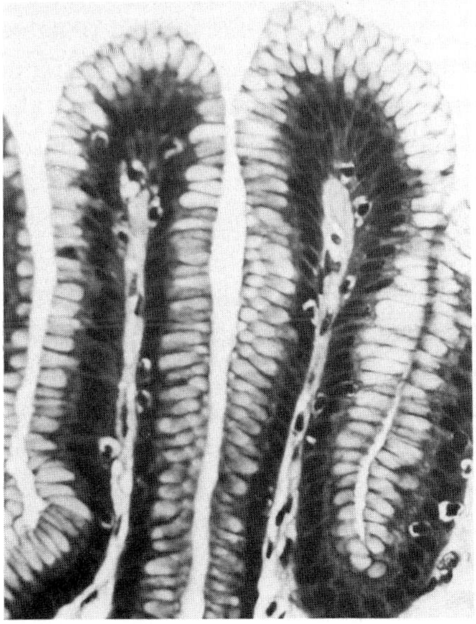

314

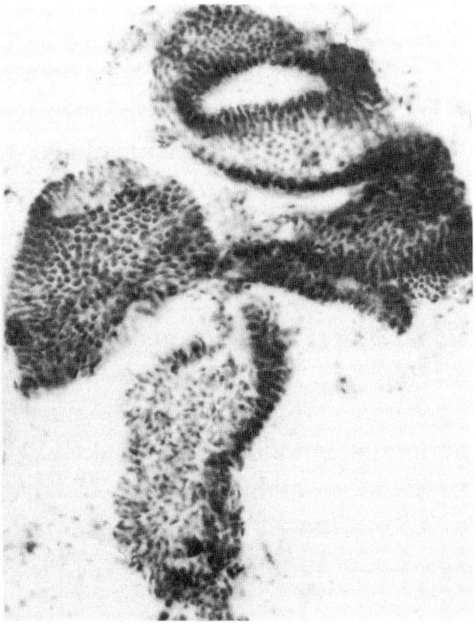

315

Fig. 314. Surface epithelium of hyperplastic polyp of gastric mucosa showing marked mucus secretion. × 275.

Fig. 315. Fragments of gastric mucosa in washings from patient with benign hyperplastic polyps. × 86.

ever, multiple adenomatous polyps are depicted. Those in the fundus and antrum of the stomach are large and sessile whilst that on the lesser curvature is clearly pedunculated. Histologically the adenomatous polyp has a papillary, villous, or tubulo-villous structure. The disorderly elongated and branched glands are lined by tall columnar epithelial cells which show varying degrees of atypia or dysplasia (fig. 317) sometimes justifying the diagnosis of carcinoma in situ.

These lesions could cause considerable diagnostic problems cytologically but fortunately the problem seldom arises in laboratory practice. The dysplastic changes may be reflected in the specimen but these will be discussed in a later chapter. Should a focus of invasive carcinoma be present then the diagnosis is essentially that of a polypoid carcinoma.

Leiomyoma

The stomach is the most common site for leiomyoma of the gastrointestinal tract. These neoplasms are frequently asymptomatic but patients may present with haemorrhage, iron deficiency anaemia or abdominal pain. If small the neoplasms remain intramural but if large they project from the mucosal surface in a pedunculated manner. The cut surface of the tumour has a whorled appearance (fig. 318) often with cystic spaces whilst histological examination of the tumour shows whorls of smooth muscle cells (fig. 319).

If uncomplicated, a leiomyoma produces no cytological manifestations. However, occasionally the mucosa overlying a large leiomyoma may ulcerate (fig. 320). This ulceration produces a characteristic radiological appearance (fig. 321). It also has considerable cytological implications, abnormal epithelial cells, as already described in association with a chronic peptic ulcer, often appearing in the gastric washings or brushings. In addition, brushing of the exposed tumour tissue may yield characteristic elongated smooth muscle cells (fig. 322, 323).

Fig. 316. Adenomatous polyps of gastric mucosa.

Fig. 317. Surface of adenomatous polyp of gastric mucosa showing marked abnormalities of epithelial cells. × 173.

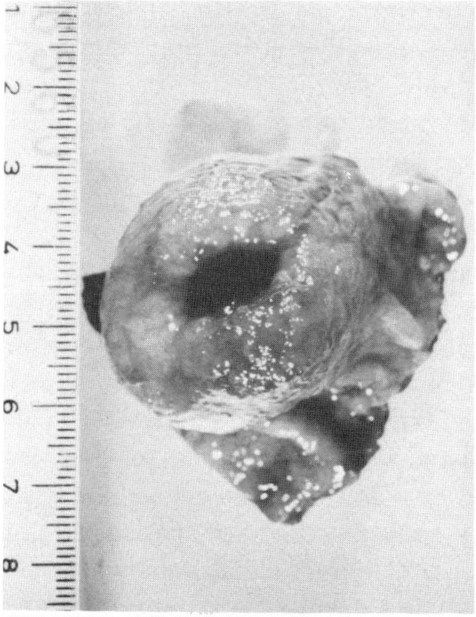

320

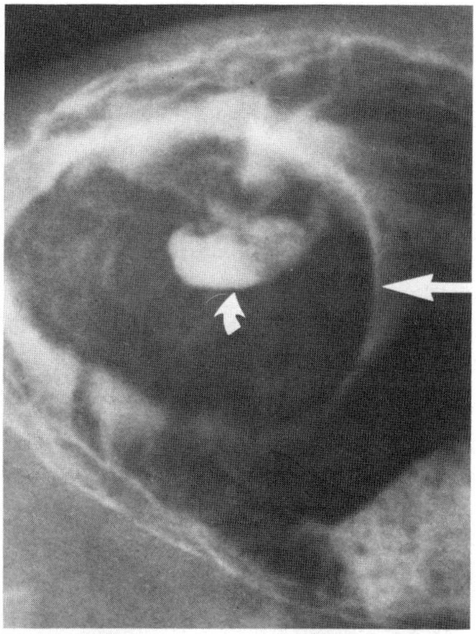

321

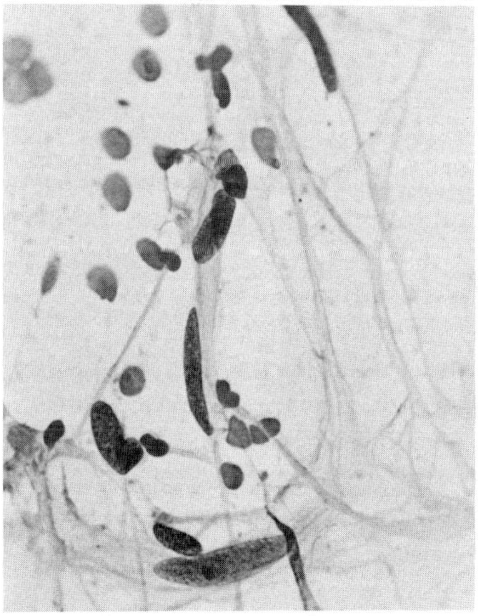

322

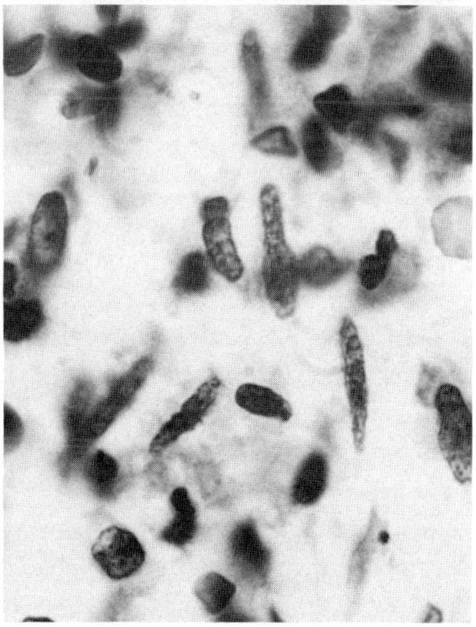

323

Fig. 322. Smooth muscle cells in gastric brushings from patient with ulcerated leiomyoma. × 435.

Fig. 323. Smooth muscle cells in imprint of gastric leiomyoma. × 860.

In summary, *Ming* distinguishes two principal growth patterns – one primarily exhibiting intraluminal growth, the other an intramural growth. Combinations of both do, of course, occur. Essentially, therefore, carcinomas of the gastric mucosa present, as do most carcinomas of body surfaces, as fungating, ulcerating and infiltrating lesions as seen in figures 324–327. These growth characteristics can be demonstrated clearly by radiological means (fig. 328–330).

Whilst the gross morphologic characteristics are not of great significance in the cytologic interpretation they may be of importance in specimen collection. Thus an extensively ulcerated tumour of the type depicted in figure 325 frequently yields a great deal of necrotic debris and inflammatory exudate. Indeed direct brushing of the floor of the ulcer may result in collection of this material only or the malignant cells, if harvested, may be very degenerate and hence difficult to assess. The diffusely infiltrative type, including the classical linitis plastica or 'leather-bottle' stomach as shown in figures 326 and 327 may not shed many cells into the gastric lumen nor may the carcinoma tissue be readily accessible to direct brushing or washing techniques. Hence again a problem of sampling may arise.

Histological Appearances

A great variety of histologic patterns are seen in gastric carcinoma and indeed *Stout* [62] has stated that 'histological classification is hopeless and unrewarding'. Nevertheless many classifications have been proposed most conforming to the traditional approach of classification according to the presence or absence of recognizable tissue patterns.

In 1965 *Lauren* [34] proposed a classification that departed from the classical descriptions and allocated gastric carcinoma into two main groups – intestinal type and diffuse type. In general terms intestinal type carcinomas have a glandular pattern, usually accompanied by papillary formation or solid components, and are made up of large pleomorphic cells with large hyperchromatic nuclei often showing mitoses (fig. 331). Intestinal metaplasia may be prominent in the gastric mucosa in the vicinity of the neoplasm. This type of neoplasm is usually well demarcated. The diffuse type of car-

---

Fig. 324. Large fungating carcinoma of stomach.

Fig. 325. Ulcerating carcinoma of stomach. Note loss of rugal pattern in region of ulcer crater.

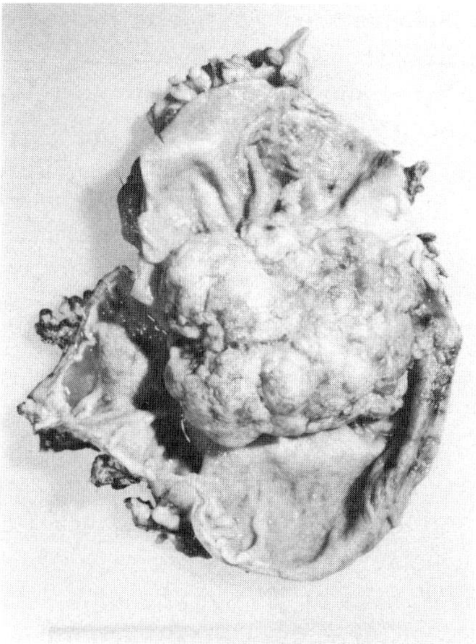
324

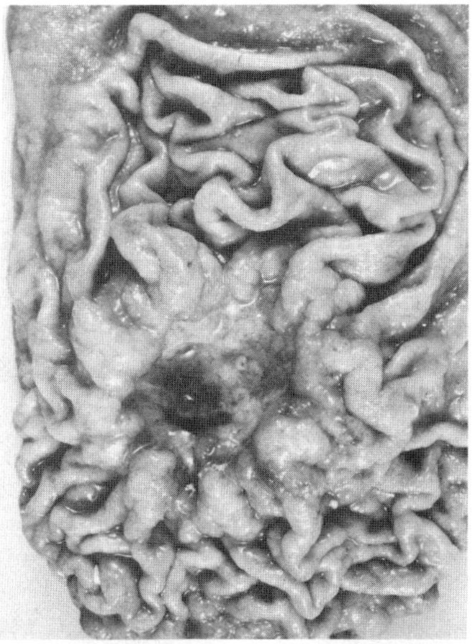
325

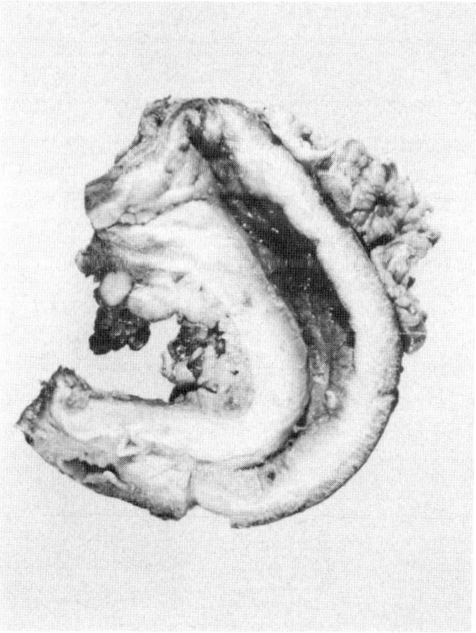
326

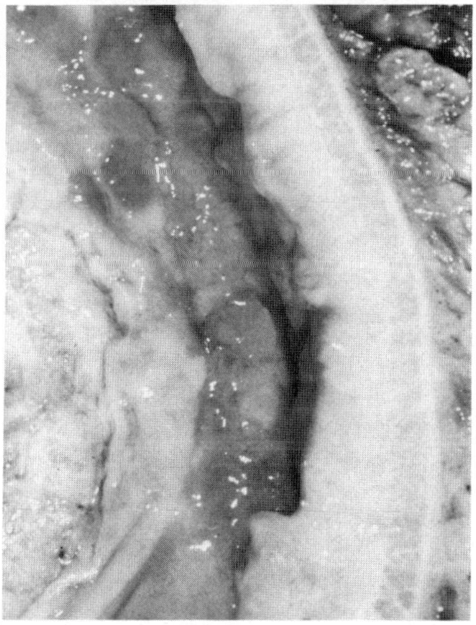
327

Fig. 326. Infiltrating carcinoma of stomach producing linitis plastica or 'leather-bottle' stomach.

Fig. 327. Infiltrating carcinoma of stomach showing marked thickening of gastric wall and rigidity of mucosal rugae.

cinoma may show some glandular arrangement, particularly in the more superficial part of the tumour, but is mainly characterized by scattered single cells or small clusters of cells, the latter showing poor cell cohesion. Many of the cells contain mucus, and typical signet-ring forms may be seen, but non-secretory cells also occur (fig. 332). This type of neoplasm is poorly demarcated with a tendency to spread widely within the gastric wall. Whilst the majority of gastric carcinomas can be subdivided into intestinal and diffuse types, a minority cannot be so classified and hence an 'indeterminate' or 'mixed' category is required. In addition, squamous cell carcinoma and adenosquamous carcinoma are separately categorized.

This type of classification would appear to be of some importance since the two major types of carcinoma of the stomach, the intestinal and the diffuse, and the less common mixed type, show differences in epidemiological factors, clinical behaviour and natural history. In addition, they are distinguishable cytologically with a considerable degree of accuracy. Thus *Rilke* [54], for example, in a study of 210 cases of gastric carcinoma with histological follow-up, achieved a cytological sensitivity of 0.935 for intestinal type carcinoma, 0.85 for the diffuse type, and 0.88 for the mixed type, the latter accounting for approximately 10% of the cases. Cytologic identification of the histologic type was based on a combination of criteria, none specific by itself, although the pattern of cellularity and the morphology of the nuclei remained the most useful diagnostic features.

Despite the theoretical advantages of the Lauren classification and its cytologic applicability, the author prefers to use a more traditional system of nomenclature, as exemplified by the World Health Organization classification [47].

WHO classification of malignant epithelial gastric tumours:

Adenocarcinoma
- Papillary
- Tubular
- Mucinous

Signet-ring cell carcinoma
Adenosquamous carcinoma

Fig. 328. Fungating and infiltrating carcinoma of stomach. Obvious thickening of the gastric wall is indicated by the straight arrow whilst the curved arrows show the very marked and irregular constriction of the gastric lumen.

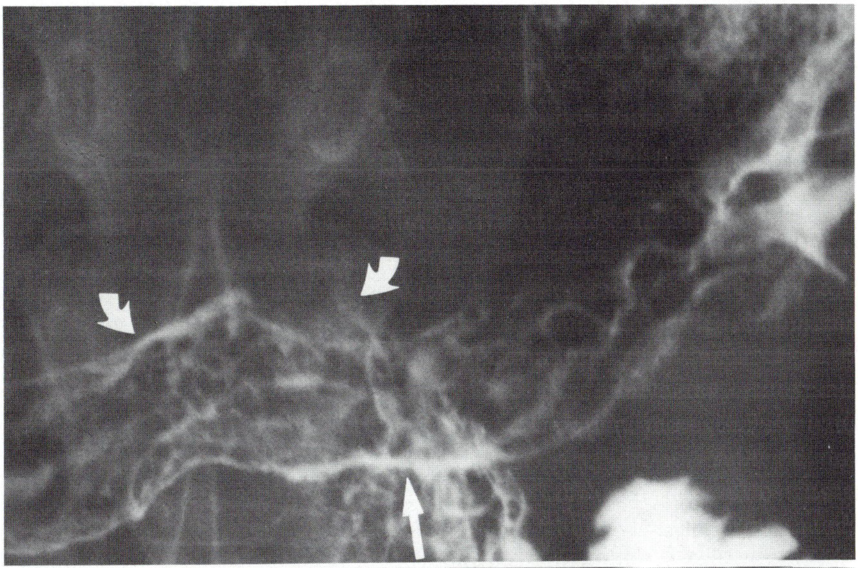

328

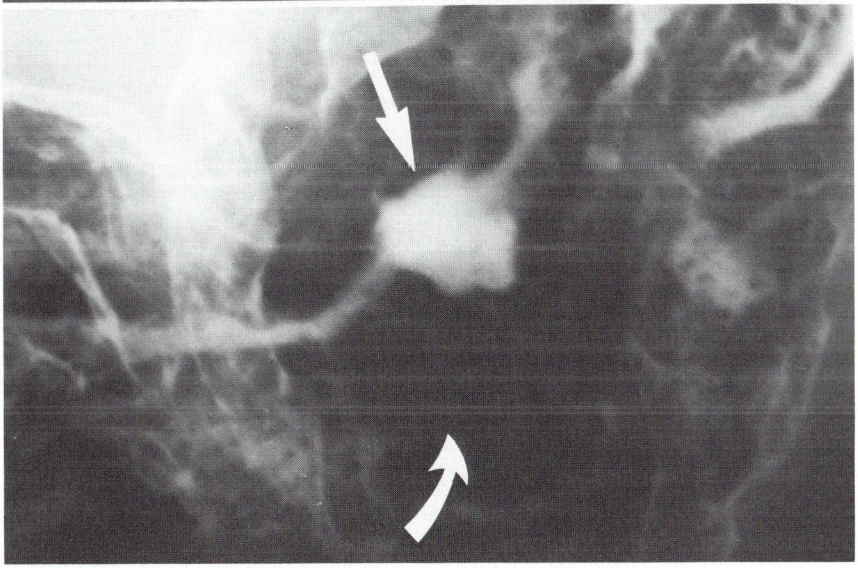

329

Fig. 329. Ulcerating carcinoma of stomach. The straight arrow indicates a central area of ulceration filled with contrast medium. Note the irregular outline of the ulcer crater. The surrounding mucosa shows a prominent loss of rugal pattern, as indicated by the curved arrow, this loss being due to malignant infiltration.

Squamous cell carcinoma
Undifferentiated carcinoma
Unclassified carcinoma

In using this classification, however, it must be appreciated that not only are there marked differences in the appearances of separate carcinomas but also often considerable variation within individual tumours, either in structure, differentiation, or both. In addition, in applying this and similar classifications to the cytodiagnosis of gastric carcinoma some confusion may be created by the variable use of the term 'differentiated'. As pointed out by *Koss* [32], in a totally different context, the word differentiation, although used freely in discussions of tumour pathology, is difficult to define. In a classical sense it is used to describe 'the emergence of a specialized cell from a primitive precursor'. Specialization in turn is a functional term denoting a specialized activity on the part of a cell or tissue (e.g. mucus secretion, keratin formation) and this activity can usually be recognized or inferred from morphologic examination. In tumour pathology a neoplasm is classified according to the tissue it resembles (histological classification) or according to the tissue from which it is derived (histogenetic classification) the former method being preferable.

We further take note of the degrees of resemblance, this feature being expressed in terms of degrees of 'differentiation'. Thus a well differentiated tumour closely resembles the normal tissue whereas a poorly differentiated neoplasm shows very little resemblance to the normal. The histopathologist depends on tissue pattern to establish the degree of differentiation. If an adenocarcinoma is characterized by clearly defined, well structured, glandular acini it is said to be well differentiated; if it is composed of sheets of cells with only occasional poorly formed glandular structures it is described as a poorly differentiated adenocarcinoma. The cytopathologist may also be guided by tissue patterns, 'malignant' acini or papillae, for example, being seen not uncommonly in cell preparations. However, the cytopathologist is much more reliant on individual cell structure, the cell showing cytoplasmic keratinization indicating origin from a squamous cell carcinoma, whilst the

---

Fig. 330. Infiltrating carcinoma of stomach. The wall of the stomach is obviously thickened as indicated, for example, by the two small curved arrows. The lumen of the stomach is markedly but fairly uniformly constricted as indicated by the straight arrow.

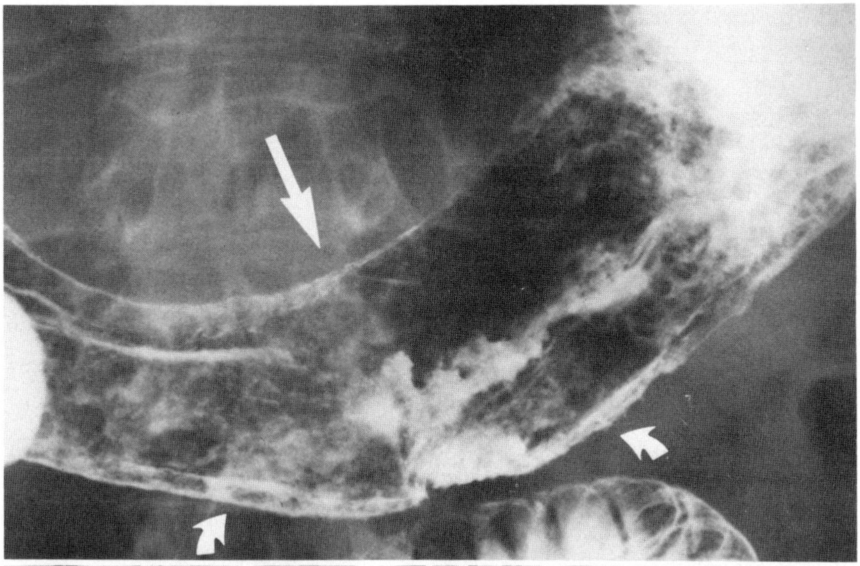

330

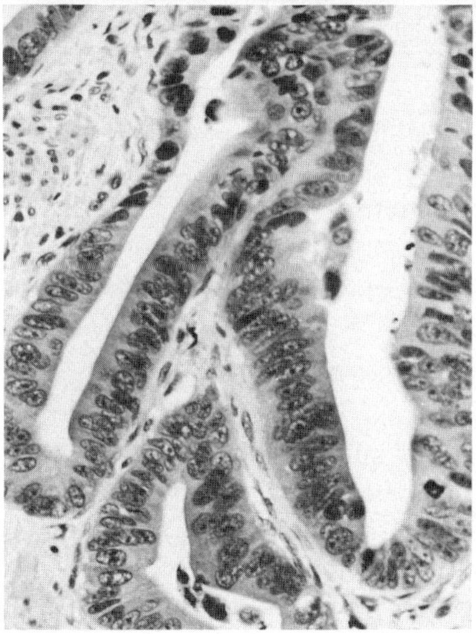

331

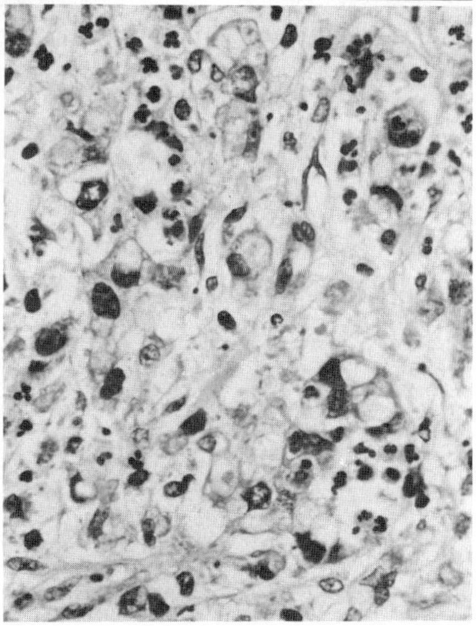

332

Fig. 331. Section of gastric carcinoma of intestinal type (Lauren). Note the well formed tubular structures, the crowded, and in places stratified, tumour cell nuclei and the suggestion of a brush border. × 205.

Fig. 332. Section of gastric carcinoma of diffuse type (Lauren). Note the scattered single cells or small clusters of cells, the latter showing poor cell cohesion. × 345.

cell whose cytoplasm shows evidence of secretory vacuolation is deduced to be arising from an adenocarcinoma.

These concepts are of considerable importance in a discussion of gastric cytology. In the classical linitis plastica, for example, the wall of the stomach is diffusely infiltrated by single cells, frequently with no evidence of gland formation. At a histological level this tumour would be regarded as being undifferentiated. Nevertheless, in many cases the infiltrating cells are clearly mucus-producing, infiltration by 'signet-ring' cells being quite common. At a cytological level such cells would be regarded as showing 'glandular' differentiation or as being derived from an adenocarcinoma. With this concept in mind the term 'undifferentiated' or 'anaplastic' should only be used to designate those cells that have relatively scanty cytoplasm and in which there is no evidence of specialized cytoplasmic function.

These problems of classification and nomenclature have been discussed in some detail since cytodiagnosis must be based on a clear understanding of the histological features of the lesions being studied. Accordingly it is suggested that the cytological diagnosis of gastric carcinoma be based on the recognition of cells that are shed from:

Adenocarcinoma – of all degrees of differentiation and including the so-called 'signet-ring' carcinoma.
Undifferentiated or 'anaplastic' carcinoma.
Adenosquamous carcinoma.
Squamous cell carcinoma.

As already indicated the differentiated forms of adenocarcinoma may be subdivided into, for example, papillary and tubular types, each in turn being further categorized by the degree of differentiation (fig. 333, 334). Within this group also are the so-called mucinous adenocarcinomas in which the individual carcinoma cells are surrounded by large amounts of mucin (fig. 335) and the signet-ring cell carcinoma, an adenocarcinoma in which the majority of cells contain within their cytoplasm a large droplet of mucin which displaces the nucleus to the periphery of the cell (fig. 336).

Similarly, although less usefully perhaps, the undifferentiated carcinoma may also be subdivided. Thus terms such as 'solid', where the cells are closely packed and the tumours have a well defined boundary (fig. 337) and 'scirrhous' where the tumour cells are few relative to the fibrous stroma (fig. 338), may be applied to an undifferentiated lesion.

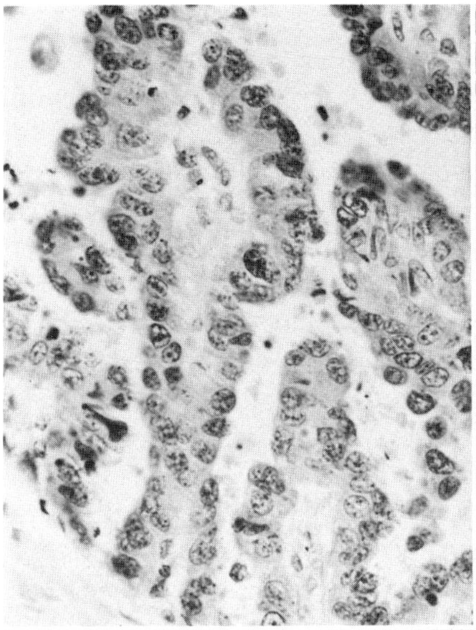

333

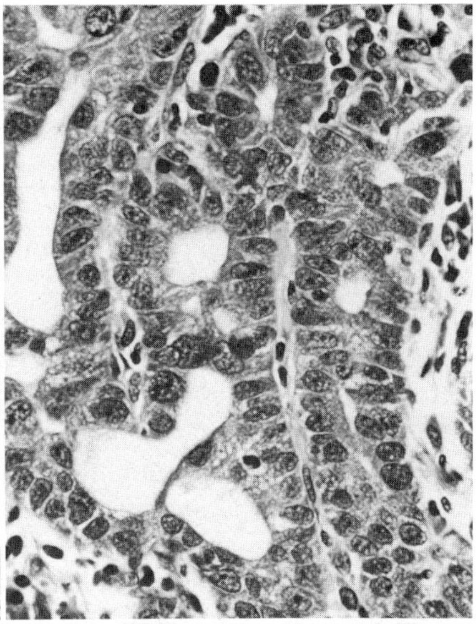

334

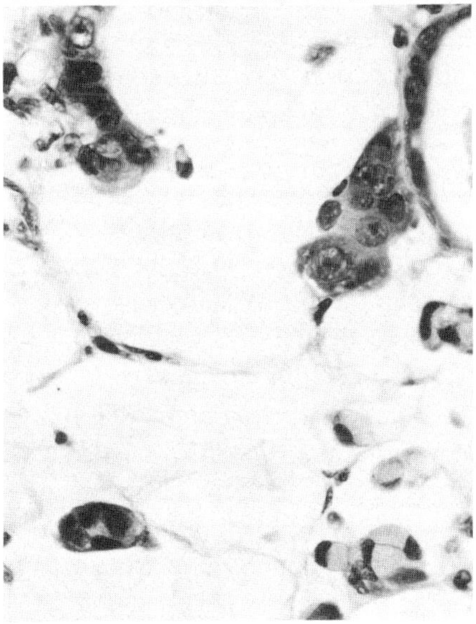

335

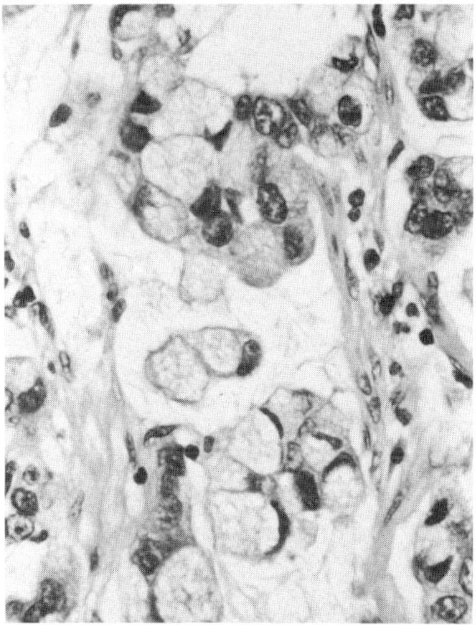

336

Fig. 333. Well differentiated papillary adenocarcinoma of stomach. ×275.
Fig. 334. Well differentiated tubular adenocarcinoma of stomach. ×275.
Fig. 335. Mucinous adenocarcinoma of stomach. ×275.
Fig. 336. Signet-ring cell carcinoma of stomach. ×345.

Cytological Diagnosis

It is important to note from the outset that malignant criteria are usually present and clearly recognizable. Indeed it is stressed that identification of such criteria is essential if false diagnoses of malignancy are to be avoided. Malignant criteria have been described and illustrated in considerable detail in an earlier chapter. As emphasized in this early discussion there is no absolute criterion of malignancy. Nevertheless gastric cancers are usually characterized by a number of malignant criteria present in numerous cells. In the diagnosis of gastric malignancy certain criteria are more useful than others. For example, as has already been emphasized, the chromatin pattern may be markedly disturbed in non-neoplastic conditions such as peptic ulceration and superficial gastritis. Hence such disturbances of chromatin must be interpreted with extreme caution. However, the cells shed from a gastric carcinoma show frequently a chromatin pattern that is characteristic. The chromatin has a very coarse or 'sharp' granularity (fig. 339) that contrasts with the 'softer' and finer appearance of the chromatin in a cell associated with gastritis and/or peptic ulceration (fig. 340). This contrast is by no means unique to gastric cytology. Thus the differentiation between the reactive or regenerative changes associated with chronic cervicitis and those of endocervical adenocarcinoma may depend on a similar contrast.

A criterion more readily assessed, perhaps, is the nuclear outline. Although the nuclei in cells associated with gastritis or peptic ulceration may be enlarged they usually retain a relatively smooth outline (fig. 341). This contrasts dramatically with the irregular outline of many gastric carcinoma cells some of which are characterized by deep notches or, alternatively, projections into the surrounding cytoplasm (fig. 342).

However, it is in small groups of cells and tissue fragments that malignant criteria are of most value. Nuclear moulding (fig. 343) is a valuable indicator of malignancy this phenomenon being seldom if ever seen in reactive or regenerative processes. Finally, pleomorphism is suggestive of a carcinoma (fig. 344), the cells from peptic ulceration and/or gastritis usually being relatively uniform in size and shape.

As already emphasized, gastric adenocarcinoma may manifest all degrees of differentiation. At the one extreme are well differentiated papillary and tubular lesions and here there is often a remarkable correlation between the cytological manifestations and the subsequent histological findings (fig. 345–348). As discussed the signet-ring carcinoma may also be regarded as well differentiated, at least at the cytological level (fig. 349,

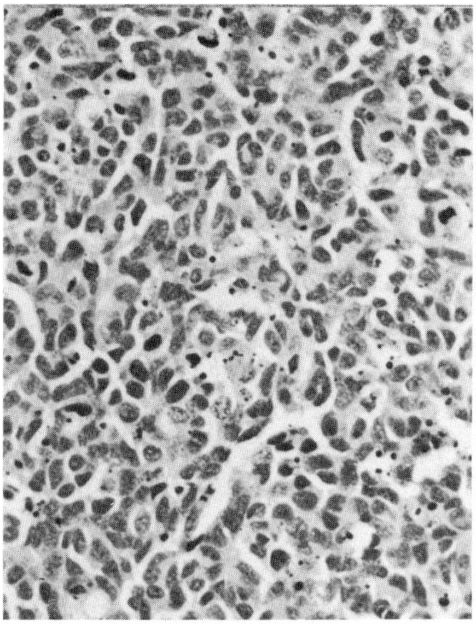

337

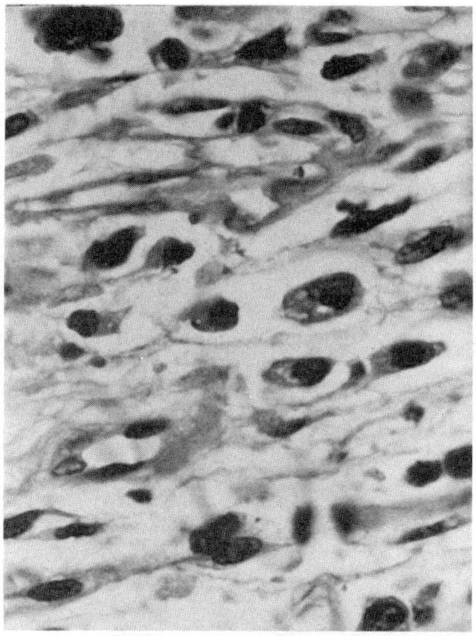

338

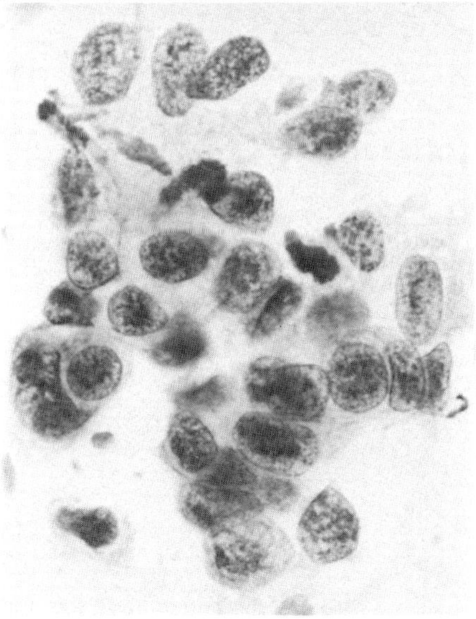

339

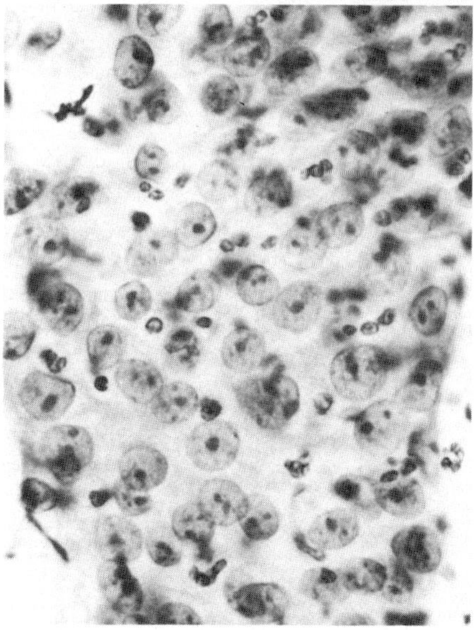

340

Fig. 337. Undifferentiated carcinoma of stomach showing 'solid' pattern. × 205.

Fig. 338. Undifferentiated carcinoma of stomach showing 'scirrhous' pattern. × 545.

Fig. 339. Gastric carcinoma cells showing characteristic 'sharp' granularity of chromatin material. × 545.

Fig. 340. Cells in gastric brushings from patient with chronic superficial gastritis showing finer, 'softer' granularity of chromatin material. × 545.

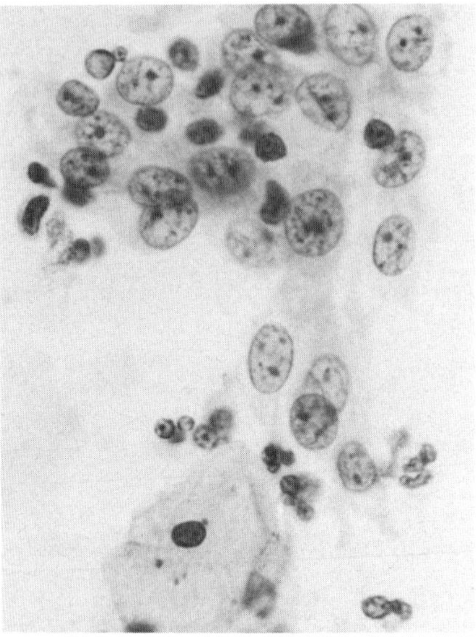

341

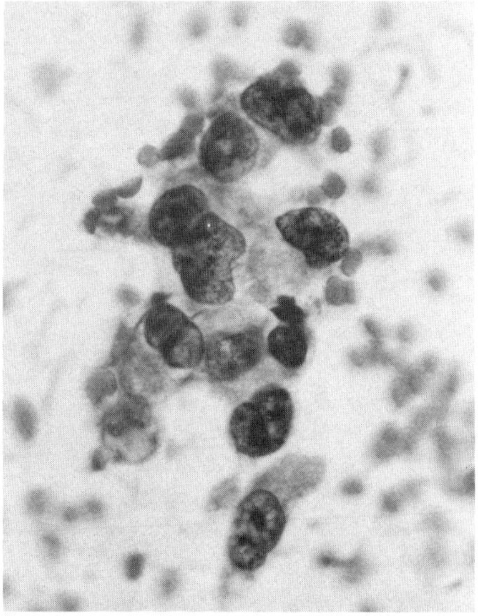

342

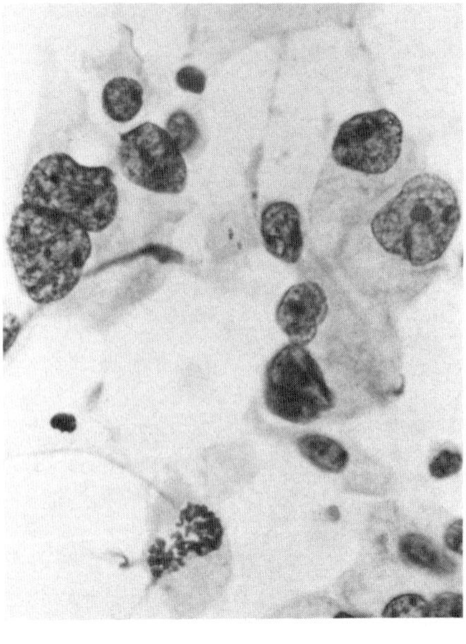

343

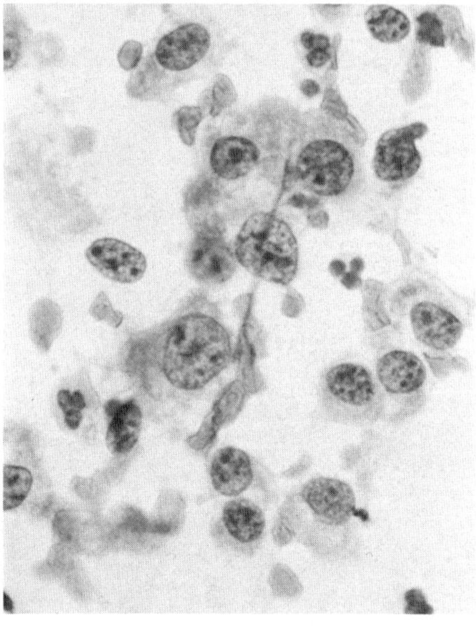

344

Fig. 341. Cells in gastric washings from patient with chronic peptic ulcer showing smooth or regular nuclear outlines. × 545.

Fig. 342. Cells in gastric washings from patient with moderately well differentiated adenocarcinoma showing marked irregularity of nuclear outlines. × 545.

Fig. 343. Cells in gastric brushing from patient with adenocarcinoma showing marked nuclear moulding. Note also abnormal mitotic figure in lower part of photograph. × 545.

Fig. 344. Cells in gastric brushing from patient with poorly differentiated adenocarcinoma showing marked pleomorphism. × 545.

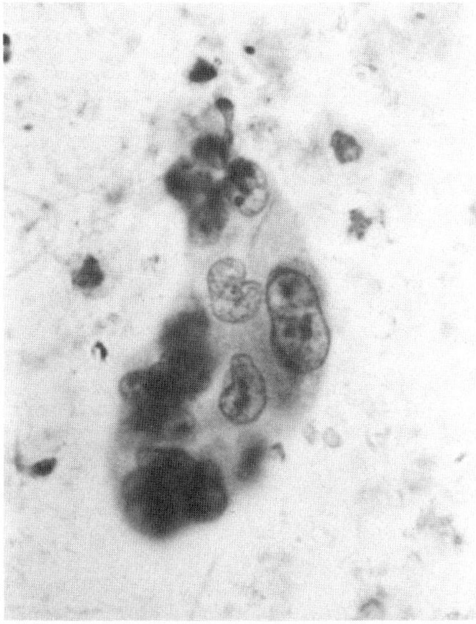

345

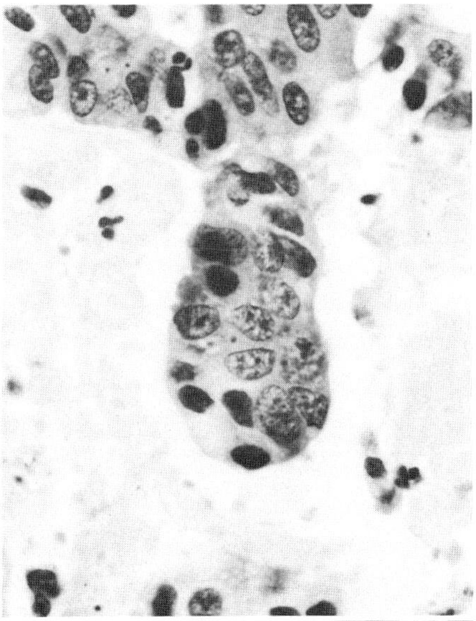

346

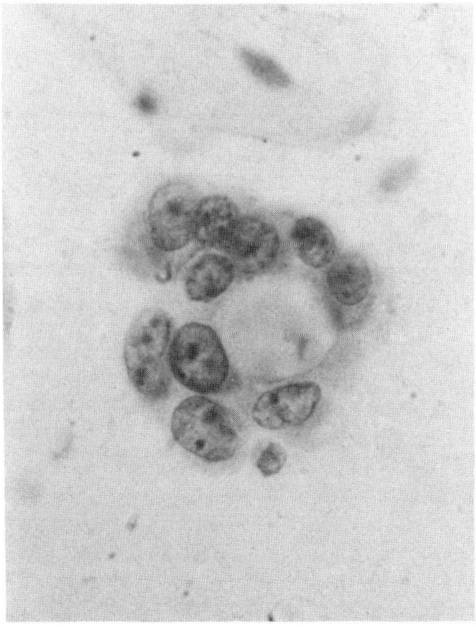

347

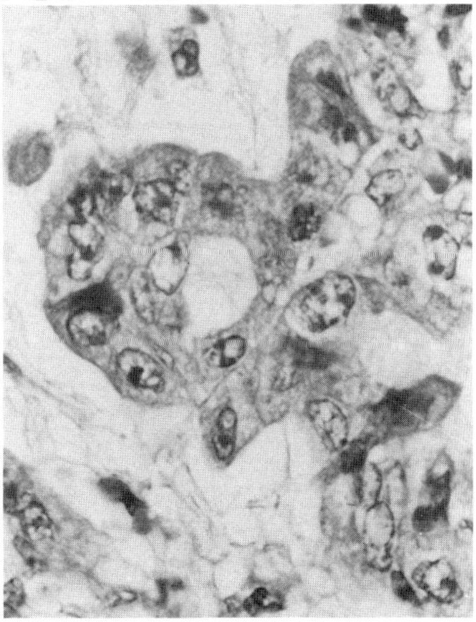

348

Fig. 345. Papillary fragment in gastric washings from patient with well differentiated papillary adenocarcinoma. × 545.

Fig. 346. Section of papillary adenocarcinoma. × 545.

Fig. 347. Acinar structure in gastric washings from patient with well differentiated tubular adenocarcinoma of stomach. × 545.

Fig. 348. Section of well differentiated tubular adenocarcinoma. × 545.

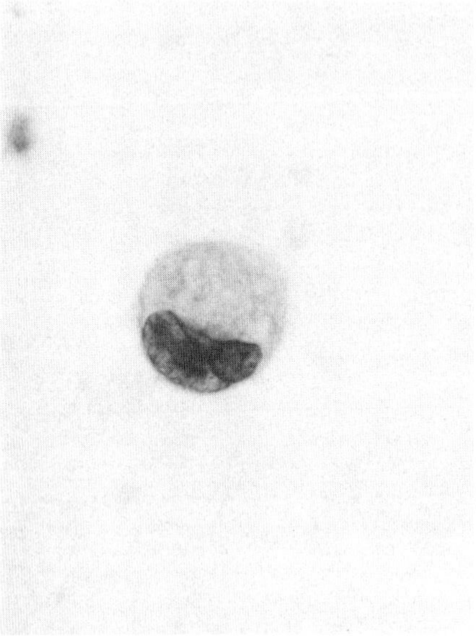

349

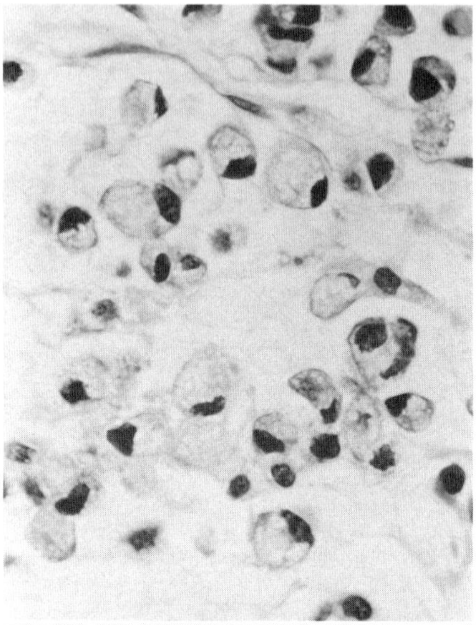

350

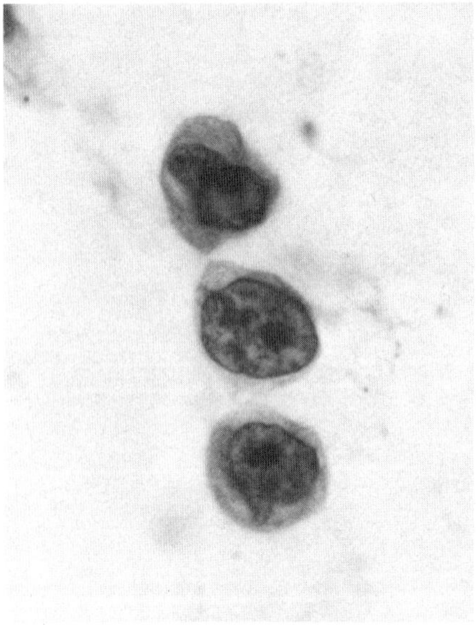

351

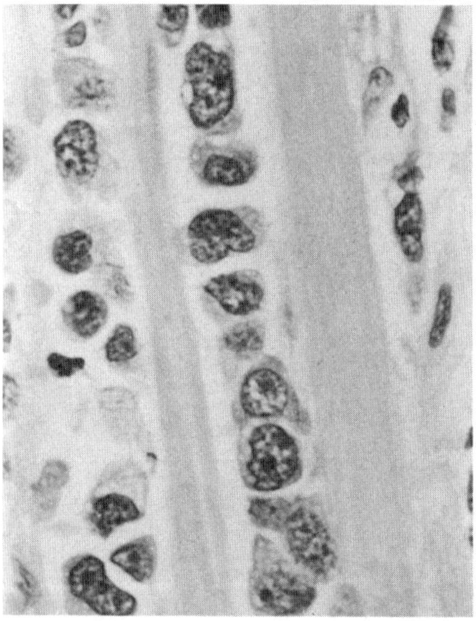

352

Fig. 349. Signet-ring carcinoma cell in gastric washings. Note abnormal pointed nucleus, with prominent malignant criteria, and abundant vacuolated cytoplasm. × 545.

Fig. 350. Section of signet-ring cell carcinoma of stomach. × 545.

Fig. 351. Cells from undifferentiated or anaplastic carcinoma of stomach showing epithelial grouping of cells. × 860.

350). At the other end of the spectrum is the completely undifferentiated or anaplastic carcinoma the undifferentiated neoplasm being reflected clearly in the cytological findings. Again the cytologic and histologic correlation may be quite striking (fig. 351, 352).

The majority of adenocarcinomas of the stomach fall between these two extremes, the carcinoma being characterized by small groups or sheets of cells with abnormal nuclei and variable degrees of cytoplasmic differentiation. These appearances are illustrated in figures 353–360. Considerable care is required in the evaluation of the so-called signet-ring cells (fig. 361) since these may be confused with the goblet cells shed from an area of intestinal metaplasia (fig. 362). However, a careful evaluation of nuclear detail usually leads to a correct diagnosis, the nucleus of the carcinoma cell showing malignant criteria whereas that of the metaplastic cell is usually bland in appearance. In addition, the nuclei of the cancer cell are frequently pointed and tend to mould around the cytoplasmic vacuole whereas the nuclei of the metaplastic cell have smoother contours and are usually not deformed by the vacuole.

Fig. 352. Section of undifferentiated or anaplastic carcinoma of stomach showing similar arrangement of malignant cells as seen in previous figure. × 545.

Fig. 353. Papillary fragment in gastric washings from patient with well differentiated papillary adenocarcinoma. Note the polymorphonuclear leucocytes within the cytoplasmic vacuoles. × 545.

Fig. 354. Tissue fragments in gastric brushings from patient with well differentiated adenocarcinoma. Note both acinar and papillary structures. × 545.

Fig. 355. Cells in gastric washings from patient with moderately well differentiated adenocarcinoma of stomach. Cytoplasmic vacuolation is prominent. × 545.

Fig. 356. Cells in gastric brushings from patient with moderately well differentiated adenocarcinoma of stomach. Cytoplasmic vacuolation is again prominent and these vacuoles contain polymorphs. × 545.

Fig. 357. Cells in gastric brushings from patient with moderately well to poorly differentiated adenocarcinoma of stomach. Note small cytoplasmic vacuoles which indicate nature of neoplasm. × 545.

Fig. 358. Cells in gastric brushings from patient with poorly differentiated adenocarcinoma of stomach. Note abundant finely vacuolated cytoplasm. × 545.

Fig. 359. Cells in gastric brushings from patient with poorly differentiated adenocarcinoma of stomach. Group in lower portion of photograph suggests a 'cell-in-cell' arrangement but polymorphs indicate clearly the presence of a cytoplasmic vacuole. × 545.

Fig. 360. Cells in gastric brushings from patient with poorly differentiated adenocarcinoma of stomach. Multinucleate cell in centre of photograph has abundant finely vacuolated cytoplasm. × 545.

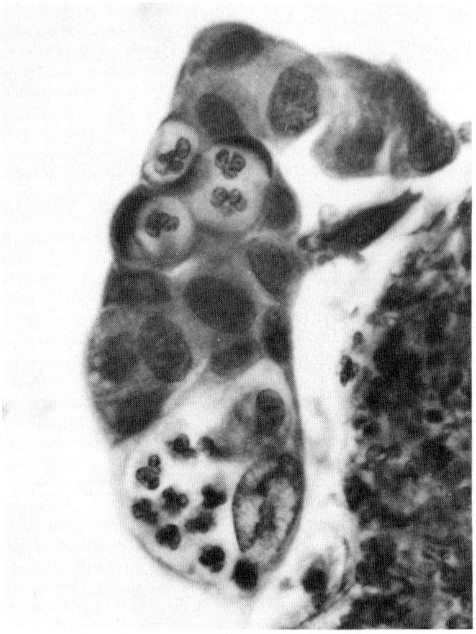

353

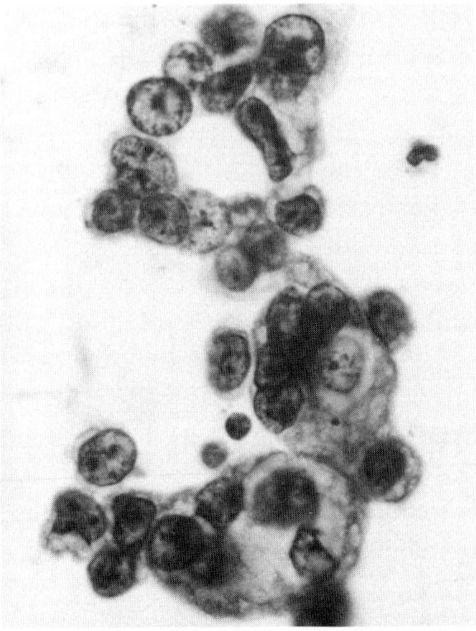

354

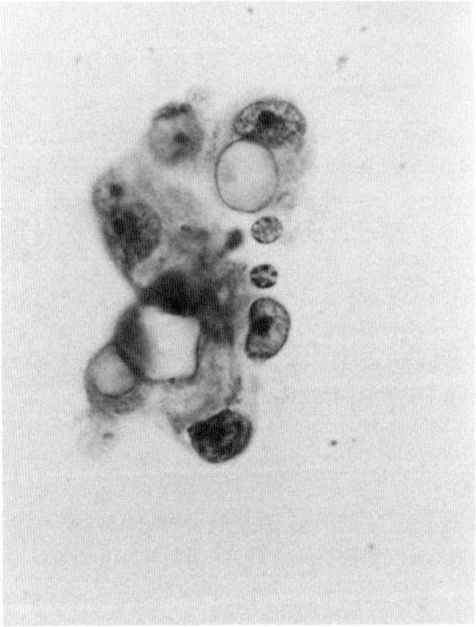

355

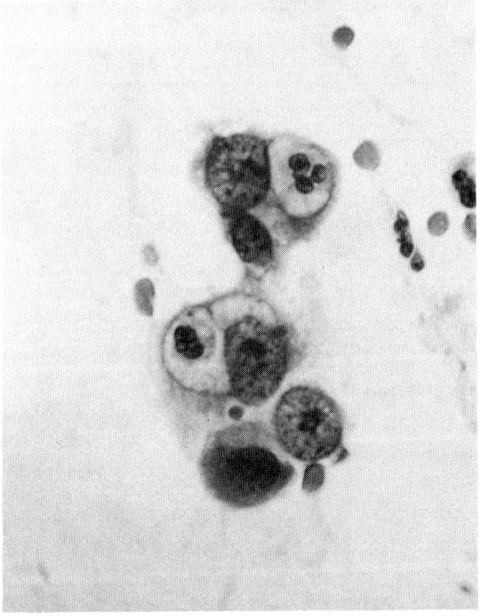

356

For legends, see p. 177.

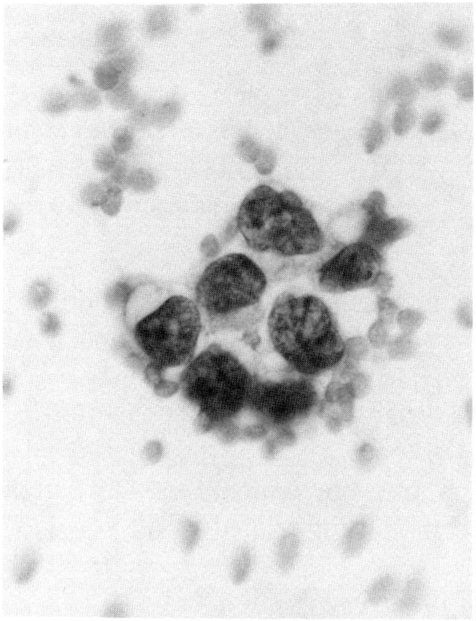

357

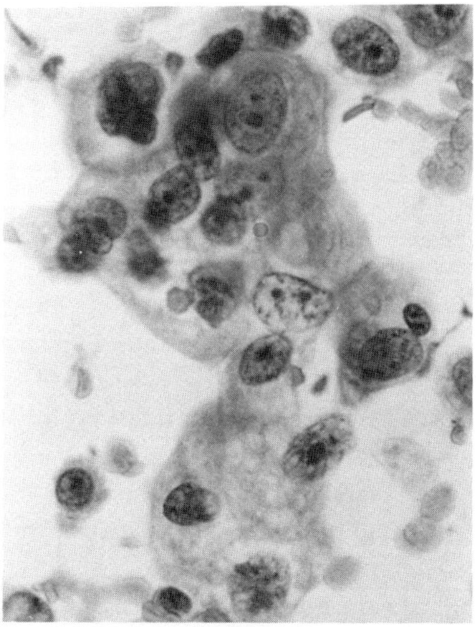

358

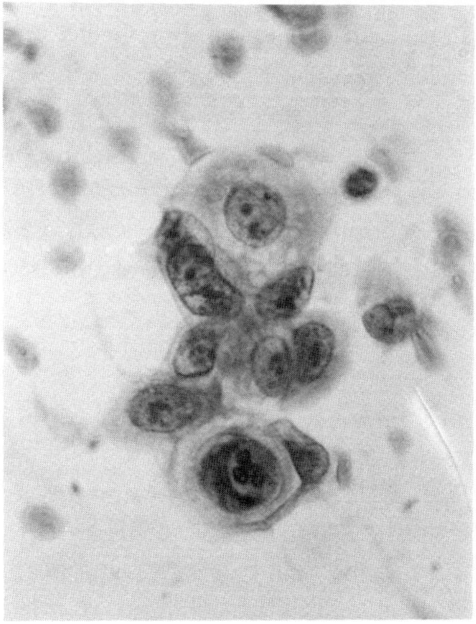

359

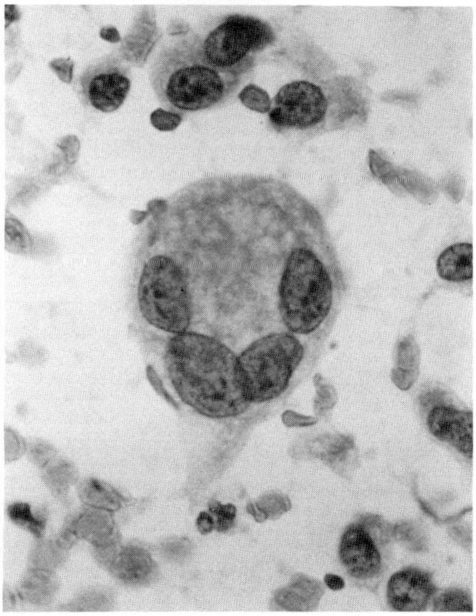

360

For legends, see p. 177.

Malignant Lymphomas

In most cases lymphomas of the stomach, particularly the small ones are local manifestations of a disseminated disease which is already evident elsewhere or will be manifest subsequently. According to *Dawson* et al. [11] the following five criteria are necessary to establish that a lymphoma in the intestinal tract is a primary one:

There must be no palpable enlargement of superficial lymph nodes.
There must be no radiological evidence of mediastinal lymph node enlargement.
The white cell count, both total and differential, must be within normal limits.
At laparotomy the alimentary lesion must predominate and any lymph node involvement must be confined to those related to the involved area of gut.
There must be no involvement of the liver or spleen.

When these criteria are applied strictly primary lymphomas of the alimentary tract are relatively uncommon. Thus *Warren and Lulenski* [67] found only 28 such cases among a total of 3,132 gastro-intestinal tumours. The stomach is more frequently affected by lymphomas than the remainder of the alimentary tract being involved in 51% of cases in one series [2]. Although uncommon, primary gastric lymphoma occurs sufficiently frequently as to be an important diagnostic problem to those who practise gastric cytology.

Problems of Nomenclature

'Nowhere in pathology has a chaos of names so clouded clear concepts as in the subject of lymphoid tumours.'

These words, written by *Rupert Willis* in 1967 [68] are unfortunately as applicable, if not more so, in 1984. Several systems of nomenclature are currently in use throughout the world and the diagnostic pathologist reporting on a specific lesion will usually use that classification that is most acceptable to him but will supplement his diagnosis with one or more synonymous terms. The original classification of malignant lymphomas into Hodgkin's disease, lymphosarcoma (well and poorly differentiated) and reticulum cell sarcoma has persisted in most contemporary texts on gastric cytology although in some the term lymphocytic lymphoma is substituted for lymphosarcoma and histiocytic lymphoma for reticulum cell sarcoma.

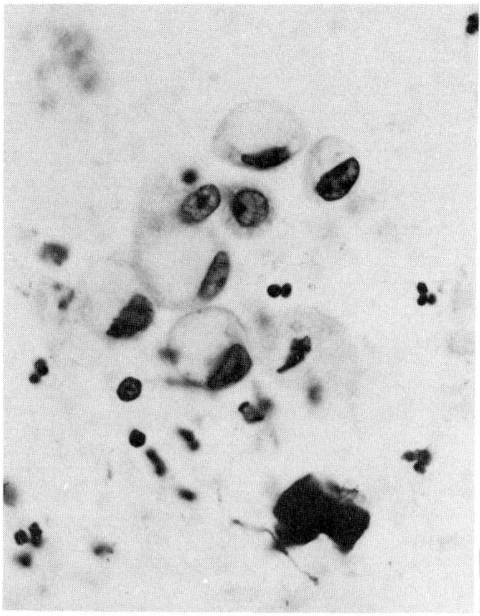

361

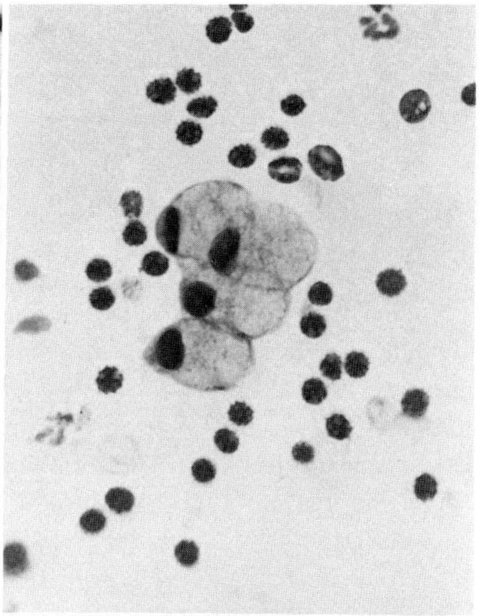

362

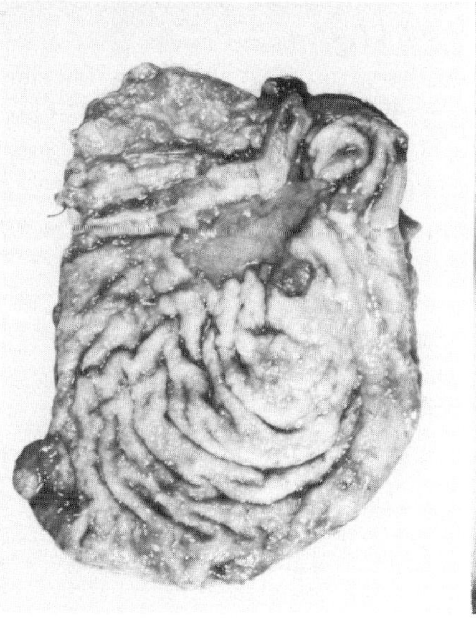

363

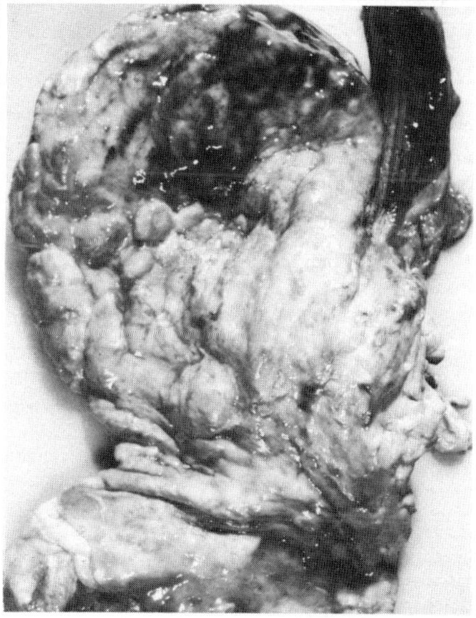

364

Fig. 361. Group of signet-ring carcinoma cells in gastric washings. Note pointed nuclei, with prominent malignant criteria, and tendency of nucleus to mould around cytoplasmic vacuole. × 545.

Fig. 362. Group of goblet cells in gastric brushings from patient with intestinal metaplasia. Note the smooth outlines and 'bland' appearance of the nuclei. × 545.

Fig. 363. Ulcerated malignant lymphoma of stomach.

Fig. 364. Extensive infiltration of gastric wall by malignant lymphoma with prominent rigid rugal pattern.

This classification, with or without modifications, is useful but it does not lend itself to clinical and therapeutic correlations nor does it pay cognizance to the current concepts of the derivation and nature of the lymphoid cell series. Many classifications have been proposed of greater or lesser complexity and of varying degrees of usefulness. In all Hodgkin's disease is regarded as a separate, reasonably non-contentious entity, the problems of nomenclature largely relating to the so-called non-Hodgkin's group. Although, as will be emphasized, a detailed knowledge of histological classification is not necessary for gastric cytodiagnosis the pathologist should be aware of these classifications so that he may more closely correlate his cytological findings with the histological diagnosis and also communicate these findings in a more meaningful way to the clinician.

Perhaps the most widely used classifications are those of *Rappaport* [53], and the Kiel classification the latter being derived from the *Lennert* classification [36]. A useful comparison of these two classifications was prepared by *Gerard-Marchant* et al. in 1974 [17].

The other classification with which the pathologist should be familiar is the 'Working Formulation of Non-Hodgkin's Lymphoma for Clinical Usage – US National Cancer Institute Non-Hodgkin's Lymphoma Study, 1980'. This is reproduced in full as it is becoming more widely used and does appear to meet some of the problems inherent in the other systems of nomenclature available currently.

A Working Formulation of Non-Hodgkin's Lymphoma for Clinical Usage (US National Cancer Institute Non-Hodgkin's Lymphoma Study, 1980)

Low Grade	*Intermediate Grade*
A. Malignant Lymphoma	D. Malignant Lymphoma, follicular
Small Lymphocytic	Predominantly large cell
Consistent with chronic lymphocytic leukaemia	Diffuse areas
Plasmacytoid	Sclerosis
B. Malignant Lymphoma, follicular	E. Malignant Lymphoma, diffuse
Predominantly small cleaved cell	Small cleaved cell
Diffuse areas	Sclerosis
Sclerosis	F. Malignant Lymphoma, diffuse
C. Malignant Lymphoma, follicular	Mixed, small and large cell
Mixed, small cleaved and large cell	Sclerosis
Diffuse areas	Epithelioid cell component
Sclerosis	

G Malignant Lymphoma, diffuse
- Large cell
  - Cleaved cell
  - Non-cleaved cell
  - Sclerosis

High Grade

H. Malignant Lymphoma
- Large cell, immunoblastic
  - Plasmacytoid
  - Clear cell
  - Polymorphous
  - Epithelioid cell component

I. Malignant Lymphoma
- Lymphoblastic
  - Convoluted cell
  - Non-convoluted cell

J. Malignant Lymphoma
- Small non-cleaved cell
  - Burkitt's
  - Follicular areas

Miscellaneous
- Composite
- Mycosis fungoides
- Histiocytic
- Extramedullar
  - Plasmacytoma
- Unclassifiable
- Other

Those who wish to consider these problems in greater depth are referred to the appendix on 'Classification and Nomenclature' in *Robb-Smith and Taylor's* excellent book entitled 'Lymph Node Biopsy' [55].

In the actual practice of gastric cytology it is possible to adopt a much simpler approach to the problems of terminology. In general a diagnosis of malignancy can usually be made cytologically in a patient with a malignant lymphoma of the stomach and the prediction that the malignancy is a lymphoma is also usually possible. It is much less common, however, to be able to state with certainty the type of lymphoma present. The final and precise classification must await the availability of imprint and histological material to enable an assessment to be made of predominant cell type and tissue architecture and to allow the application of surface marker and enzyme techniques. For the gastric cytologist, therefore, the following simplified classification is adequate depending, as it does, on cell identification and descriptive morphology:

Hodgkin's Lymphoma – No subdivision is desirable or necessary.

Non-Hodgkin's Lymphoma
- Small cell ('lymphocytic' lymphoma or lymphosarcoma).
- Large cell ('histiocytic' lymphoma or reticulum cell sarcoma).

Macroscopic Appearances

The lymphoma develops in aggregates of lymphoid tissue that are commonly present in the adult gastric mucosa. Their mode of growth is thus similar to that of a carcinoma, the neoplasm commencing in the mucosa and subsequently infiltrating the submucosa and muscle coats. Many of the tumours form a plaque-like mass the surface of which is ulcerated (fig. 363). The edges of the ulcer are raised and everted. Some tumours are polypoid (fig. 364) whilst others cause marked thickening of the gastric wall. Occasionally extensive infiltration of the mucosa will result in a rigid giant rugal pattern. Multiple nodules of lymphoma tissue are not uncommon. These various features may be demonstrated radiologically (fig. 365). A gastric lymphoma, therefore, resembles closely the gross characteristics of a carcinoma and indeed this is usually the provisional clinical diagnosis when the patient is referred for cytological investigation. The investigation is thus of major importance in these cases.

Histological Features

Histological examination shows a dense infiltration of the mucosa by tumour cells (fig. 366). The distribution may be diffuse or nodular and, as already indicated, this is an important feature in the classification of the lesion. The mucosal glands may be absent but those that are present usually retain their normal architecture (fig. 367). It is this persistence of apparently normal tissue structures surrounded by neoplastic lymphoma cells that contrasts vividly with the destruction of similar tissue structures by a carcinomatous infiltration. Nevertheless, the glandular epithelial cells in continuity with the lymphoma tissue frequently show morphologic abnormalities (fig. 368) and these may confuse the cytologic diagnosis. Low power examination suggests a uniformity or monotony of the neoplastic cell infiltrate. However, a closer inspection reveals a variability of cell pattern and indeed, as also discussed, the recognition of the predominant cell type is fundamental to the classification of the lesion present. Thus the cells may be small (fig. 369) or large (fig. 370) or a mixture of both types may be seen. The characteristic Reed-Sternberg (fig. 371) cell, to be described below, is essential for the diagnosis of Hodgkin's disease.

Cytological Diagnosis

On the basis of the simplified classification suggested above there are three main cell types that may be recognized in a cytological preparation:

(1) The small cell, or 'lymphocytic' cell, is a round cell that is usually larger than the normal lymphocyte. It has a round nucleus and a prominent

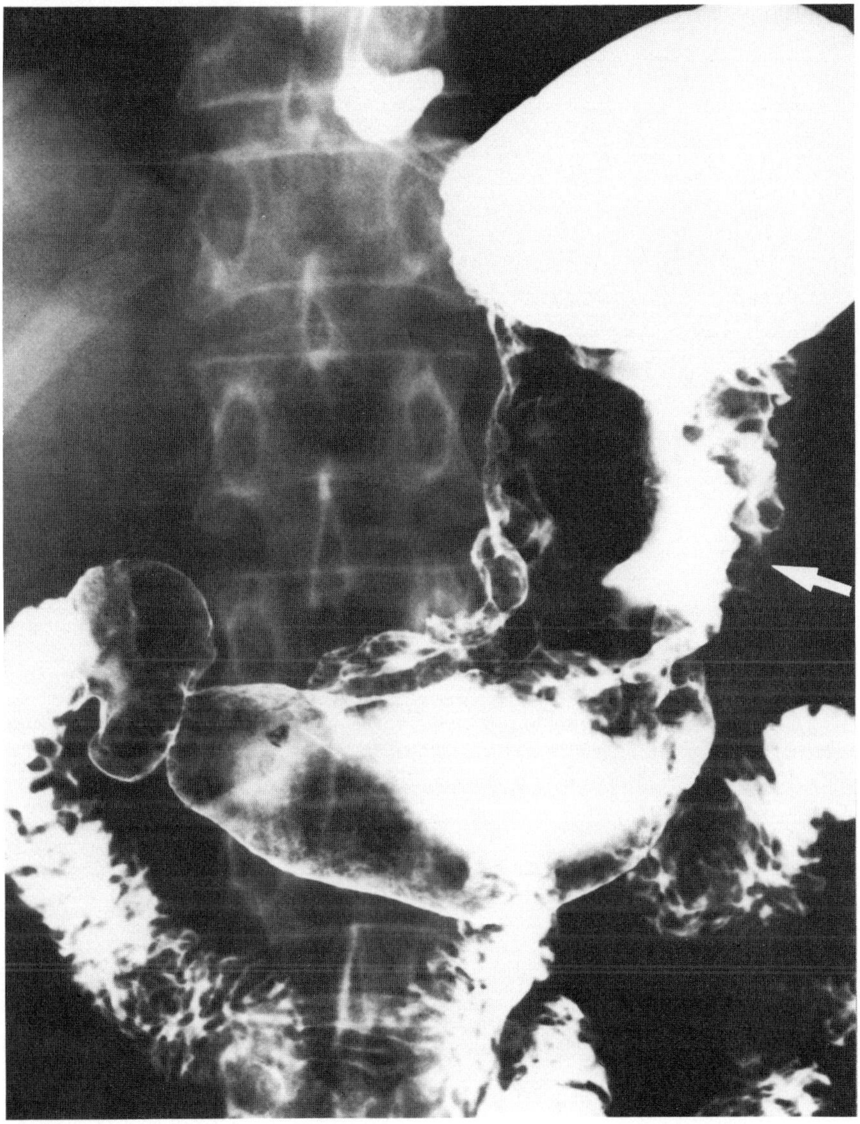

Fig. 365. Radiological appearances of extensive primary gastric lymphoma. There is a large ulcer on the posterior wall of the body of the stomach, the ulcer crater being filled with contrast medium as indicated by arrow. The wall of the stomach is extensively infiltrated by tumour tissue.

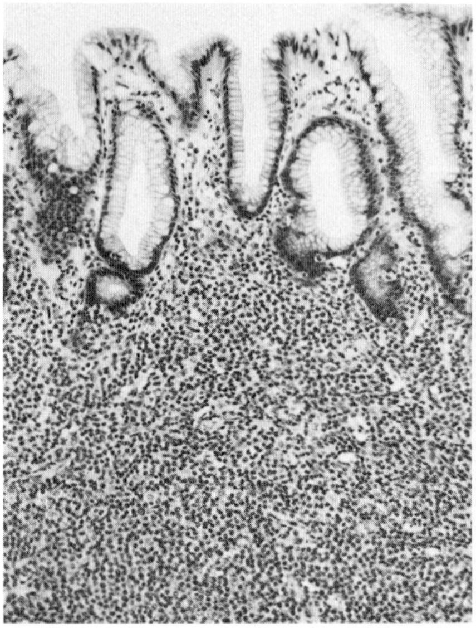

366

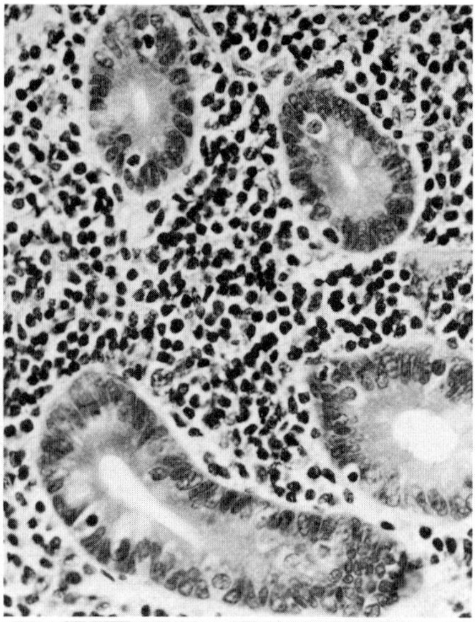

367

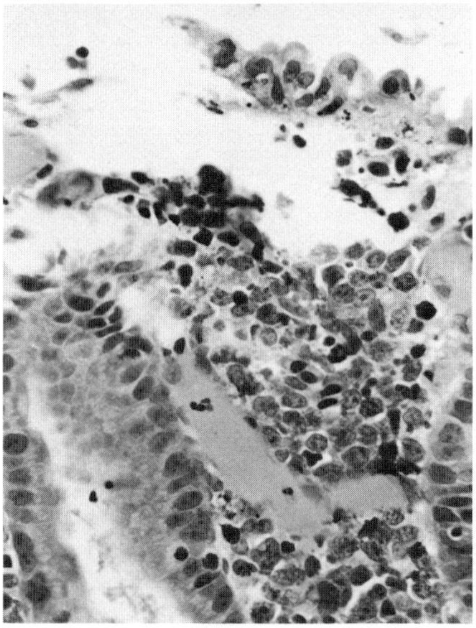

368

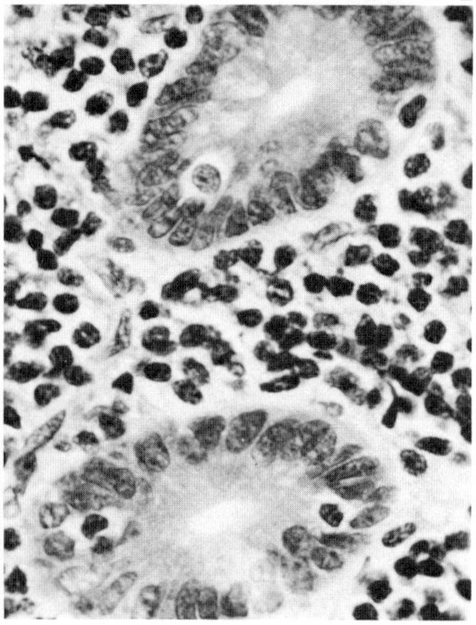

369

Fig. 366. Malignant lymphoma of stomach showing dense infiltration of the mucosa by tumour cells. × 86.

Fig. 367. Gastric glands surrounded by dense infiltrate of malignant lymphoma cells. × 275.

Fig. 368. Marked nuclear abnormalities of glandular epithelial cells in association with lymphomatous infiltration. × 275.

Fig. 369. Malignant lymphoma of small-cell type. × 545.

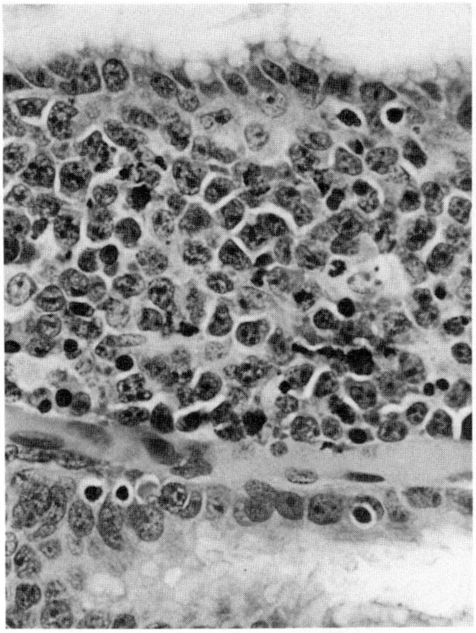

370

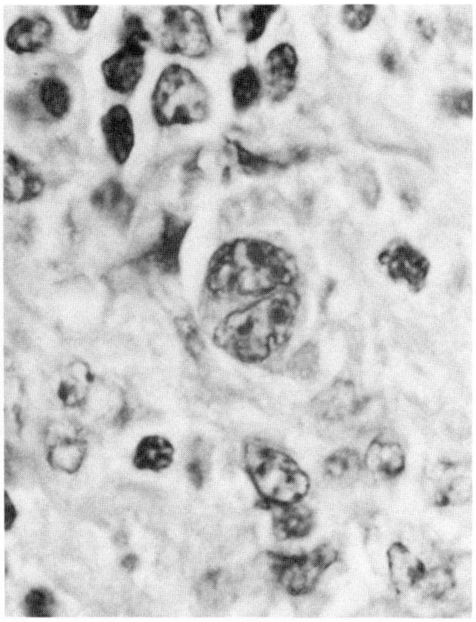

371

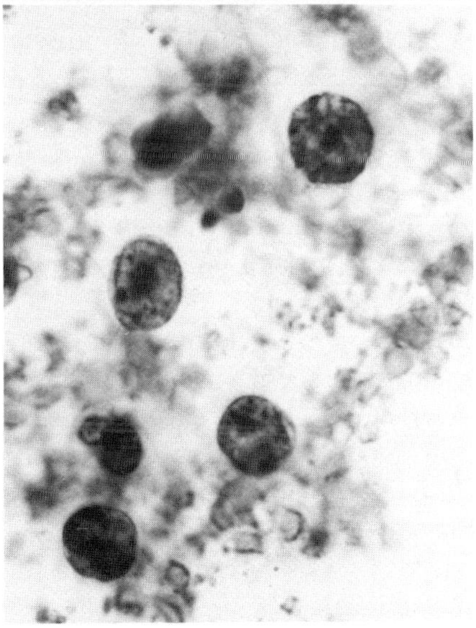

372

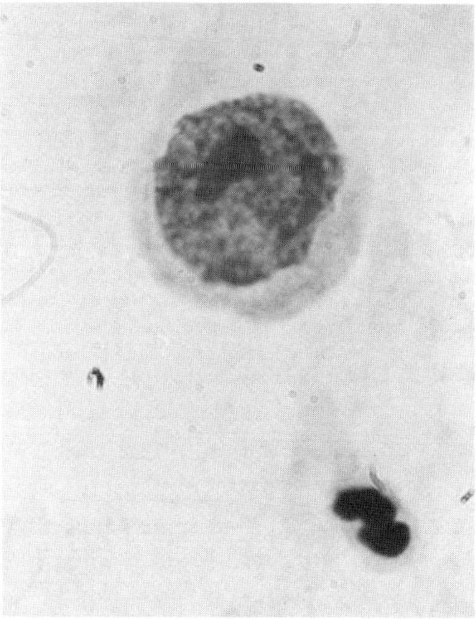

373

Fig. 370. Malignant lymphoma of large-cell type. × 345.

Fig. 371. Reed-Sternberg cell characteristic of Hodgkin's disease. × 860.

Fig. 372. Cells of small-cell lymphoma in gastric brushings. Note large nucleus, scanty cytoplasm, and prominent nucleolus. × 860.

Fig. 373. Lymphoma cell of large-cell type. Note the large irregular nucleus, more abundant cytoplasm, and very prominent irregular nucleolus. × 860.

nucleolus. The cytoplasm is scanty being seen usually as a narrow rim surrounding the nucleus (fig. 372).

(2) The large cell ('histiocyte or reticulum cell') is, as the name implies, a larger cell although considerable size variation is common. The cell is usually round but may be irregular in shape. The nucleus is usually irregular in outline, there are usually one or more prominent nucleoli, and the cytoplasm, although variable in amount is usually more abundant than in the small cell (fig. 373). Sometimes the large cell may have more than one nucleus.

(3) The Reed-Sternberg cell is a large cell with a nucleus that is large in proportion to the size of the cell. Usually the cell contains two or more nuclei and in the classical 'mirror-image' cell two similar nuclei are opposed to one another in the same cell (fig. 374). The nuclear membrane is thick and the nucleus tends to be rather vesicular in appearance. A large nucleolus is a prominent feature of each nucleus.

The cytological diagnosis of malignant lymphoma of the stomach depends largely on the recognition and interpretation of these various cell types. However, in addition, the following morphologic features are emphasized:

(1) As already discussed in an earlier chapter, the cytological preparation from a washing or brushing specimen of a gastric lymphoma frequently has a distinctive pale blue hazy background. The nature of this is uncertain but its presence may suggest the diagnosis of malignant lymphoma before diagnostic cells are seen – indeed it may lead to a more careful scrutiny of the material for the sometimes elusive lymphoma cells. In a paper published in 1973 [56] the authors commented upon 'a heretofore undescribed proteinaceous material' that they observed within the neoplastic nodules of malignant lymphomas. Special histochemical procedures failed to reveal the nature of the precipitate and electron micrographs prepared from one patient showed extracellular clusters of a fine fibrillar material of unknown nature. Whether this report has any relevance to our own observations is uncertain but there is no doubt that these observations are useful in the diagnosis of malignant lymphoma of the stomach.

Fig. 374. Reed-Sternberg cell in gastric washings. Note the 'mirror-image' nuclei. $\times$ 860. Specimen courtesy of Dr. *Elaine Waters,* Royal Perth Hospital, Western Australia.

Fig. 375. Cells from large-cell malignant lymphoma of stomach showing 'brainlike convolutions' of nucleus. $\times$ 860.

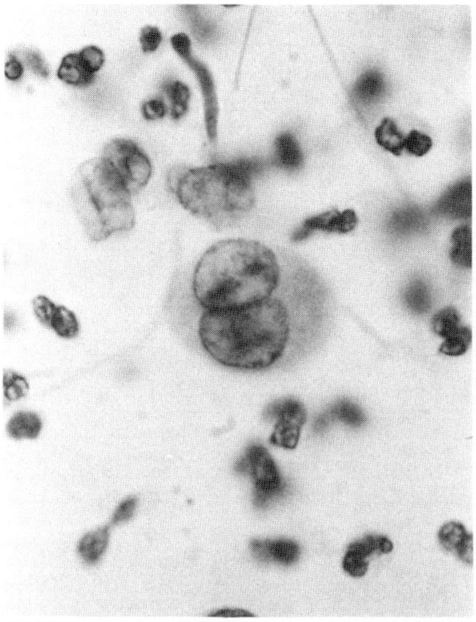

374

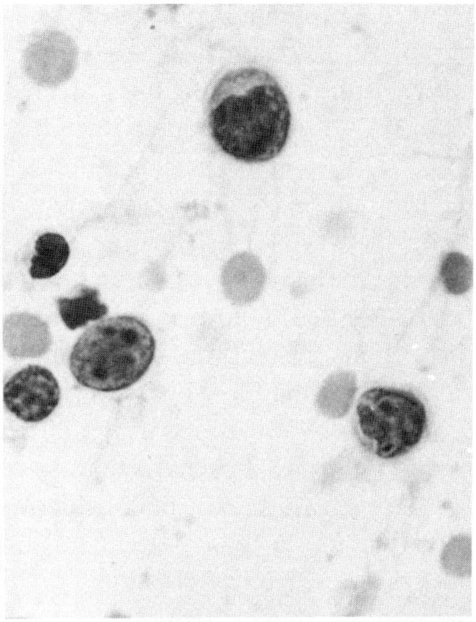

375

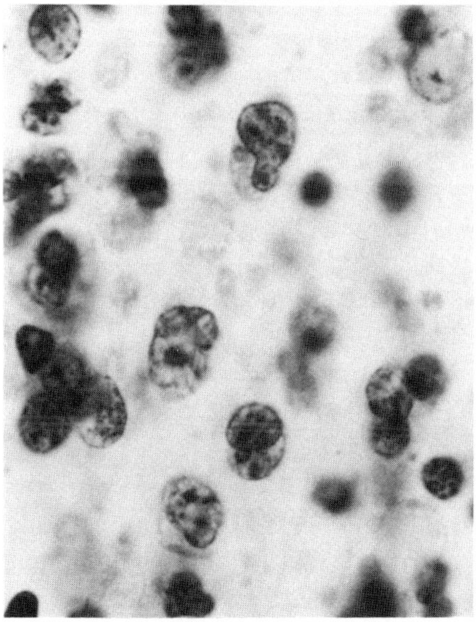

376

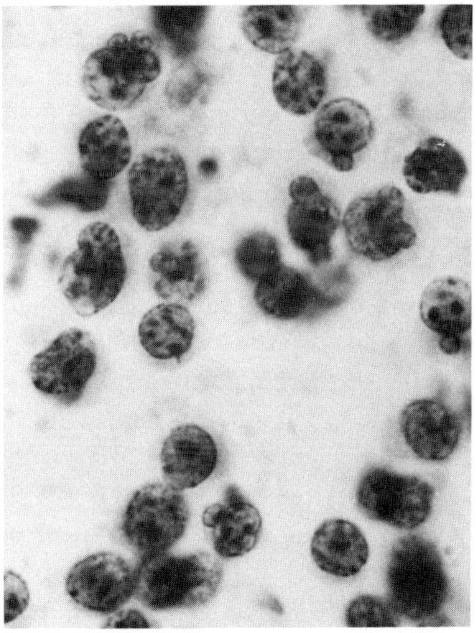

377

Fig. 376. Cells from large-cell malignant lymphoma of stomach showing prominent nuclear protrusions. × 860.

Fig. 377. Cells from large-cell malignant lymphoma of stomach showing nuclear protrusions. × 860.

(2) *Takeda* [63] describes the nuclei of the cells of 'histiocytic lymphoma or reticulum cell sarcoma' as having 'brainlike convolutions' and indicates that similar features are seen in the nuclei of cells associated with Hodgkin's disease. Whilst this description may be apt with some cells (fig. 375) it is our experience that a more characteristic irregularity of the nuclear outline is the presence of small, large or multiple nipple-like protrusions. This characteristic was described by *Melamed* [40] in association with the malignant lymphoma he characterized as the 'variable lymphocyte type' observed by him in malignant effusions. Thus he described 'nodular protrusions, resembling the budding of yeast' as being prominent in some cases of lymphoma. This feature was emphasized by *Koss* [31] who states that 'nuclear protrusions in the form of small, tongue-like projections, single or multiple, are characteristic of malignant tumours'. We have also found this feature to be of considerable value in the identification of malignant lymphoma cells, particularly of the large-cell type, and examples are illustrated in figures 376 and 377. Curiously this abnormality is not prominent in histological sections.

(3) Cleaved cells are clearly recognized histologically and, as already indicated, their presence is utilized in some systems of nomenclature. Cleaved nuclei are also seen in cytological preparations being evident in both small and large cells. Their significance should not be overestimated as cleaved nuclei merely indicate a follicular centre origin of the cell and may be seen in both normal and abnormal lymphoid cells. However, the presence of a cleaved nucleus as depicted in figures 378 and 379 may be helpful in deciding that an individual malignant cell has been derived from a lymphoma.

(4) In the paper already referred to, *Melamed* [40] also described the phenomenon of 'nuclear fragmentation'. He regarded this as one of the least common cell patterns but 'one of the most striking and specific'. He suggested that this change may represent the effects of therapy since it was never seen in effusions from untreated patients. *Koss* [31] referred to this phenomenon as a form of 'massive karyorrhexis' and, whilst agreeing that on occasions it may represent the effect of treatment, he stresses that it is

---

Fig. 378. Cell from large-cell malignant lymphoma of stomach showing prominent cleaved nucleus. × 860.

Fig. 379. Cells from large-cell malignant lymphoma of stomach one of which has a cleaved nucleus. × 860.

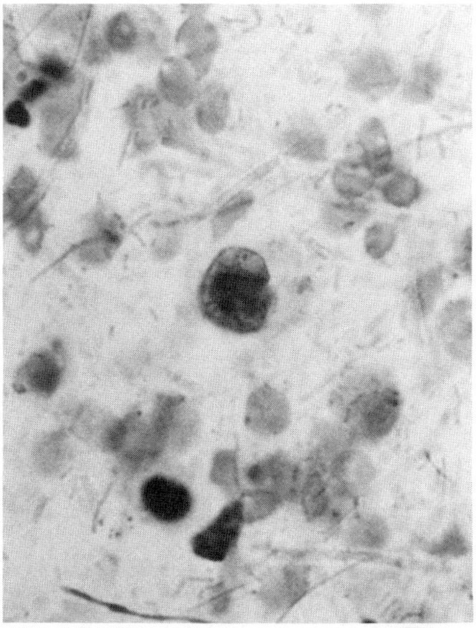

378

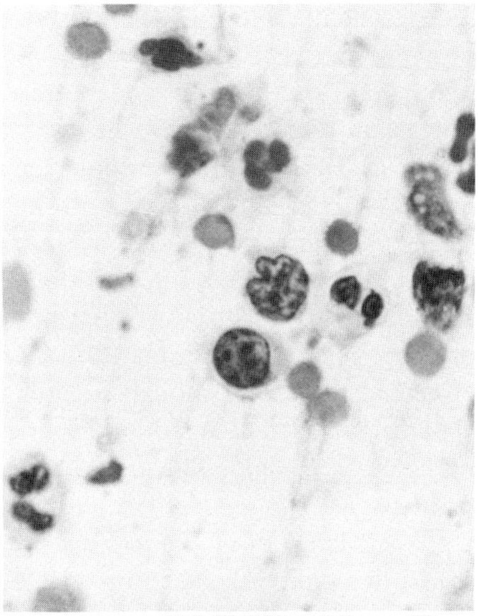

379

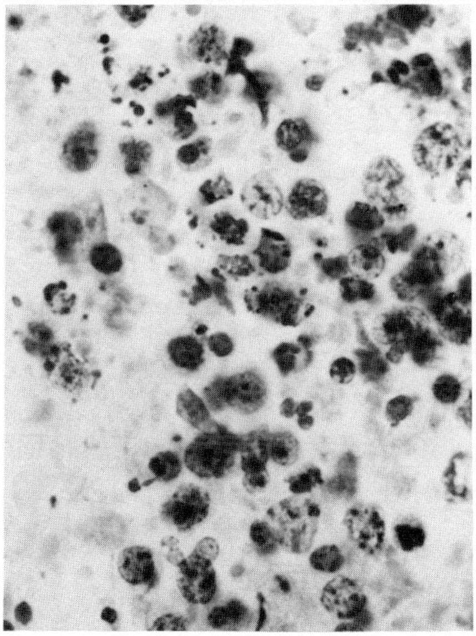

380

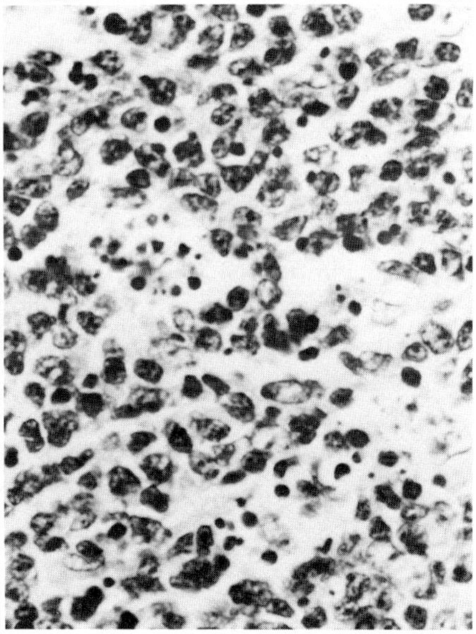

381

Fig. 380. Cells in gastric brushings from patients with untreated large-cell malignant lymphoma of stomach. Nuclear fragmentation is prominent. × 545.

Fig. 381. Section of malignant lymphoma of stomach showing nuclear fragmentation. × 545.

seen frequently in the absence of treatment. He further states that he has not observed massive karyorrhexis in any tumours other than malignant lymphoma and that the phenomenon in effusions appears to be diagnostic of this group of diseases. We have also found this a useful criterion in the diagnosis of malignant lymphoma of the stomach observing the changes in both washings and brushings from both treated and untreated patients (fig. 380). These changes are also seen quite frequently in histological sections of malignant lymphomas (fig. 381).

(5) Hodgkin's disease of the stomach is rare. In a review of 58 cases of primary gastric lymphoma *Brooks and Enterline* [4] found no cases of Hodgkin's disease although 5 cases with this original diagnosis were reclassified on review. A similar experience was reported by *Filippa* et al. [14], no cases of Hodgkin's disease being reported in a detailed analysis of 60 primary gastro-intestinal lymphomas. *Stobbe* et al. [60], however, reported a 4% incidence of Hodgkin's disease in their series of gastric lymphomas. When this type of lymphoma does occur in the stomach the most striking feature in the cytological preparation is the mixed cell population, large or 'histiocytic' lymphoma cells being prominent as is the presence of numerous eosinophils (fig. 382). As already emphasized, the identification of Reed-Sternberg cells is mandatory for the diagnosis to be made (fig. 383).

Diagnostic Problems

There is no doubt that difficulties do exist in the cytological diagnosis of gastric lymphoma and this is reflected in the results of gastro-esophageal cytology to be discussed in a subsequent chapter. The difficulties are due partly to technical problems and partly to problems of interpretation.

Technical Problems

Many lymphomas, at least in their early stages of development, are intramucosal and submucosal in site and hence the malignant cells do not present at the mucosal surface. Nevertheless ulceration is common and, if

---

Fig. 382. Lymphoma cells of large-cell type and numerous eosinophils in gastric washings from patient with Hodgkin's disease of stomach. × 860. Specimen courtesy of Dr. *Elaine Waters,* Royal Perth Hospital, Western Australia.

Fig. 383. Reed-Sternberg cell in gastric washings from patient with Hodgkin's disease of stomach. × 545. Specimen courtesy of Dr. *Elaine Waters,* Royal Perth Hospital, Western Australia.

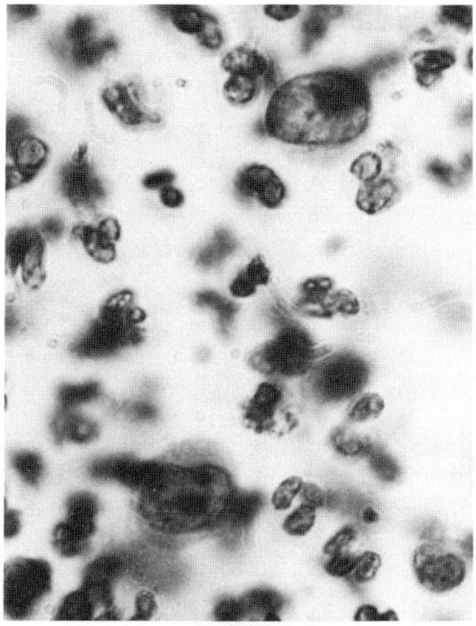

382

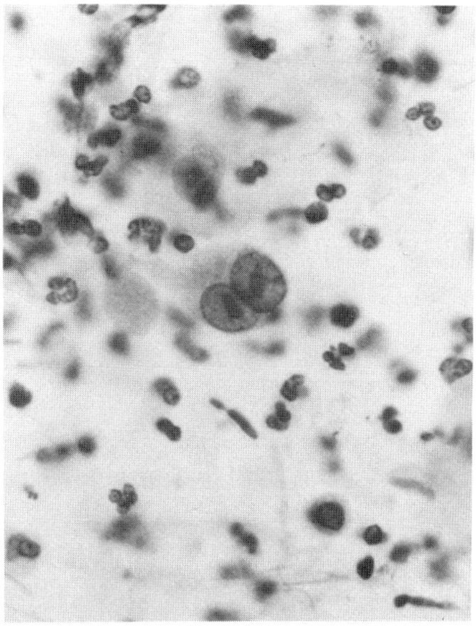

383

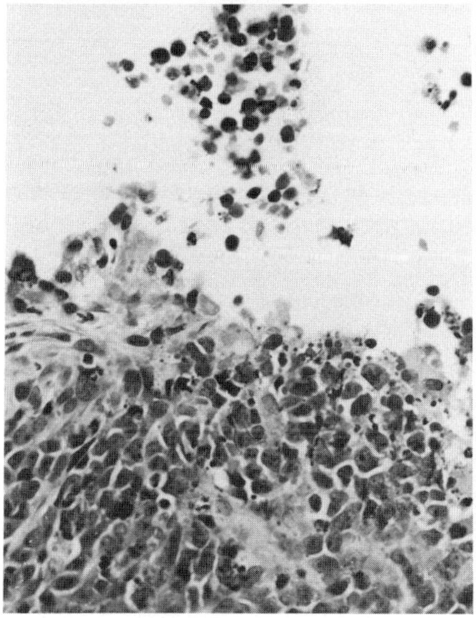

384

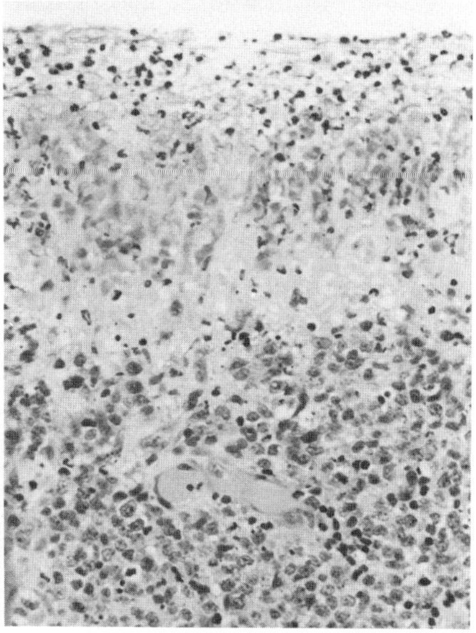

385

Fig. 384. Ulceration of gastric mucosa with exposure of viable well preserved lymphoma cells. × 220.

Fig. 385. Lymphomatous infiltration of gastric mucosa with surface ulceration. Floor of ulcer is covered by inflammatory exudate and necrotic debris. × 173.

acute and superficial, well preserved lymphoma tissue is exposed to the gastric lumen (fig. 384). In more chronic ulcers, however, although the ulcer base is composed of lymphoma tissue, the floor of the ulcer is covered by a thick layer of inflammatory exudate and necrotic debris (fig. 385). Consequently the lymphoma cells, if harvested, are few in number and may be largely concealed by cell debris. This problem of cell detection has been alluded to already in chapter 3. There is no doubt that cell yield has been improved by the advent of brushing techniques but lavage methods, if adequately performed, yield diagnostic material.

Problems of Interpretation

The Normal Lymphocyte. It may be extremely difficult to differentiate between small cell lymphoma and normal lymphocytes. The latter are usually seen in gastric preparations and may be present in large numbers in association with chronic gastritis and peptic ulceration (fig. 386). The diagnosis depends on a careful evaluation of cell size and nuclear detail (fig. 387) but one must accept that a confident distinction is not always possible.

Benign Lymphoid Hyperplasia (Pseudolymphoma). A prominent lymphocytic reaction is usually seen at the base and margins of a chronic peptic ulcer. Occasionally the lymphoid tissue is massive forming a nodular, often ulcerated, mass that may be mistaken for a malignant lymphoma. Histologically the mucosa and submucosa is infiltrated by lymphocytes with conspicuous follicle formation (fig. 388). The cytological specimen may contain numerous lymphocytes of varying degrees of maturation (fig. 389) the appearances being reminiscent of a chronic follicular cervicitis. *Prolla and Kirsner* [52] found the presence of plasma cells mixed with the lymphocytes helpful in the correct interpretation of the cellular material as benign in origin as did *Kobayashi* in association with these two workers [30]. However, they stress that plasma cells can also be seen in association with malignant lymphoma and illustrate this phenomenon. We would support

---

Fig. 386. Non-neoplastic lymphocytes in gastric brushings from patient with chronic superficial gastritis. × 545.

Fig. 387. Cells in gastric brushings from patient with malignant lymphoma of small-cell type. Note slight nuclear enlargement and chromatin irregularity. A cleaved nucleus is visible centrally. × 860.

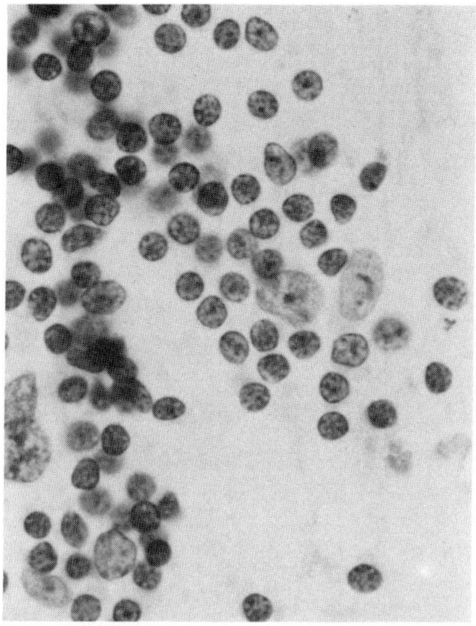

386

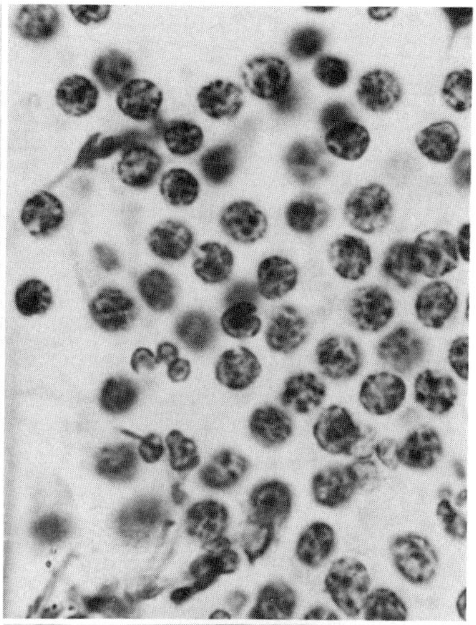

387

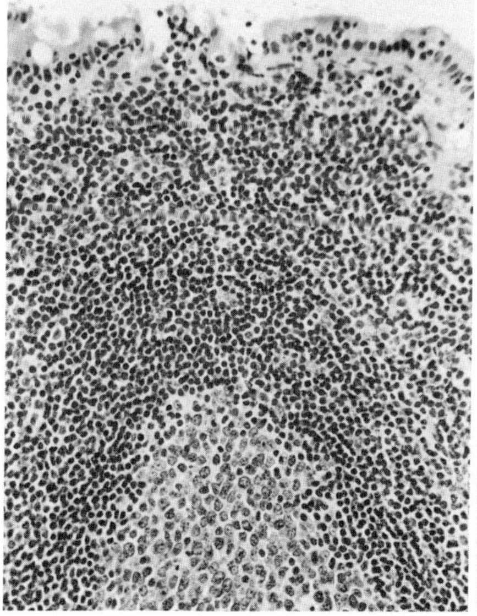

388

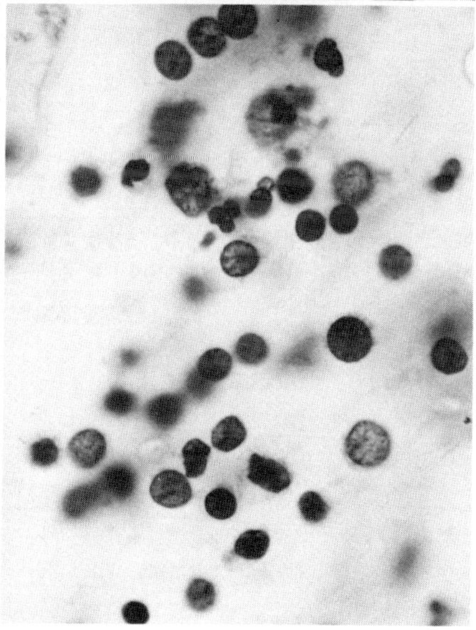

389

Fig. 388. Lymphoid follicle in mucosa of patient with benign lymphoid hyperplasia or pseudolymphoma. Note the prominent germinal centre. × 140.

Fig. 389. Lymphoid cells showing varying degrees of maturation in gastric washings from patient with benign lymphoid hyperplasia of stomach. × 545.

this latter statement most strongly and caution against reliance on plasma cells as an aid in differential diagnosis. Histological examination of gastric lymphomas almost always discloses a superficial zone of plasma cells admixed with inflammatory cells, including eosinophils, situated between the main mass of tumour tissue and the surface epithelial cells (fig. 390, 391). Monoclonal antibody studies have revealed in some cases an identity of these cells with the tumour cells [39] and hence they presumably represent an expression of differentiation.

Again the diagnostic problem may be insoluble but meticulous evaluation of nuclear detail will enable a correct diagnosis to be made in some cases. Fortunately a fully established 'pseudolymphoma' is uncommon but follicular gastritis occurs more frequently.

Undifferentiated or Anaplastic Carcinoma. This is perhaps the most frequently encountered diagnostic problem the large lymphoma cells being mistaken for carcinoma cells or vice versa. The most useful criteria in the differential diagnosis of these two conditions are:

Cell Presentation. This is undoubtedly the most useful feature. Although undifferentiated carcinoma cells present frequently as single cells a careful examination of the whole specimen almost invariably reveals the epithelial grouping characteristic of a carcinoma (fig. 392). Conversely malignant lymphoma cells are always shed singly and never present as cohesive groups or clumps (fig. 393). However, some caution is necessary in the interpretation of brushing specimens when the neoplastic lymphoid cells may be so crowded as to assume an epithelial appearance (fig. 394). Indeed, as depicted in figure 395 the apparent moulding of nuclei and the relatively abundant cytoplasm presumably moulded into a columnar shape by the pressure of contiguous cells may produce an appearance indistinguishable from that of carcinoma.

Nuclear Features. The nuclear outline of lymphoma cells tends to be more uniform than that of carcinoma cells although, as already discussed and illustrated, the former may have very irregular nuclei with prominent protrusions. The chromatin material of the lymphoma cell nucleus is

---

Fig. 390. Lymphomatous infiltration of gastric mucosa showing obvious lymphoma cells in deeper zone and numerous plasma cells more superficially. × 275.

Fig. 391. Higher power view of previous figure showing predominance of plasma cells in superficial mucosa. × 545.

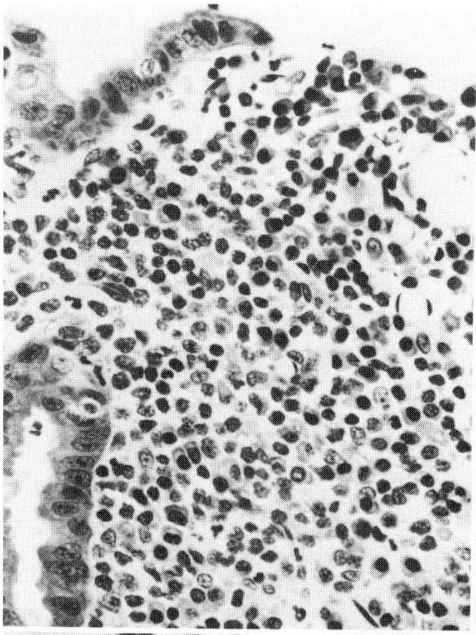

390

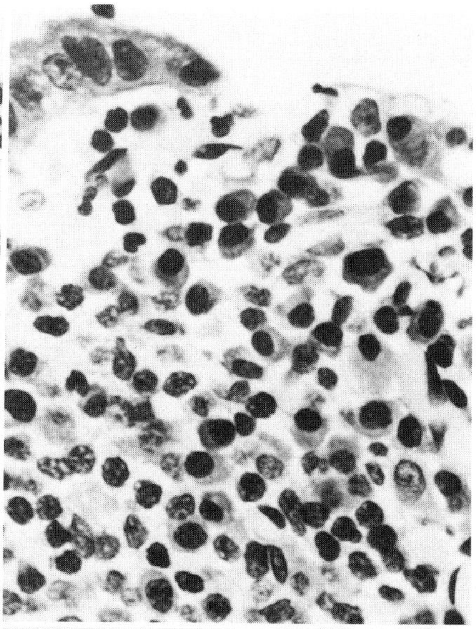

391

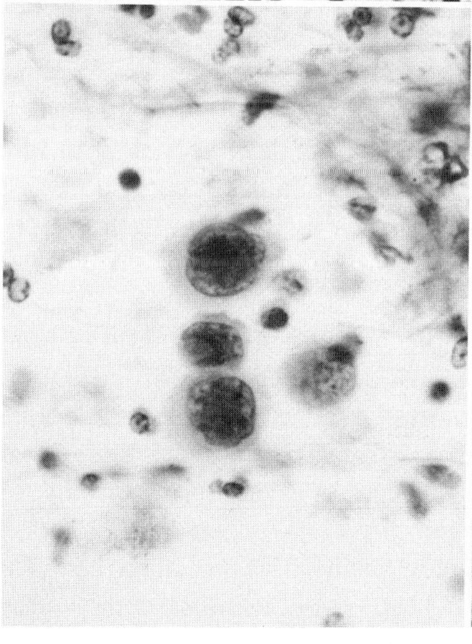

392

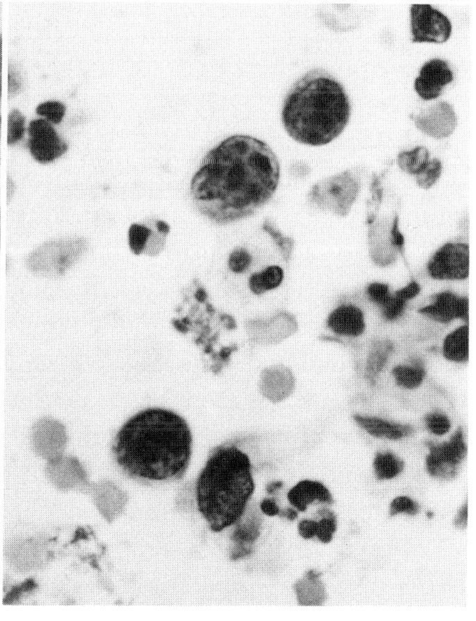

393

Fig. 392. Undifferentiated or anaplastic carcinoma cells in gastric washing showing characteristic epithelial grouping. × 545.

Fig. 393. Malignant lymphoma cells of large-cell type showing characteristic separation of individual cells. × 860.

usually less coarsely clumped than that of the carcinoma cell and the former lack the areas of nuclear clearing so characteristic of a malignant epithelial cell (fig. 396, 397).

Cytoplasmic Features. Although there may be a suggestion of vacuolation of the cytoplasm of the lymphoma cell vacuolation is never a prominent feature and vacuoles, if present, tend to be small and somewhat ill-defined (fig. 398). However, occasionally they may be more prominent (fig. 399). Conversely the carcinoma, being essentially an 'undifferentiated adenocarcinoma' may show, at least in some of its cells, cytoplasmic vacuolation. The vacuole may be extremely small (fig. 400) or readily recognizable (fig. 401).

Newer Techniques. The various means of differential diagnosis described above are all based on the morphologic assessment of specimens stained by the routine Papanicolaou method. Other techniques are available to assist with specific diagnostic problems and these undoubtedly will become standard procedures in all diagnostic cytology laboratories.

The use of *monoclonal antibodies* for the diagnosis of malignancy and, in particular, the resolution of difficult problems of differential diagnosis has been described both for tissue sections [15, 50] and cytological preparations [8, 13, 21, 45, 64, 65, 69]. *Gatter* et al. [15] used a panel of seven monoclonal antibodies, selected to include reagents reactive with both epithelial and lymphoid cells, to distinguish between anaplastic carcinoma and lymphoma. *Pizzolo* et al. [50] described a similar approach to this diagnostic problem although this group used only one antibody – the anti-human leucocyte antibody. This agent detects a human leucocyte antigen that is expressed strongly on B and T lymphoid cells but is not found on normal mesenchymal and epithelial tissues. Both these studies were carried out on tissue sections but there seems no reason why similar methods could not be used with cytological preparations derived from the stomach. The use of monoclonal antibodies to distinguish between undifferentiated carcinoma and malignant lymphoma is illustrated in plate 1.

In recent years the *electron microscope* has become an established tool in the routine histopathology laboratory [23]. One of its most valuable fea-

---

Fig. 394. Group of malignant lymphoma cells of large-cell type in gastric brushings. Note the crowding of cells with pseudoepithelial grouping. × 545.

Fig. 395. Higher power view of cells depicted in previous figure. Note apparent columnar configuration due presumably to pressure of contiguous cells. × 860.

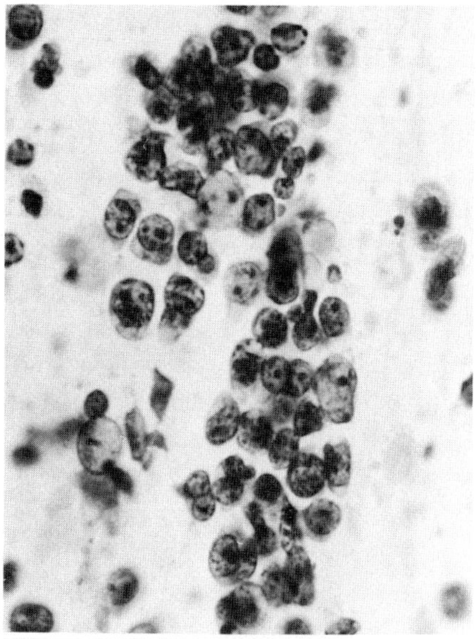

394

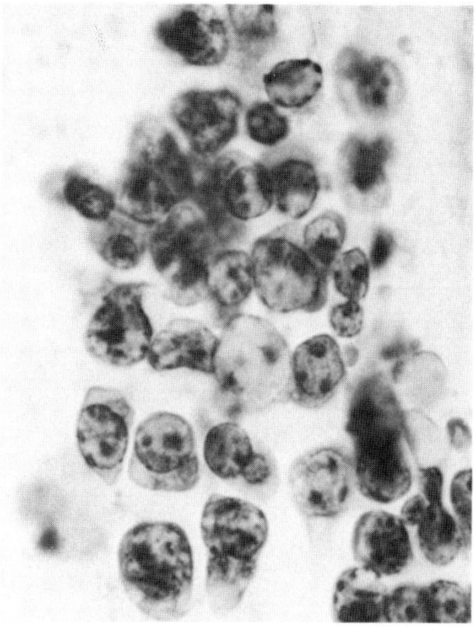

395

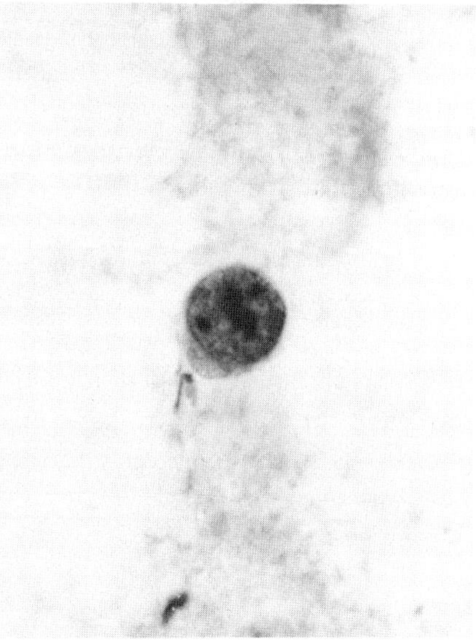

396

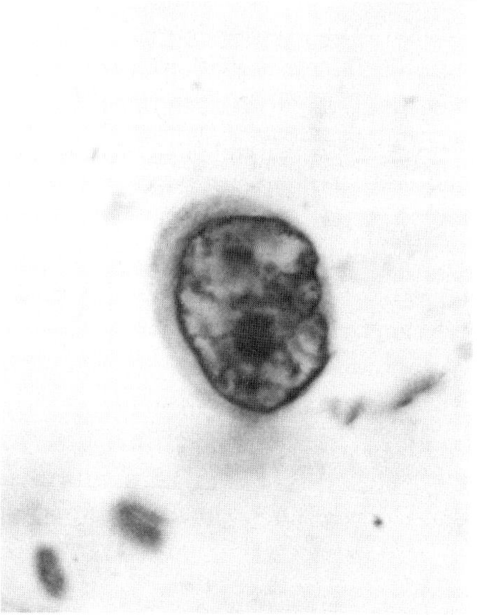

397

Fig. 396. Single malignant lymphoma cell of small-cell type from gastric washings showing characteristic uniform nuclear outline. ×860.

Fig. 397. Single undifferentiated or anaplastic carcinoma cell in gastric washings showing irregular nuclear membrane, chromatin clumping, and nuclear clearing. ×860.

tures is its ability to distinguish between different types of malignant neoplasms that may be morphologically inseparable on routine examination by light microscopy. This is particularly so in the differential diagnosis of anaplastic carcinomas and malignant lymphomas. The distinction ultrastructurally between these two entities is usually achieved largely by the demonstration in carcinoma cells of the various indicators of epithelial differentiation such as intercellular junctions (fig. 402, 403), basal lamina, acinus formation and cytoplasmic tonofilaments or, conversely, by the absence of such features in lymphoma cells (fig. 404). In addition, electron microscopy may reveal specific features of lymphoma cells, such as the cleaved nuclei of some lymphomas, thus aiding in the establishment of a diagnosis and the classification of the lymphoma present. The author is indebted to Dr. *Elaine Waters* of Western Australia for providing details of, and materials from, the case depicted in figures 405–407 which illustrates the use of electron microscopy as an aid in the cytological diagnosis of malignant lymphoma. The patient was a 47-year-old man who presented with an ulcer on the lesser curvature of the stomach. Gastric brushings showed numerous lymphoma cells predominantly of small-cell or lymphocytic type (fig. 405). 3 months later the patient presented with a right sided pleural effusion. Examination of this fluid again revealed lymphoma cells similar to those seen in the gastric brushings (fig. 406). Degenerative changes, with prominent karyorrhexis, were evident in this second specimen, possibly reflecting the effects of treatment given in the interim period.

Examination by electron microscopy of thin sections of a cell pellet prepared from the pleural effusion showed numerous abnormal lymphoid cells (fig. 407). The electron microscopist, Dr. *J.A. Armstrong,* described the lymphoid elements as being ‘of medium size, with large activated nuclei in which most of the chromatin is dispersed and the nucleoli are large and complex’. He concluded that the lymphoid cells were neoplastic and that the features were consistent with the diagnosis of non-Hodgkin’s lymphoma. Cleaved nuclei are a prominent feature.

Fig. 398. Cells from malignant lymphoma of large-cell type showing ill-defined cytoplasmic vacuolation. × 860.

Fig. 399. Cells from malignant lymphoma of large-cell type of stomach showing clearly defined cytoplasmic vacuoles. × 860.

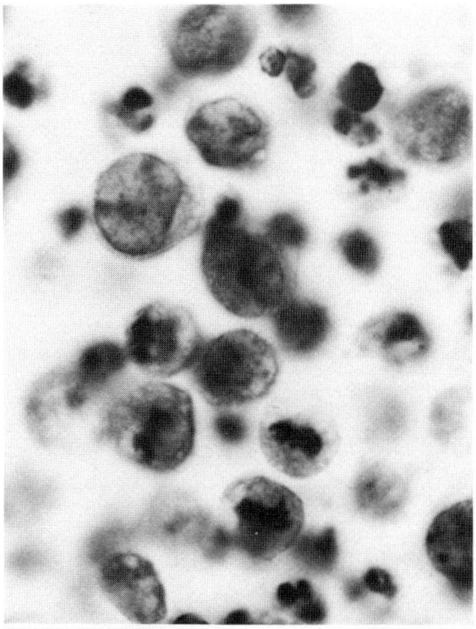

398

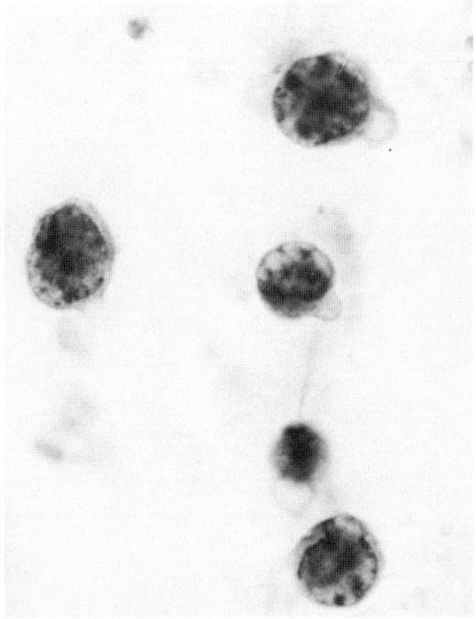

399

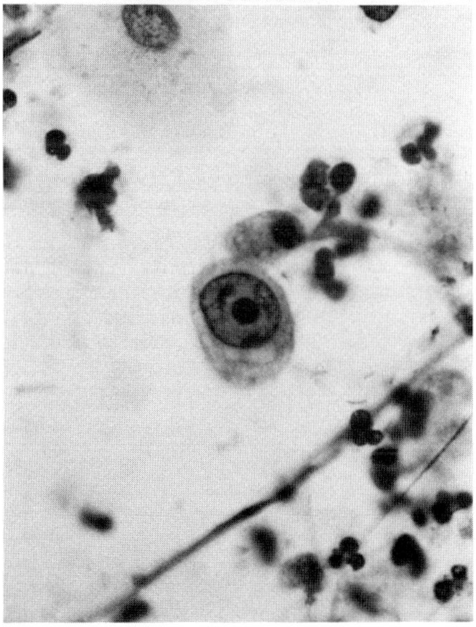

400

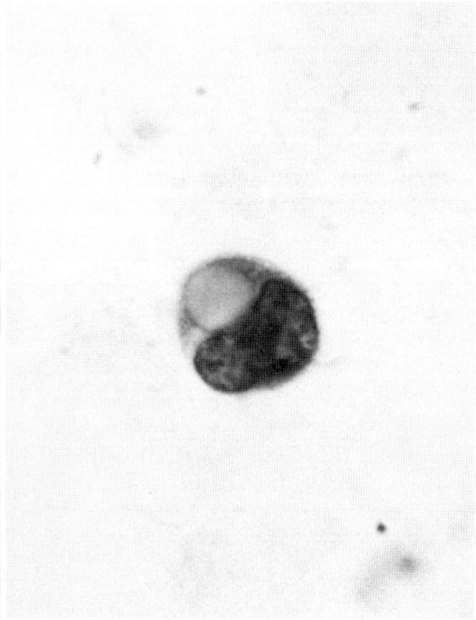

401

Fig. 400. Cells from undifferentiated or anaplastic carcinoma of stomach showing small but clearly defined cytoplasmic vacuole. × 545.

Fig. 401. Cells from undifferentiated or anaplastic carcinoma of stomach showing prominent cytoplasmic vacuole. × 860.

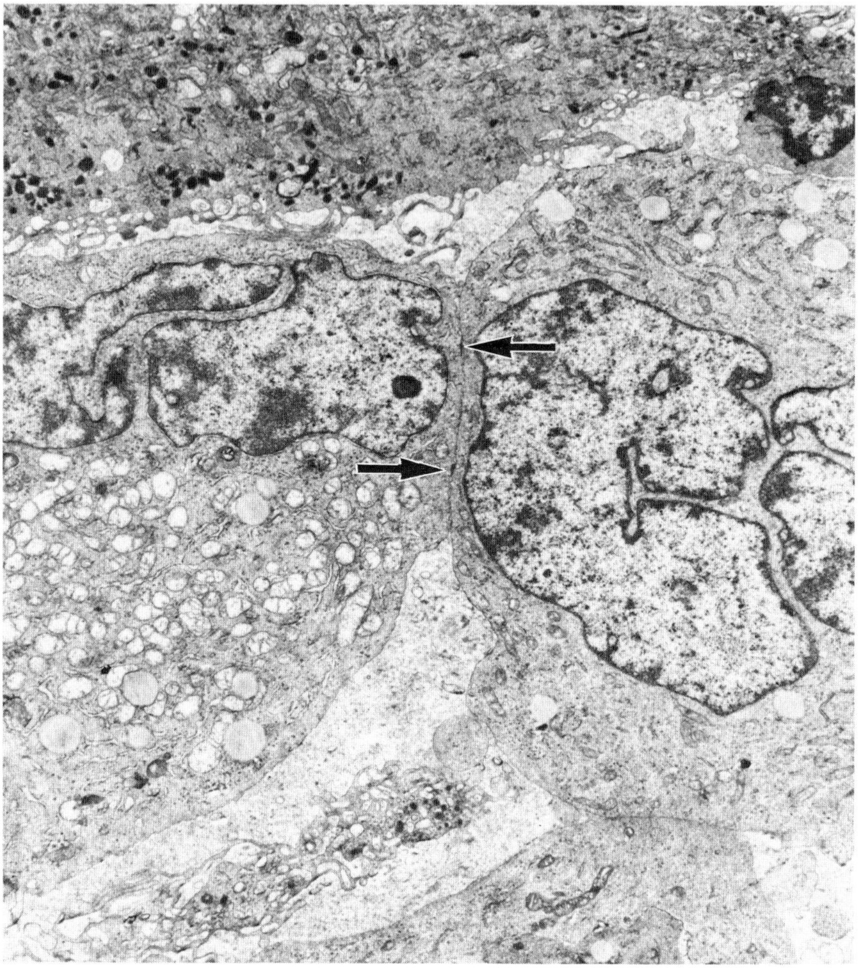

Fig. 402. Electron micrographs of carcinoma showing intercellular junctions as indicated by arrows. $\times$ 6,000.

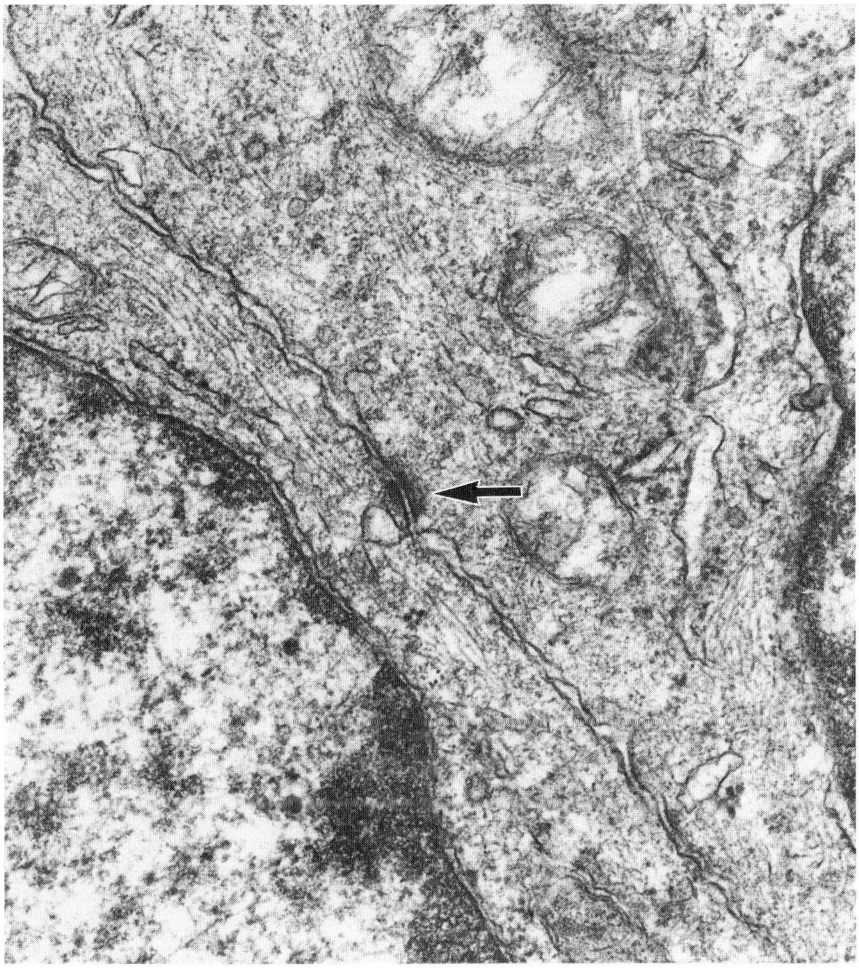

Fig. 403. Electron micrograph of carcinoma cells showing intercellular junction (arrow). ×47,000.

404
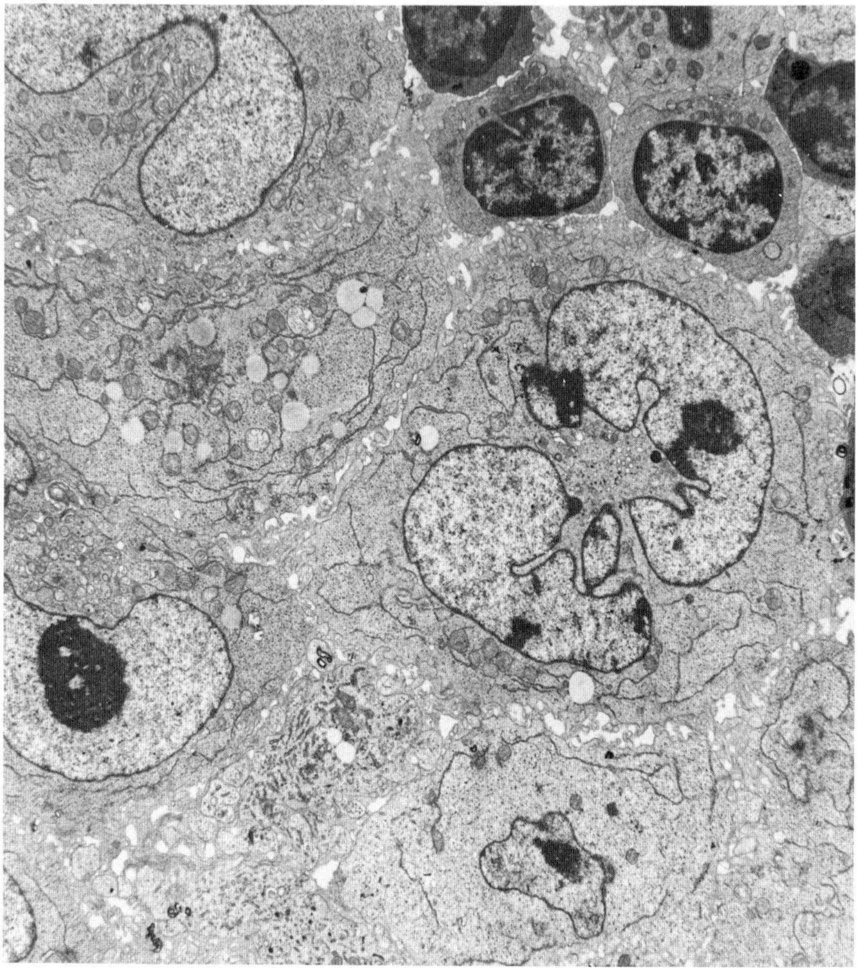

Fig. 404. Electron micrograph of malignant lymphoma tissue showing absence of intercellular junctions. $\times$4,700.

Fig. 405, 406. Malignant lymphoma cells of small-cell type in gastric brushings (fig. 405) and pleural effusion (fig. 406) of same patient. $\times$860. Specimens courtesy of Dr. *Elaine Waters,* Royal Perth Hospital, Western Australia.

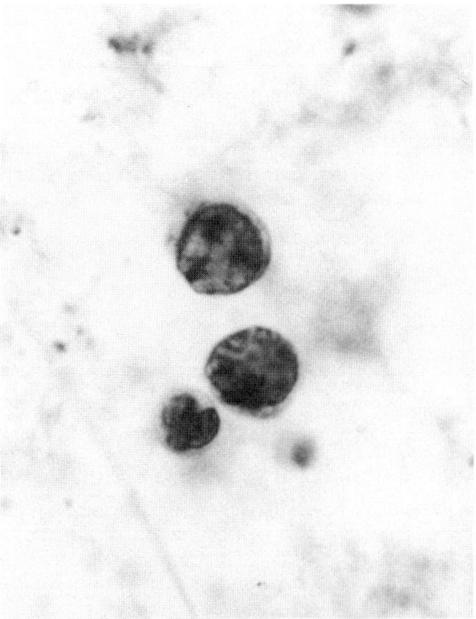

405

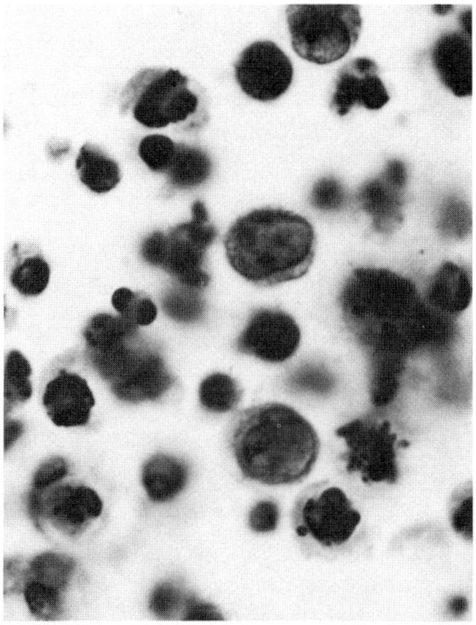

406

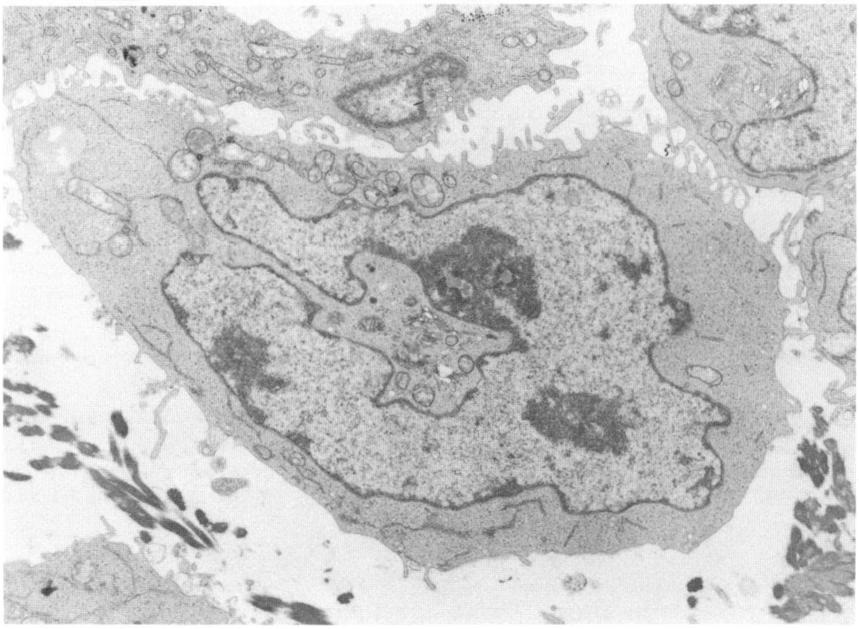

407

Fig. 407. Electron micrograph of malignant lymphoma cells from pleural effusion. × 5,200. Photograph courtesy of Dr. *J. A. Armstrong,* Royal Perth Hospital, Western Australia.

Secondary Neoplasms

The term 'secondary' is used in a very loose sense to describe the occurrence, in gastric specimens, of malignant cells derived from neoplasms of sites other than the stomach. There are three ways in which this may occur:

(1) The abnormal cells may be swallowed. This may be seen with carcinoma of the mouth, esophagus or bronchus (fig. 408).

(2) The stomach may be involved by direct extension from a contiguous or adjacent organ such as the esophagus (fig. 409), transverse colon or pancreas (fig. 410).

(3) True metastasis may occur but is uncommon. *Davis and Zollinger* [10] reported a series of 23,019 autopsies in which there were 67 metastatic malignancies of the stomach these including 13 lymphoid tumours, 3 fibrosarcomas, and 51 carcinomas. *Choi* et al. [7] reported 28 cases of breast carcinoma metastatic to the stomach 21 of which had gross lesions, 6 showing a thickening of the gastric wall that resembled linitis plastica. This feature was also noted by *Graham and Goldman* [20]. In most series published breast is one of the commoner primary sites and an example of this type of metastatic lesion is seen in figure 411. We have not diagnosed true metastatic malignancy of the stomach by cytological means.

References

1 Bennington, J.F.; Porus, R.; Ferguson, B.; et al.: Cytology of a gastric sarcoidosis. Report of a case. Acta cytol. *12:* 30–36 (1968).

2 Berg, J.W.: Primary lymphomas of the human gastro-intestinal tract. Natn. Cancer Inst. Monogr. *32:* 211–220 (1969).

3 Brandborg, L.L.; Taniguchi, L.; Rubin, C.E.: Exfoliative cytology in non-malignant conditions of the upper intestinal tract. Acta cytol. *5:* 187–190 (1961).

4 Brooks, J.J.; Enterline, H.T.: Primary gastric lymphomas: a clinicopathologic study of 58 cases with long-term follow-up and literature review. Cancer *51:* 701–711 (1983).

Fig. 408. Squamous carcinoma cells in gastric washings from patient with primary bronchogenic carcinoma. × 545.

Fig. 409. Squamous carcinoma cells in gastric washings from patient with esophageal carcinoma. × 545.

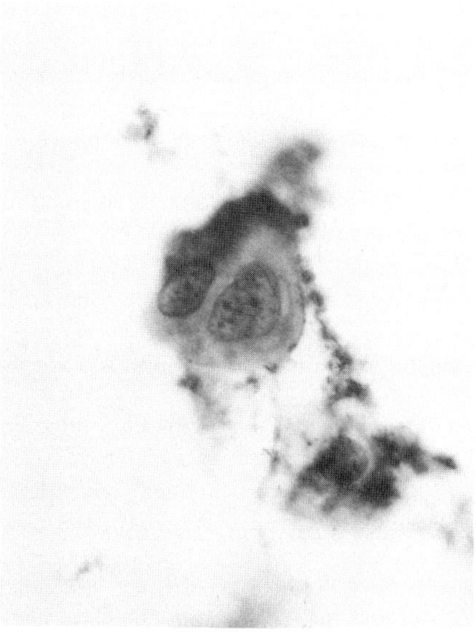

408

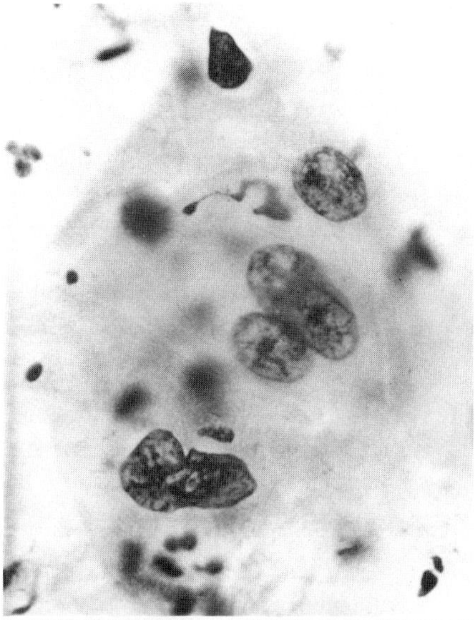

409

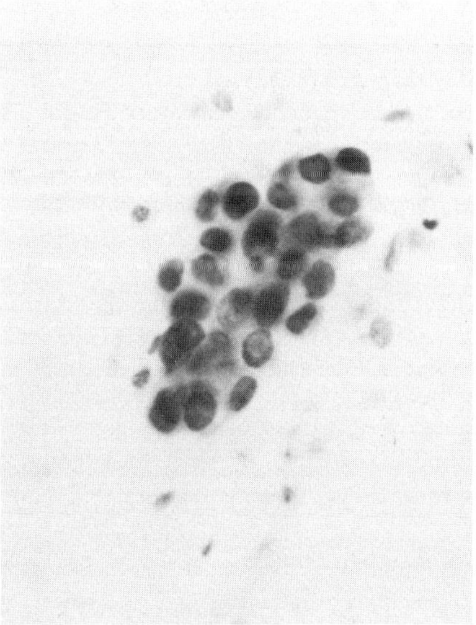

410

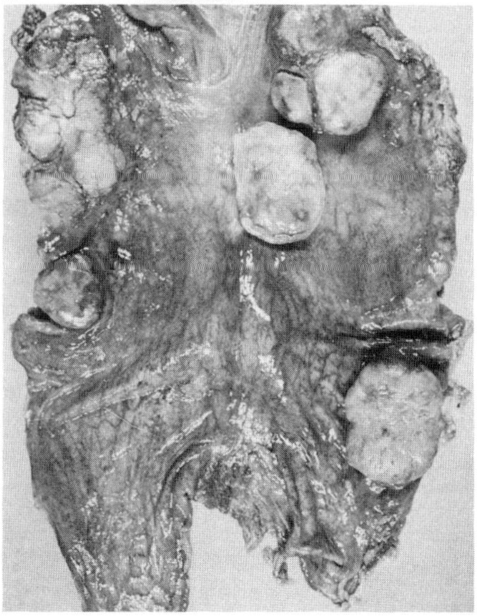

411

Fig. 410. Carcinoma cells in gastric washings of patient with carcinoma of pancreas with direct extension to stomach. × 375.

Fig. 411. Multiple metastases of primary carcinoma of breast to stomach. Photograph courtesy of Professor *P. Bhathal,* Royal Melbourne Hospital.

5 Cheli, R.; Carli, C.; Ciancamerla, G.: Cytology of the gastric mucosa in the development of chronic gastritis. Am. J. dig. Dis. *19:* 161–166 (1974).

6 Cheli, R.; Giacosa, A.: in Sherlock, Morson, Barbara, Veronesi, Precancerous lesions of the gastrointestinal tract (Raven Press, New York 1983).

7 Choi, S.H.; Sheehan, F.R.; Pickren, J.W.: Metastatic involvement of the stomach by breast cancer. Cancer *17:* 791–797 (1964).

8 Coleman, D.V.; Ormerod, M.G.; Dearnaley, D.P.: Immunoperoxidase staining in tumor marker distribution studies in cytologic specimens. Acta cytol. *25:* 205–206 (1981).

9 Cook, A.R.: Gastric damage by drugs and the role of the mucosal barrier. Aust. N.Z. J. Med. *6:* suppl. 1, pp. 26–29 (1976).

10 Davis, G.H.; Zollinger, R.W.: Metastatic melanoma of the stomach. Am. J. Surg. *99:* 94–96 (1960).

11 Dawson, I.M.P.; Cornes, J.S.; Morson, B.C.: Primary malignant lymphoid tumours of the intestinal tract. Br. J. Surg. *49:* 80 (1961).

12 Eker, R.; Efskin, J.: The pathology and prognosis of gastric carcinoma. Based on 1,314 partially and totally resected cases. Acta chir. scand., suppl. 264, pp. 1–182 (1960).

13 Epenetos, A.A.; Canti, G.; Taylor-Papadimitriou, J.; et al.: Use of two epithelium-specific monoclonal antibodies for diagnosis of malignancy in serous effusions. Lancet *ii:* 1004–1006 (1982).

14 Filippa, D.A.; Lieberman, P.H.; Weingrad, D.N.; et al.: Primary lymphomas of the gastrointestinal tract. Am. J. surg. Pathol. *7:* 363–372 (1983).

15 Gatter, K.C.; Abdulaziz, Z.; Beverley, P.; et al.: Use of monoclonal antibodies for the histopathological diagnosis of human malignancy. J. clin. Path. *35:* 1253–1267 (1982).

16 Gibbs, D.D.: Exfoliative cytology of the stomach (Butterworths, London 1968).

17 Gerard-Marchant, R.; Hamlin, I.; Lennert, K.; et al.: Classification of non-Hodgkin's lymphoma. Lancet *ii:* 406–408 (1974).

18 Goldman, H.; Ming, S.-C.: Mucins in normal and neoplastic gastrointestinal epithelium. Archs Path. *85:* 580–586 (1968).

19 Graham, R.M.; Rheault, M.H.: Characteristic cellular changes in epithelial cells in pernicious anemia. J. Lab. clin. Med. *43:* 235–254 (1954).

20 Graham, W.P.; Goldman, L.: Gastro-intestinal metastases from carcinoma of the breast. Ann. Surg. *159:* 477–480 (1964).

21 Hancock, W.W.; Medley, G.: Monoclonal antibodies to identify tumour cells in CSF. Lancet *ii:* 739–740 (1983).

22 Hattori, T.; Fujita, T.: Cellular kinetics of the gastric mucosa. Taisha *14:* 877–891 (1977).

23 Henderson, D.W.; Papadimitriou, J.: Ultrastructural appearances of tumours (Churchill Livingstone, Edinburgh 1982).

24 Henning, N.; Witte, S.: Atlas of gastrointestinal cytodiagnosis (Thieme, Stuttgart 1970).

25 Imai, T.; Kubo, T.; Watakabe, H.: Chronic gastritis in Japanese with reference to high incidence of gastric carcinoma. J. natn. Cancer Inst. *47:* 179–195 (1971).

26 Jarvi, O.; Lauren, P.: On the role of heterotopias of the intestinal epithelium in the pathogenesis of gastric cancer. Acta path. microbiol. scand. *29:* 26–44 (1951).

27 Kaplan, H.E.; Rigler, L.G.: Pernicious anaemia and carcinoma of the stomach – Autopsy studies concerning their interrelationship. Am. J. med. Sci. *209:* 339 (1945).
28 Kawachi, T.; Kurisu, M.; Numanyu, N.; et al.: Precancerous changes in the stomach. Cancer Res. *36:* 2673–2677 (1976).
29 Kawachi, T.: Intestinal metaplasia and its relation to gastric cancer. Taisha *14:* 919–926 (1977).
30 Kobayashi, S.; Prolla, J.C.; Kirsner, J.B.: Reactive lymphoreticular hyperplasia of the stomach. Archs intern. Med. *125:* 1030–1035 (1970).
31 Koss, L.G.: Diagnostic cytology and its histopathologic bases; 3rd ed. (Lippincott, Philadelphia 1979).
32 Koss, L.G.: Dysplasia, a real concept or a misnomer? Obstet. Gynec. *51:* 374–379 (1978).
33 Kuster, G.G.R.; ReMine, W.H.; Dockerty, M.B.: Gastric cancer in pernicious anemia and in patients with and without achlorhydria. Ann. Surg. *175:* 783 (1972).
34 Lauren, P.: The two histological main types of gastric carcinoma: diffuse and so-called intestinal type carcinoma. Acta path. microbiol. scand. *64:* 31–49 (1965).
35 Lawrence, J.C.: Gastro-intestinal polyps. Statistical study of malignancy incidence. Am. J. Surg. *31:* 499 (1936).
36 Lennert, K.; Mori, N.; Stein, H.; Kaiserling, E.: The histopathology of malignant lymphoma. Br. J. Haematol. *31:* suppl., pp. 193–203 (1975).
37 Leonards, J.R.: Are all aspirins alike? Aust. N.Z. J. Med. *6:* suppl. I, pp. 8–13 (1976).
38 Massey, B.E.; Rubin, C.E.: Stomach in pernicious anemia; cytologic study. Am. J. med. Sci. *227:* 481–492 (1954).
39 Medley, G.: Monoclonal antibody studies in gastric lymphoma (personal commun.).
40 Melamed, M.R.: The cytological presentation of malignant lymphomas and related diseases in effusions. Cancer *16:* 413–431 (1963).
41 Ming, S.-C.: Tumors of the esophagus and stomach. Atlas of tumor pathology, fasc. 7 (Armed Forces Institute of Pathology, Washington 1973).
42 Morson, B.C.; Dawson, I.M.P.: Gastrointestinal pathology, 2nd ed. (Blackwell, Oxford 1979).
43 Morson, B.C.: Carcinoma arising from areas of intestinal metaplasia in the gastric mucosa. Br. J. Cancer *9:* 377–385 (1955).
44 Mosbech, J.; Bidebaek, A.: Mortality from and risk of gastric carcinoma among patients with pernicious anemia. Br. med. J. *ii:* 390 (1950).
45 Nadji, M.: The potential value of immunoperoxidase techniques in diagnostic cytology. Acta cytol. *24:* 442–447 (1980).
46 Nieburgs, H.E.; Glass, G.B.J.: Gastric cell maturation disorders in atrophic gastritis, pernicious anemia and carcinoma. Histologic site of origin and diagnostic significance of abnormal cells. Am. J. dig. Dis. *8:* 135–159 (1963).
47 Oota, K.; Sobin, L.H.: Histological typing of gastric and oesophageal tumours. International Histological Classification of Tumours, No. 18 (World Health Organization, Geneva 1977).
48 Perez-Mota, A.; Casanova, A.; Hidalgo, A.: Gastroscopic cytology in intestinal metaplasia. Endoscopy *19:* 1–6 (1977).

49 Peters, M.; Weiner, J.; Whelan, G.: Fungal infection associated with gastroduodenal ulceration: endoscopic and pathologic appearances. Gastroenterology *78:* 350–354 (1980).

50 Pizzolo, G.; Sloane, J.; Beverley, P.; et al.: Differential diagnosis of malignant lymphoma and nonlymphoid tumors using monoclonal anti-leucocyte antibody. Cancer *46:* 2640–2647 (1980).

51 Prolla, J.C.; Kobayashi, S.; Yoshi, Y.; et al.: Diagnostic cytology of the stomach in gastric syphilis. Report of two cases. Acta cytol. *14:* 333–337 (1970).

52 Prolla, J.C.; Kirsner, J.B.: Handbook and atlas of gastrointestinal exfoliative cytology (University of Chicago Press, Chicago 1972).

53 Rappaport, H.: Tumours of the haemopoietic system. Atlas of tumour pathology, sect. 3, fasc. 8 (Armed Forces Institute of Pathology, Washington 1966).

54 Rilke, F.: Malignant lesions of the stomach: cytohistologic correlations. Acta cytol. *23:* 517–518 (1979).

55 Robb-Smith, A.H.T.; Taylor, C.R.: Lymph node biopsy (Miller Heyden, London 1981).

56 Rosas-Uribe, A.; Variakojis, D.; Rappaport, H.: Proteinaceous precipitate in nodular (follicular) lymphomas. Cancer *31:* 534–542 (1973).

57 Schade, R.O.K.: Gastric cytology. Principles, methods and results (Edward Arnold, London 1960).

58 Schell, R.F.; Dockerty, M.B.; Comfort, M.W.: Carcinoma of the stomach associated with pernicious anemia. Surgery Gynec. Obstet. *98:* 710 (1954).

59 Shearman, D.J.C.; Finlayson, N.D.C.; Wilson, R.; et al.: Carcinoma of the stomach and early pernicious anemia. Lancet *ii:* 403 (1966).

60 Stobbe, J.A.; Dockerty, M.B.; Bernatz, P.E.: Primary gastric lymphoma and its grades of malignancy. Am. J. Surg. *112:* 10–19 (1966).

61 Stewart, M.J.: Observations on the relation of malignant disease to benign tumours of the intestinal tract. Br. med. J. *ii:* 567 (1929).

62 Stout, A.P.: Tumours of the stomach. Atlas of tumor pathology, sect. 6, fasc. 21 (Armed Forces Institute of Pathology, Washington 1953).

63 Takeda, M.: Atlas of diagnostic gastrointestinal cytology (Igaku-Shoin, New York 1983).

64 To, A.; Coleman, D.V.; Dearnaley, D.P.; Ormerod, M.G.; Steele, K.: Use of antisera to epithelial membrane antigen for the cytodiagnosis of malignancy in serous effusions. J. clin. Path. *34:* 1326–1332 (1981).

65 To, A.; Dearnaley, D.P.; Ormerod, M.G.; Canti, G.; Coleman, D.V.: Indirect immunoalkaline phosphatase staining of cytologic smears of serous effusions for tumour marker studies. Acta cytol. *27:* 109–113 (1983).

66 Videbaek, A.; Mosbeck, J.: The aetiology of gastric carcinoma elucidated by a study of 302 pedigrees. Acta med. scand. *149:* 137 (1954).

67 Warren, S.; Lulenski, C.R.: Primary solitary lymphoid tumours of the gastro-intestinal tract. Ann. Surg. *115:* 1–12 (1942).

68 Willis, R.A.: Pathology of tumours; 4th ed. (Appleton-Century-Crofts, New York 1967).

69 Woods, J.C.; Spriggs, A.I.; Harris, H.; McGee, J.O'D.: A new marker for human cancer cells. 3. Immunocytochemical detection of malignant cells in serous fluids with the Ca 1 antibody. Lancet *ii:* 512–514 (1982).

6. Early Cancer and Precancer of Esophagus and Stomach

The cytological diagnosis of superficial or 'early' carcinoma of the stomach and esophagus has much in common with that of the more advanced lesions of both organs and could well have been considered in the appropriate sections of the two previous chapters. However, the topic has been allotted its own chapter as it raises some separate issues of major importance, both in a practical and in a theoretical sense. As already discussed, the diagnosis of gastric and esophageal cancer in a minimally invasive, or even preinvasive stage, offers the only real hope of improving the results of treatment of these diseases. However, another aspect of considerable interest is that the study of these early lesions may shed some light on the histogenesis of both carcinoma of the esophagus and of the stomach. An analogy could be drawn with cancer of the uterine cervix where the application of the techniques of diagnostic cytology to the detection of this condition and its precursors has completely revolutionized our understanding of its development.

Early Cancer and Precancer of the Esophagus

The esophagus, being lined for the greater part by stratified squamous epithelium with a well defined basement membrane, lends itself readily to a study of preinvasive lesions referred to elsewhere as dysplasia and carcinoma in situ. The occurrence of these conditions within the esophageal mucosa has been recognized for some considerable time and both can be demonstrated frequently in esophagectomy specimens removed for invasive carcinoma by examination of the mucosa adjacent to the main lesion. However, until recently descriptions of these conditions and their cytological manifestations have been few and somewhat sporadic [2, 8, 10, 14, 28, 30, 32, 35].

In 1973 the Co-ordinating Group for the Research of Esophageal Cancer, Chinese Academy of Medical Science and Honan Province [3] reported the results of an extensive study designed to detect carcinoma of the esophagus at an early stage by cytological means. A double lumen rubber tube with an abrasive balloon at the distal end was used to collect the specimens. The fasting patient was asked to swallow the tube until the collapsed balloon had passed through the cardia. The balloon was then inflated and slowly withdrawn. An adequate specimen of cells was thus collected on the surface of the balloon.

Two groups of subjects were examined in this way. The first group, comprising 7,686 cases with suspected esophageal cancer, was examined between 1963 and 1969. Carcinoma of the esophagus or gastric cardia was found in 510 cases and, of these, early carcinoma was present in 86 – an incidence of 16.3%. In the second group, studied between 1970 and 1972, 11,564 persons over 30 years of age were studied. Carcinoma of the esophagus or gastric cardia was found in 136 of the subjects and of these 96 cases were in the early stage – an incidence of 70.6%.

The authors of this paper divided the cytological findings into four grades as follows:

(1) Normal – the majority of the cells were intermediate in type with about 10% from the superficial layers. Parabasal cells were very rare.

(2) Mild dyskaryosis – hyperchromasia was seen in both intermediate and superficial type cells. The nuclei were enlarged but not more than three times that of the normal cell.

(3) Marked dyskaryosis – intermediate type cells were present with nuclei three or more times as large as the normal cell. Hyperchromasia was more marked. Numerous dyskaryotic parabasal type cells were present.

(4) Carcinoma – there were many dyskaryotic and squamous carcinoma cells present. The latter were usually single, polygonal, and comparatively uniform in shape with malignant characteristics in their nuclei. In carcinoma in situ of small size, the carcinoma cells were few in number, whilst the larger lesions, particularly when accompanied by early infiltration, shed large numbers of carcinoma cells. In the more advanced carcinomas the malignant cells often clustered together, were pleomorphic, and showed varying degrees of differentiation.

58 surgical specimens were examined in considerable detail. All specimens showed a squamous cell carcinoma with, in one, an additional glandular component. 31 cases were carcinoma in situ whilst the remaining 27

were early or superficially invasive lesions, the infiltration being limited to the submucosa.

The authors concluded from their observations that carcinoma in situ is a definite stage through which the epithelium of the esophagus must pass in the course of development of an established invasive esophageal cancer. Subsequently the results of several other studies for the detection of early esophageal cancer and precancer have been reported.

In 1979 *Crespi* et al. [7] reported a study carried out in Northern Iran in which 430 subjects were examined and in 1982 the same workers [17] reported a similar study in Northern China where 527 individuals were investigated. In both studies all subjects were submitted to an endoscopic examination and guided cytology and at least two biopsies were taken from each participant.

In addition, to revealing abnormal cells indicative of dysplasia and carcinoma in situ a very high incidence of chronic esophagitis was seen in patients from these two acknowledged high-risk areas for esophageal cancer. The esophagitis was characterized by the presence of an irregular, friable mucosa with varying degrees of oedema, hyperemia and leukoplakia but without ulceration. These changes affected the middle and lower thirds of the esophagus but the precardial region was spared. This type of chronic esophagitis contrasted with that seen in low-risk areas, such as most European countries and North America, where the changes are usually characterized by erosions and ulcerations, mostly involve the precardial region, and are almost invariably due to reflux.

The esophagitis was often accompanied by mucosal atrophy and limited follow-up studies showed a significant progression to dysplasia and cancer. On the basis of these findings the following esophageal cancer model has been proposed [18]:

Chronic esophagitis → atrophy → dysplasia → cancer.

In 1981 *Berry* et al. reported on a similar study carried out in the Republic of South Africa [1]. Using an inflatable balloon catheter with an abrasive, absorbent surface, cytologic material was collected from the total esophageal mucosa of 500 patients none of whom was suspected of having esophageal cancer. In summary, malignant cells were identified in 15 patients or 3% of the total number examined, and dysplastic cells in another 26. All lesions were confirmed by endoscopically directed biopsies and 10 of the 15 cancer patients described as having 'early stage lesions' are tabulated, 5 being described as early invasive carcinoma, 2 as microinvasive squamous carcinoma and 3 as squamous carcinoma in situ. The 5 remaining cases of invasive carcinoma are categorized as asymptomatic advanced squamous carcinoma.

For the purposes of this text the most valuable part of this paper is the morphologic description of the various lesions, particularly those designated as dysplasia and carcinoma. Dysplastic squamous cells were assessed as in cervical smears and were usually seen singly or in small groups. The presence of keratin and pearl formation was noted. The background was clean in some cases with dysplasia but many showed co-existent inflammatory change. The cytologic details were identical to those seen in the uterine cervix.

A similar study was reported in 1982, also from the Republic of South Africa [19]. In this second study 101 patients were studied, a suction-abrasive tube being used to sample the esophagus. This sampling procedure was followed immediately by endoscopy, directed biopsies and brushings being obtained from visible lesions. Positive cytology for malignancy was obtained in 44 of 47 patients with carcinoma of the esophagus. In 54 patients with benign conditions of the esophagus, 44 had normal or very mild inflammatory changes on cytology, and 10 had features of marked inflammatory atypia. There were three false-negative and no false-positive cytologic results in the series. The false negative results occurred in a group of patients with such advanced carcinoma that complete passage of the tube was prevented.

The author is deeply indebted to Dr. *Ann Berry* who, through the courtesy of Dr. *Gladwyn Leiman,* allowed him to review the cytologic material used in the preparation of her paper. Dr. *Leiman,* a participant in the second study, also very kindly allowed the author to review some of her own cases. This review indicated that the criteria used in gynaecological cytology can be applied directly to the diagnosis of dysplasia and carcinoma of the esophagus. This is illustrated in figures 412–415 which depict some of the cells seen in the South African material.

Fig. 412. Cell indicating mild to moderate dysplasia. × 545.

Fig. 413. Cells indicating mild to moderate dysplasia. × 345.

Fig. 414. 'Pearl' formation probably indicative of moderate to severe dysplasia. × 545.

Fig. 415. Abnormal squamous cell indicating esophageal carcinoma in situ. × 545. Material courtesy of Dr. *Ann Berry* and Dr. *Gladwyn Leiman,* Republic of South Africa.

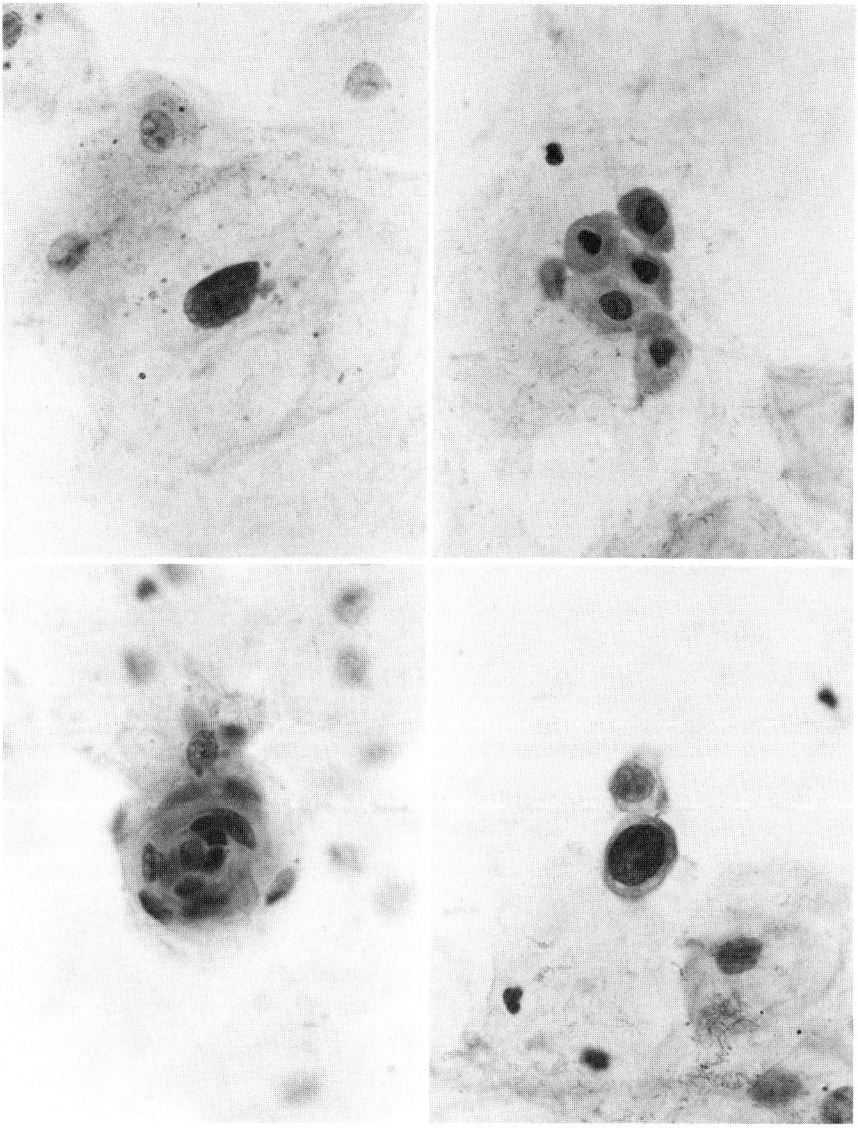

412 413

414 415

Fig. 412–415. Cells from South African esophageal cancer screening programme.

For those who do not have access to material of this sort a re-examination of washings and brushings from established cases of invasive esophageal cancer is strongly recommended. As already discussed, the esophageal mucosa adjacent to an invasive lesion frequently displays the changes of dysplasia and carcinoma in situ. Thus figure 416 depicts cells that indicate clearly the mild dysplastic changes seen in figure 417. Similarly figure 418 shows changes indicative of the moderate dysplasia apparent in figure 419. Finally severe dysplasia and/or carcinoma in situ may be diagnosed cytologically as seen in figures 420–423.

In summary it would appear that there is a close analogy between the development of squamous cell carcinoma of the esophagus and that of the uterine cervix. Although there are serious defects in the current systems of nomenclature it would seem advisable to apply the terms dysplasia and carcinoma in situ to the microscopically demonstrable developmental lesions until a more satisfactory system of nomenclature is adopted universally. This relatively recent area of diagnostic cytology throws considerable light on the histogenesis of esophageal cancer and provides a mechanism whereby earlier diagnosis, and hence more effective treatment, of the disease is possible.

Surface or Superficial Cancer of the Stomach

Basic Concepts and Problems of Terminology

Early gastric cancer has been defined as a carcinoma which is confined to the mucosa, or mucosa and submucosa, regardless of the presence of lymph node metastases [19]. However, it must be emphasized that this is a definition based on the gross or macroscopic appearances of the lesion. In addition, the use of the word 'early' implies a relationship between the size of the lesion and the time that it has taken to develop. Whilst this may be so it is open to question. However, the word 'early' in this context is not meant to imply a stage in the genesis of the cancer but rather to describe the potential curability of the lesion. Thus it has been suggested [13] that a carcinoma in the early stage may be defined clinically as 'a carcinoma in the stage with a high possibility of permanent cure by means of simple gastrectomy'. By contrast the term 'carcinoma in the initial stage' may be used for a case in which 'only a minute, histologically barely detectable, yet undebatable malignant lesion is present'. Such a lesion would not be manifest

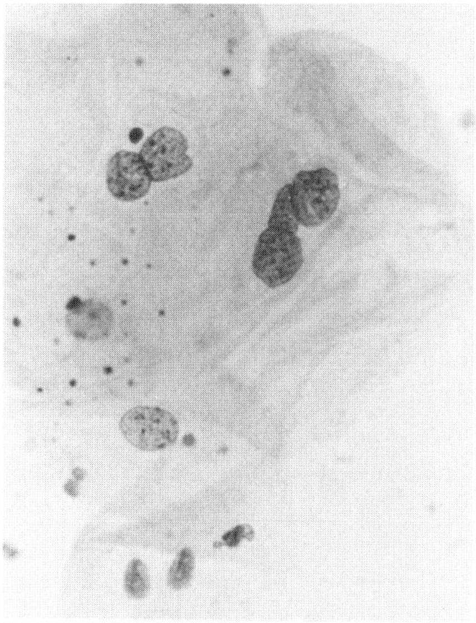

416

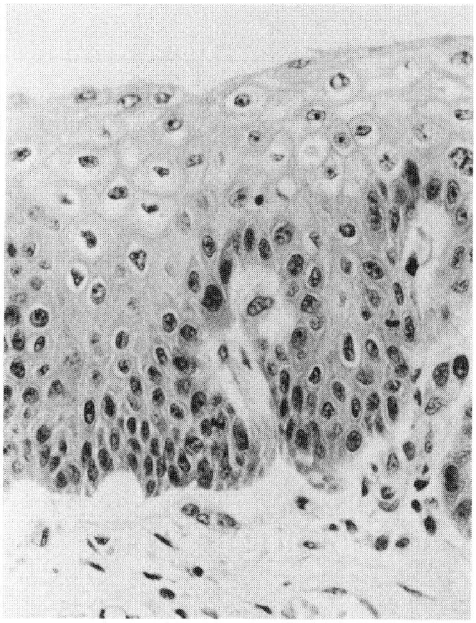

417

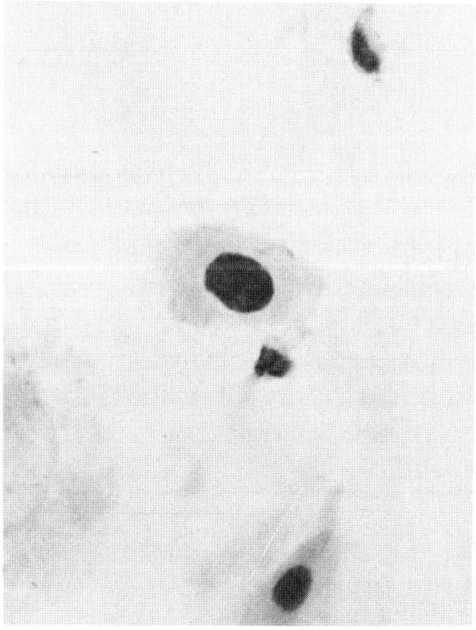

418

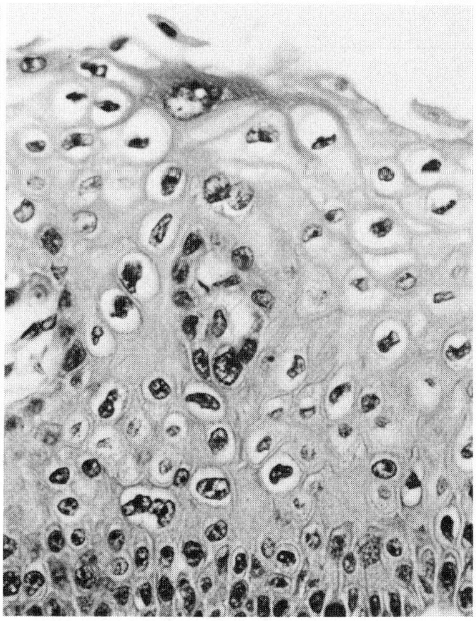

419

Fig. 416. Cells from esophageal washings showing evidence of mild dysplasia. × 545.
Fig. 417. Biopsy of esophagus confirming presence of mild dysplasia. × 205.
Fig. 418. Abnormal cells in esophageal washings indicating moderate dysplasia. × 545.
Fig. 419. Biopsy of esophagus confirming presence of moderate dysplasia. × 275.

clinically but may be detected if surgery was carried out for an associated benign lesion and the specimen examined meticulously. Alternatively, of course, its presence may be revealed by cytological examination.

The histopathologist and cytopathologist, both dependent on cell and tissue morphology, are better served by a descriptive nomenclature and it is generally accepted that the terms *superficial* or *surface* carcinoma are best used for these small, and presumably 'early', gastric cancers. In turn superficial or surface carcinoma can be subdivided microscopically into intramucosal carcinoma, where the connective tissues of the lamina propria have been infiltrated, and submucosal carcinoma, where there is invasion also of the submucosal tissues but the muscularis remains intact.

True carcinoma in situ of the stomach undoubtedly exists. However, the unequivocal exclusion of early invasion is extremely difficult, if not impossible, and hence it is probably preferable not to apply the term 'carcinoma in situ' to the stomach except as a theoretical concept. In order to overcome some of these difficulties of nomenclature *Nagayo* [21] coined the term 'borderline lesion' and this term is used widely in the Japanese literature. However, it has not gained wide acceptance in Europe and North America.

Macroscopic Features

In 1962 a meeting of the Japan Gastroenterological Endoscopic Society resolved to adopt a uniform classification of early gastric cancer [20].

Early gastric cancer was divided into three main groups and three subgroups on the basis of the macroscopic appearances at endoscopy and on gastrectomy specimens. These various groups are depicted diagrammatically in figure 424. The lesions are described as follows:

Type I	The Protruded Type The tumour projects clearly into the lumen. All polypoid, nodular and villous tumours are included in this group.
Type II	The Superficial Type This is further subdivided into three subgroups.
Type II(a)	Elevated In carefully prepared gastrectomy specimens this is seen as a flat, plaque-like lesion, well circumscribed, and only raised above the surrounding mucosa by a few millimetres.

Type II(b)	Flat No abnormality is visible macroscopically, although some colour change may be seen endoscopically and in very carefully prepared gastrectomy specimens.
Type II(c)	Depressed The surface is slightly depressed below the adjacent mucosa for not more than the thickness of the submucosa. A thin covering of exudate may indicate surface erosion.
Type III	The Excavated Type This is essentially ulceration of variable depth into the gastric wall. This is rarely seen in pure form but may be combined with any of the other types.

In describing a particular lesion the symbols are used and when, as is often the case, more than one type is seen in combination, the symbol for each type is given. In this case the dominant macroscopic feature is placed first, e.g. I + II(c).

In making a plea for all pathologists to become familiar with the Japanese classification of early gastric cancer, *Morson* [15] suggests that it may be more acceptable if only the descriptive terms are used rather than the symbols. He further suggests that the words protuberant or polypoid may be preferable to 'protruded' in Type I.

Histological Features

Histologically cases of surface or superficial carcinoma are essentially the same as the more advanced lesions and hence the same principles of classification apply. Thus the lesions can be classified in accordance with the World Health Organization Classification preferred in chapter 5 or, alternatively, the classification of Lauren into intestinal and diffuse forms can be applied. Indeed classification of these early forms of gastric cancer is much less complex than that of the more advanced forms with their frequently bewildering variety of tissue patterns. The depth of penetration of the neoplasm is the most critical feature in the histological assessment and this has already been alluded to. The neoplasm may be confined completely to the gastric mucosa, the so-called intramucosal carcinoma (fig. 425, 426).

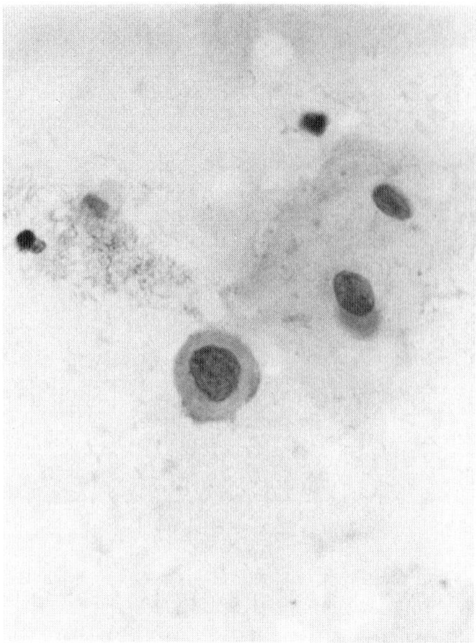
420

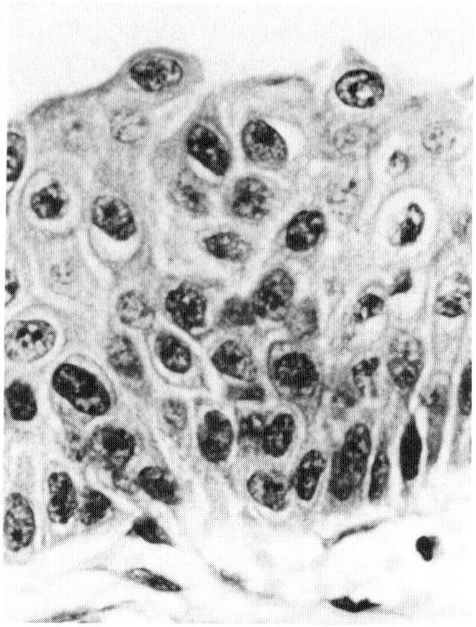
421

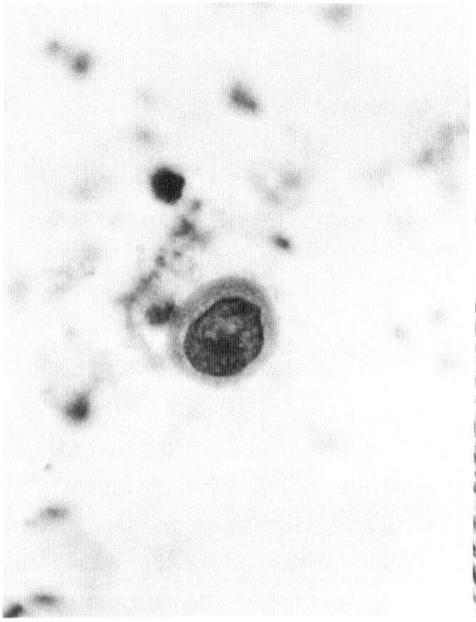
422

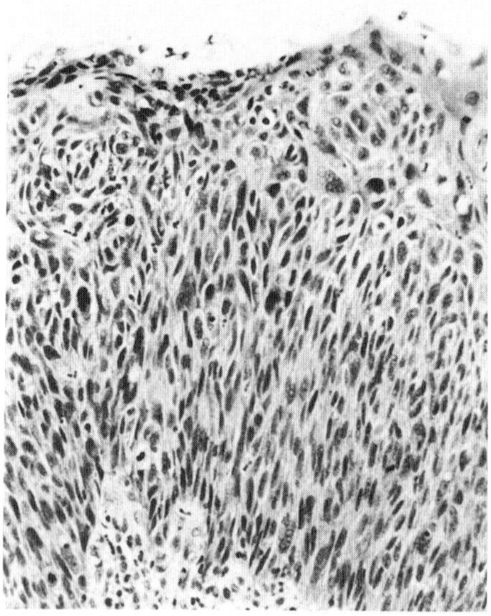
423

Fig. 420. Cells in esophageal brushings indicating moderate to severe dysplasia. × 545.

Fig. 421. Biopsy of esophagus showing moderate to severe dysplasia or possibly carcinoma in situ. × 345.

Fig. 422. Cell in esophageal brushings indicative of carcinoma in situ. × 545.

Fig. 423. Biopsy of esophagus showing area of carcinoma in situ. × 140.

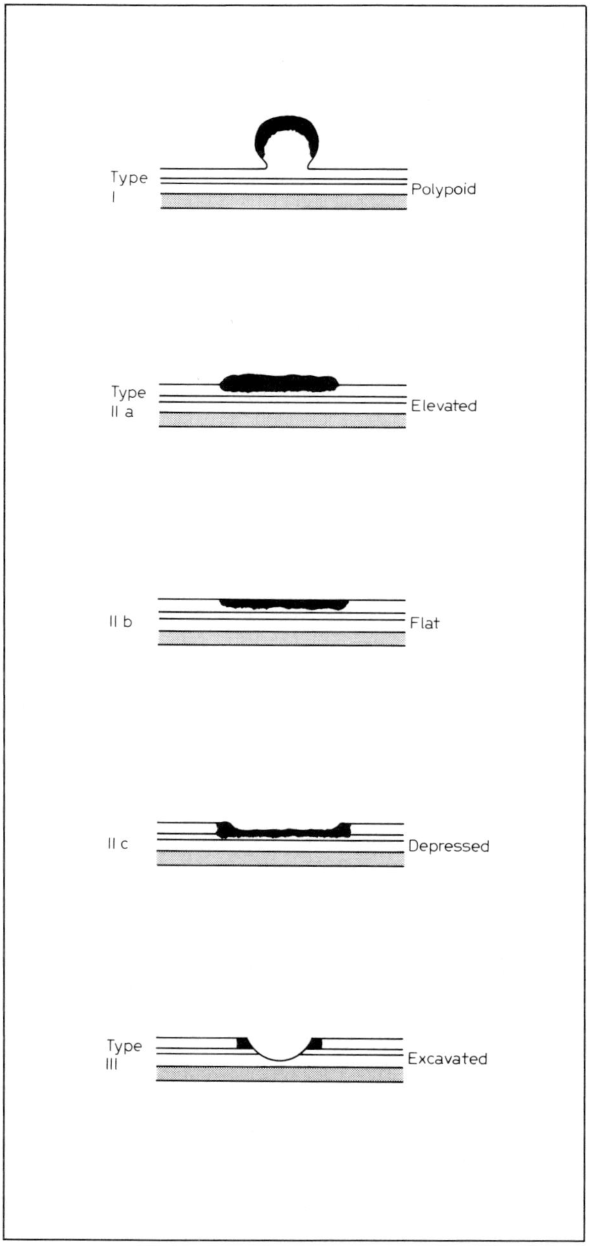

Fig. 424. Diagrammatic representation of classification of early gastric cancer – Japan Gastroenterological Endoscopic Society.

Alternatively it may penetrate the submucosa but, by definition, it must not infiltrate the muscle coats. As already emphasized, the diagnosis of carcinoma in situ is always suspect in the stomach.

Cytologic Diagnosis

It is generally stated that early cancer of the stomach and advanced cancer are not distinguishable cytologically or that the differences when present, are relatively insignificant. In his excellent monograph on gastric cytology, published in 1968, *Gibbs* [9] states categorically that cells from surface carcinoma of the stomach are identical to those obtained from macroscopically obvious carcinomas. This statement is reiterated equally categorically by *Prolla and Kirsner* [26], and either stated or inferred in the more recent publications by the many Japanese workers in this field. *Takeda* [33] describes in some detail the cytological features of early carcinoma of both the 'intestinal' and 'gastric' types but emphasizes that 'no cytologic differences are noted between early gastric carcinoma and advanced cancer in these two types'. *Schade* [27], who must be regarded as the pioneer in the cytological diagnosis of early gastric cancer, adopted a similar stance although he did emphasize that a surface carcinoma sheds cells more abundantly than more advanced tumours and that the cellular material obtained in the former cases was not contaminated by debris. *Koss* [12] also comments on the freedom of the specimen from necrotic debris and the good preservation of the cancer cells. In addition, he describes the tendency of cells from well differentiated superficial gastric adenocarcinomas to form more compact clusters than in the more advanced cancers. Undoubtedly the cytological features of many cases of superficial gastric cancer are identical to those seen in the more advanced lesions. However, in our laboratory considerable stress is laid on the features already described namely, the clean background with a conspicuous absence of tumour debris, and the small compact cell clusters formed by cells from surface lesions (fig. 427, 428). In addition, the cells from these superficial cancers quite frequently display abnormalities that appear to lie between those of extreme regenerative activity, such as may be seen in a florid chronic superficial gastritis, in the active phase, and unequivocal malignancy. Thus there is a tendency to increased pleomorphism, some irregularity of nuclear membranes (fig. 429) and an increasing disorder in the relationship of one cell to another (fig. 430).

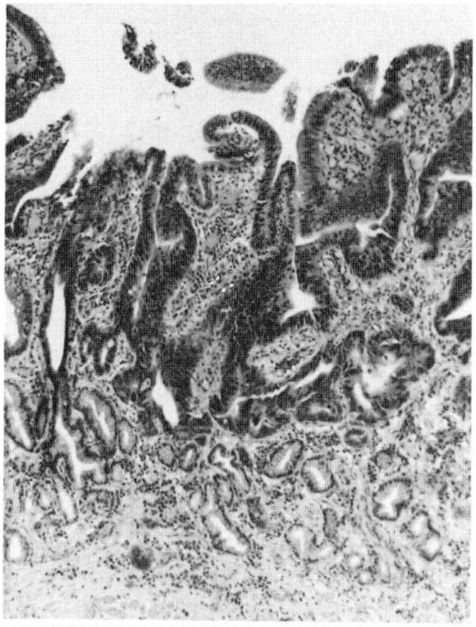

425

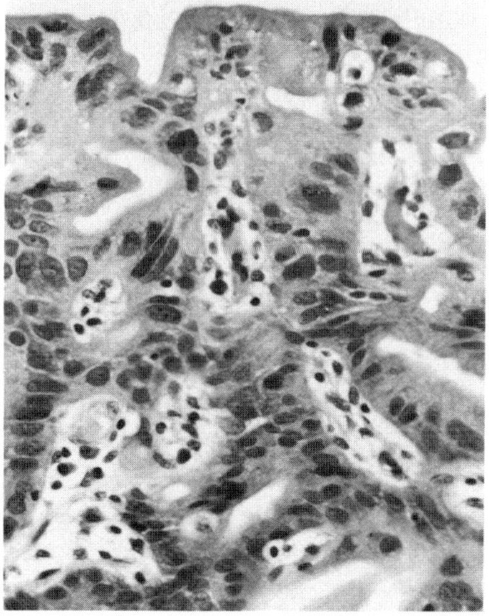

426

Fig. 425. Intramucosal carcinoma. Abnormal glands are recognized clearly by their hyperchromasia. × 55.

Fig. 426. Intramucosal carcinoma. Abnormalities are largely limited to the glands but there is some penetration of the lamina propria. × 205.

Precursor and Precancerous Lesions of the Stomach

As in other areas of the body, cytological findings have acted as a catalyst to a great deal of investigation into the development of gastric cancer and this, coupled with more sophisticated endoscopic and histologic techniques, has resulted in the accumulation of an impressive body of knowledge concerning the histogenesis of carcinoma of the stomach.

Several studies of the gastric mucosa in high risk populations have now been carried out and reported in considerable detail [4, 5]. These studies would suggest that a spectrum of change from the normal gastric mucosa to established carcinoma can be observed the situation being similar to that which has been so extensively documented in relation to the uterine cervix. Some of these conditions and their cytological manifestations have already been discussed in an earlier chapter but the salient points are reiterated for completeness of the current discussions.

The first observable lesion of significance would appear to be chronic superficial gastritis a condition manifested, as already discussed, by an infiltration of the lamina propria by lymphocytes, plasma cells and polymorphonuclear leucocytes usually accompanied by necrosis of epithelial cells and regenerative changes in the glandular neck region.

In the earlier discussion chronic gastritis was considered as a single lesion although it was indicated that both acute and chronic phases may occur. In considering this condition in relation to gastric cancer precursors, however, it is relevant to note the several different types of chronic gastritis undoubtedly occur and that the following three types have been fairly well documented [6]:

(1) *Autoimmune chronic gastritis (ACG)* accompanies the precursor pernicious anaemia syndrome and was referred to by *Strickland and Mackay* [31] as type A gastritis. It is believed to be due to injury by autoantibodies against intrinsic factor and parietal cells and its characteristic distribution is in keeping with this supposed pathogenesis. Thus the antrum is usually spared the changes being present diffusely throughout the corpus and fundus of the stomach. Autoimmune chronic gastritis is usually associ-

---

Fig. 427. Small compact cluster of carcinoma cells from surface carcinoma of stomach. Note 'clean' background. × 545.

Fig. 428. Small compact cluster of carcinoma cells from surface carcinoma of stomach. Background is again 'clean'. × 545.

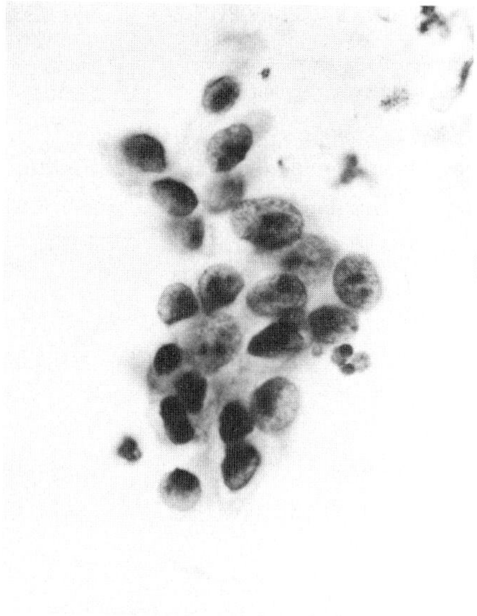

427

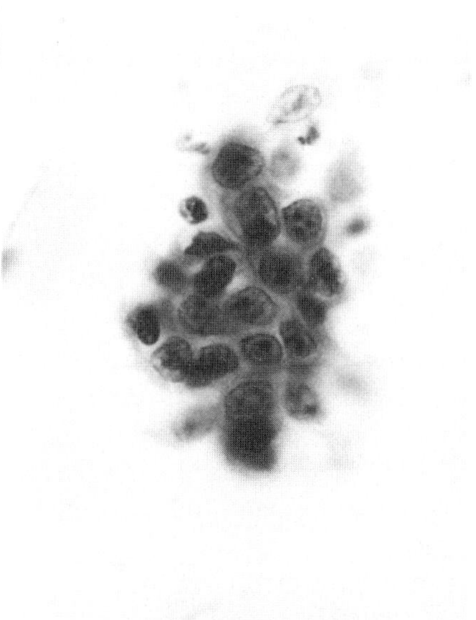

428

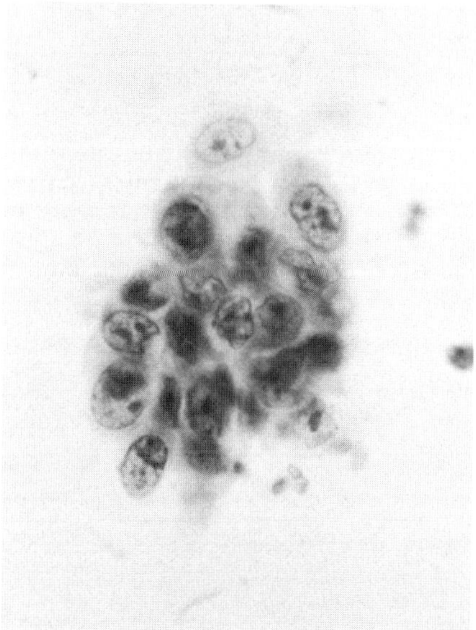

429

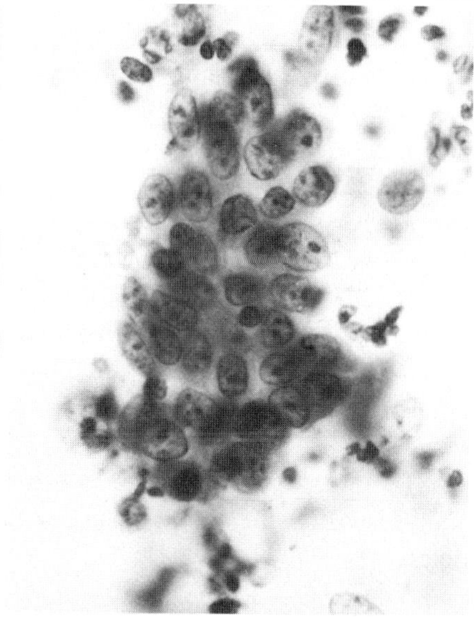

430

Fig. 429. Cells from surface carcinoma of stomach. Some features suggest chronic superficial gastritis but note irregularity of nuclear membranes and some pleomorphism. ×545.

Fig. 430. Cells from surface carcinoma of stomach. Note pleomorphism and cell disarray. ×545.

ated with atrophy and intestinal metaplasia and the patient is exposed to an increased risk of gastric cancer.

(2) The second type, which is seen most commonly in populations with an elevated gastric cancer risk, is *environmental chronic gastritis (ECG)*. The condition is characteristically multifocal involving the antrum and body of the stomach. This lesion is also associated with atrophy and intestinal metaplasia and is considered to be the most common precursor of gastric cancer.

(3) *Hypersecretory chronic gastritis (HCG)* is seen in patients with peptic ulcers of the duodenum or gastric antrum and it is assumed that similar mechanisms operate. In the presence of a duodenal ulcer the gastritis is characteristically limited to the gastric antrum but may extend to the body if the ulcer is localized in the stomach. Atrophy and intestinal metaplasia are not prominent and there does not appear to be an increased risk of gastric cancer.

It can thus be seen that at least two types of chronic gastritis may be associated with atrophy and intestinal metaplasia. Chronic atrophic gastritis has also been discussed previously as has intestinal metaplasia. As indicated in this earlier discussion the metaplastic cells may not reach full maturity. This immaturity is reflected functionally by a lack of the complete set of intestinal enzymes and morphologically by abnormalities of nuclear morphology the latter characterized by increased nuclear size, hyperchromasia and irregular shape. In pathological terms these changes are known as 'dysplasia', the term also implying distortion of the glandular architecture.

In 1978 a World Health Organization Expert Committee agreed that dysplasia was the most suitable term for these lesions [16, 29]. The condition was defined as having the following three histological features: (1) cellular atypia, (2) abnormal differentiation, (3) disorganized mucosal architecture.

In an analysis of a large number of cases of gastric dysplasia *Nagayo* [22] described the histological features of this condition in considerable detail:

---

1 Cell atypia
  - a Dense distribution of hyperchromatic, slender, and elongated nuclei in tall columnar epithelia
  - b irregular arrangement of nuclei leading to pseudostratification
  - c Increased nucleocytoplasmic ratio
  - d Nuclear pheomorphism
  - e Disturbances of cell polarity

2 Evidence of abnormal differentiation
  a Decrease in the number of cells with secretory granules or of goblet cells
  b Disappearance of surface cells on the surface of the mucosa
  c Atrophy of pyloric glands leading to appearances of intestinal metaplasia and/or pseudopyloric gland formation of the fundic glands
  d Loss of Paneth cells or their irregular distribution
  e Increase in the layer of generative cells
3 Evidence of disorganized mucosal architecture
  a Irregularities in the form and structure of foveolae, such as elongation, distortion, dilatation, branching, or fusion
  b Cystic dilatation of glands with or without irregular contour
  c Diffuse or sporadic glandular heterotopia with or without proliferative change
  d Irregularity of muscularis mucosae and/or fibrous or scar formation at the base of the lesion
  e Loss of smoothness or regularity of the mucosal surface

As is the case with squamous dysplasia, gastric dysplasia was classified into three grades – mild, moderate and severe – according to the severity of the changes. It would be inappropriate to dwell at length on the histological diagnosis of gastric dysplasia and its cytological manifestations. However, it must be stressed that the more severe degrees of dysplasia are very difficult to distinguish cytologically from early carcinoma. This diagnostic dilemma is illustrated in figures 431–434. The abnormal cells (fig. 431, 432) were obtained from a patient who on repeated biopsy had histological evidence of dysplasia only (fig. 433, 434).

Jass [11] also observed a spectrum of dysplasia but described two principal types. Type I resembled the epithelium lining colonic adenomas whereas Type II was composed of eosinophilic cells with basally situated vesicular nuclei. A link was demonstrated between incomplete intestinal metaplasia, Type II dysplasia and the more poorly differentiated cancers of intestinal type.

In summary, therefore, during the development of a gastric cancer from an apparently normal gastric mucosa the following stages can be identified:

1 Chronic superficial gastritis
2 Chronic atrophic gastritis
3 Immature intestinal metaplasia
4 Dysplasia – mild, moderate and severe.

Severe dysplasia has been equated with carcinoma in situ but this latter term should be used with care as it may lead to confusion with invasive superficial or surface carcinoma – a condition already discussed.

The Histogenesis of Gastric Carcinoma

The gastric cancer model outlined in the previous section is supported by a vast amount of evidence, both epidemiological and histological, and would appear to be a valid explanation of the development of many cases of carcinoma of the stomach. However, although many gastric cancers arise from an area of intestinal metaplasia, thus conforming to the cancer model, there is abundant evidence that not all do so. Some cancers arise in a gastric mucosa that shows no evidence of intestinal metaplasia and indeed it has been suggested that such origin may occur from 'normal' or 'ordinary' gastric mucosa. It is in this context that the classification of Lauren, already referred to, is particularly useful. Thus, as indicated in chapter 5, Lauren classified carcinoma of the stomach into the intestinal type, which represented 53% of all cases, and the diffuse type which accounted for 33%. 14% of cases could not be so classified. The available evidence suggests that the more frequently occurring intestinal type of carcinoma arises on a basis of intestinal metaplasia, and hence conforms to the cancer model discussed above in some detail, whereas the diffuse type arises in a gastric mucosa that is usually free of intestinal metaplasia. Indeed it is this latter type that is said by some to arise from 'ordinary' gastric epithelium [23–25]. As suggested by *Morson* [15] the study of early gastric carcinoma should provide information about alternative mechanisms of histogenesis.

In a recent paper *Takeda* et al. [34] described a study of 119 cases of early gastric carcinoma in which two distinctly different cellular types were identified. One was described by *Takeda,* in conformity with Lauren's classification, as the 'intestinal' type, whereas the other, a diffusely infiltrating mixed cell type, was designated the 'gastric' type. In supporting this proposed terminology *Takeda* states that the term 'intestinal type' indicates the

---

Fig. 431. Cells from case of gastric mucosal dysplasia. Cells are difficult to differentiate from carcinoma cells although good cohesion and retention of cell polarity is helpful. × 545.

Fig. 432. Cells from case of gastric mucosal dysplasia. × 545.

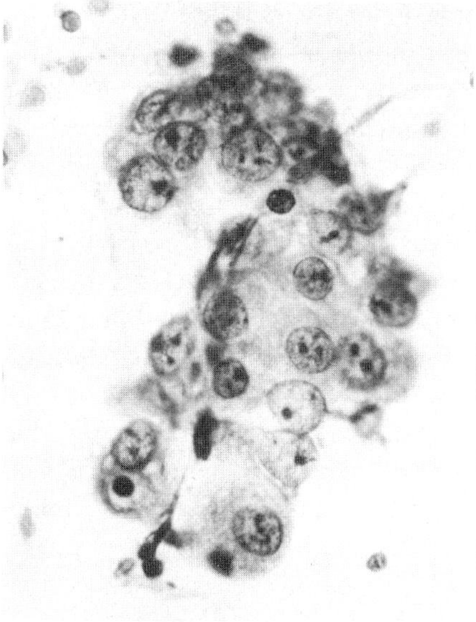

431

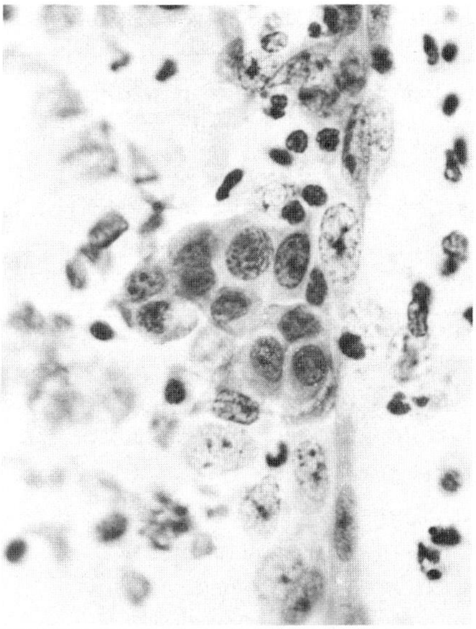

432

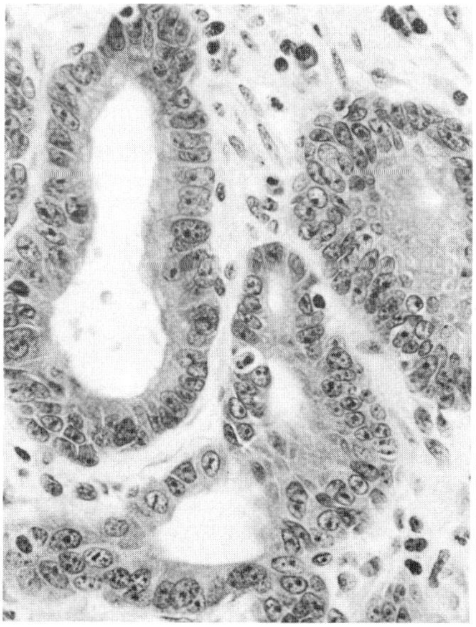

433

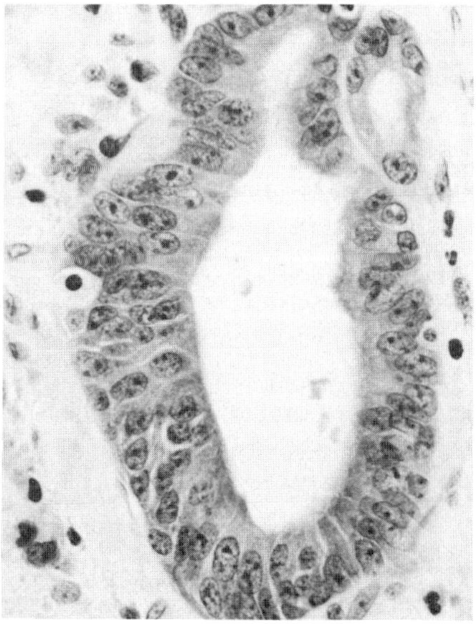

434

Fig. 433. Biopsy of stomach from which were derived the cells depicted in previous two figures. ×275.

Fig. 434. Further biopsy of same stomach again showing moderate to severe mucosal dysplasia. ×345.

origin of the cancer in intestinal metaplastic epithelium and that its cytologic features mimic that of colonic carcinoma being characterized by a uniform cellular pattern of columnar or ovoid carcinoma cells with granular cytoplasm. Conversely the 'gastric type' indicates the origin of the cancer cells in the ordinary gastric mucosa with differentiation changing undifferentiated cells into surface cell type mucus producing cells.

In our laboratory two types of early gastric cancer have also been demonstrated. The observations on these two types agree closely with those of *Takeda* in many respects but differ most significantly in others. Thus there is no doubt that many cases of gastric cancer develop in a stomach displaying evidence of intestinal metaplasia and the resultant carcinoma has the histological structure of Lauren's intestinal type of cancer. However, the other type observed does not, in our experience, arise in an 'ordinary' gastric mucosa, as described by *Takeda,* but rather appears to arise against a background of an intense chronic superficial gastritis, in an active phase, with both histological and cytological evidence of extreme regenerative activity. The resultant carcinoma has the histological pattern of Lauren's diffuse type or *Takeda's* gastric type. These points can probably be best illustrated by reference to typical cases from each group.

A case representative of the first group was a man aged 64 who presented with a long history of esophageal reflux. Endoscopy showed a hiatus hernia and there was mucosal ulceration immediately distal to the gastric cardia. Gastric washings revealed cells that were characteristic of intestinal metaplasia (fig. 435). In addition there were metaplastic cells with abnormal nuclei. Some of these were seen as goblet-type cells (fig. 436), others were tall columnar cells with abundant finely vacuolated cytoplasm and a suggestion of a brush border (fig. 437). These abnormal metaplastic cells were seen singly, in small groups, and in strips. Finally there were many clumps of unequivocally malignant cells although some of these also showed some evidence of pre-existing intestinal metaplasia (fig. 438–440).

Examination of the operative specimen showed a spectrum of changes that reflected accurately the cytological features. Intestinal metaplasia was a prominent feature in many areas (fig. 441). In many of the glands abnormal nuclei were associated with the metaplastic process and these were seen both in rounded goblet cells (fig. 442) and in taller columnar epithelium

---

Fig. 435. Goblet cells from gastric washing indicating intestinal metaplasia. × 545.

Fig. 436. Goblet cells from gastric washings. Note marked nuclear abnormalities suggestive of malignancy. × 545.

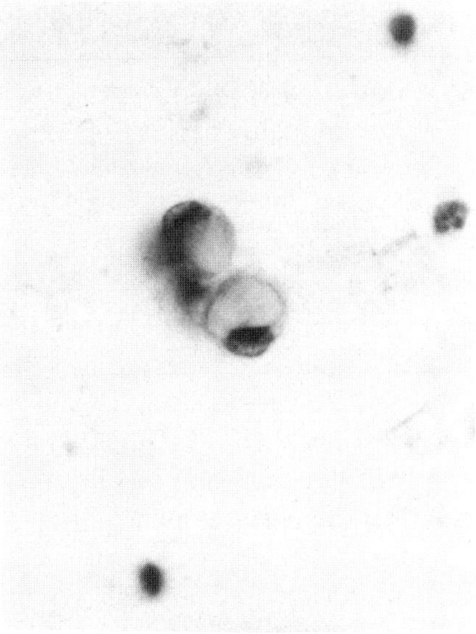

435

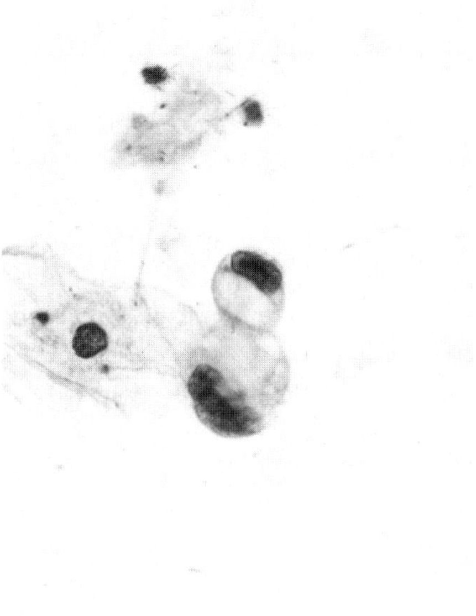

436

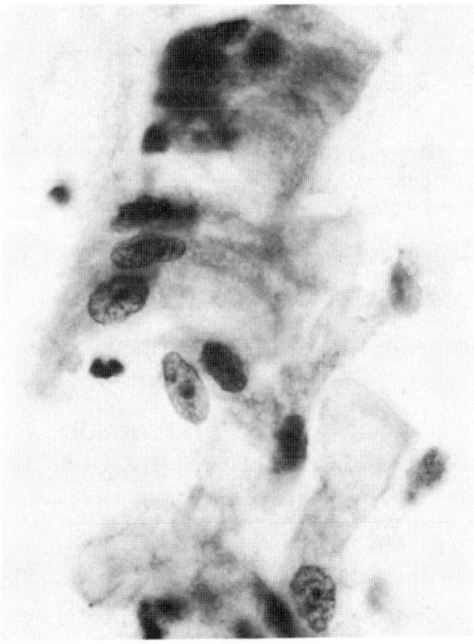

437

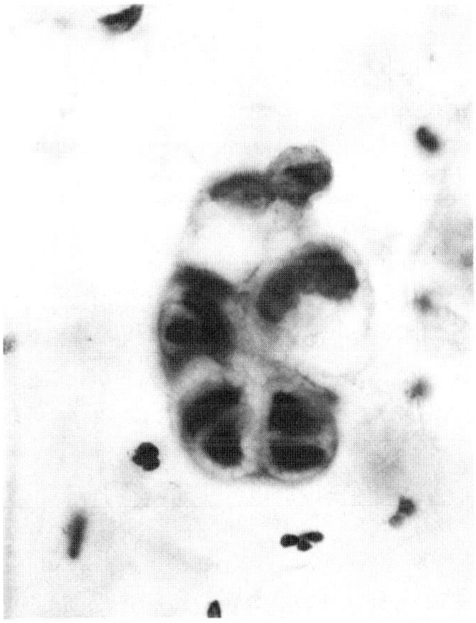

438

Fig. 437. Tall columnar cells in gastric washing, some with abundant vacuolated cytoplasm and others with suggestion of brush border. × 545.

Fig. 438. Unequivocal carcinoma cells from gastric washings with prominent cytoplasmic vacuolation. × 545.

(fig. 443). In many areas a clear transition from metaplastic epithelium to carcinoma was apparent (fig. 444, 445) whilst fully established, but superficial, carcinoma was evident elsewhere. As in the cytology specimen, there was some persistence of goblet cells in the carcinomatous areas (fig. 446).

The illustrative case of the second histogenetic type of gastric cancer is that of a woman aged 52 years who had had no less than two laparotomies over a period of 7 years. The indications for repeated surgery were persistently abnormal cytological findings, continued gastric symptoms and, on one occasion, radiological abnormalities. Gastric biopsies taken at the time of each operation were reported as showing 'dysplasia'.

Gastric washings showed numerous groups of abnormal cells (fig. 447–450). Tumour debris was conspicuously absent but there were many polymorphonuclear leucocytes present. Although the appearances in most groups were reminiscent of an intense gastritis there was evidence of pleomorphism, irregularity of nuclear membranes and a disturbed relationship of one cell to another. This cellular disarray was evident in most cell groups. Mitotic figures were also a prominent feature and these were frequently abnormal (fig. 451, 452).

In view of these marked abnormalities a total gastrectomy was performed. Examination of the specimen showed an exaggerated rugal pattern with a diffuse nodularity (fig. 453). The individual nodules had an umbilicated appearance, this appearance being due to focal superficial ulceration (fig. 454).

Multiple sections showed very extensive superficial or intramucosal carcinoma with, in one area, minimal penetration of the muscularis mucosae. Tissue fragments were being shed profusely from the abnormal mucosal surface and the cells comprising these reflected accurately the cytologic features noted in the lavage specimens (fig. 455, 456). Within the mucosa there was evidence of an intense active chronic superficial gastritis, the lamina propria being densely infiltrated by inflammatory cells (fig. 457). Many of the mucosal glands appeared to be fragmenting (fig. 458) with their lining epithelial cells spreading diffusely into the lamina propria (fig. 459, 460). A characteristic feature was the apparent 'budding off' of malignant undifferentiated epithelial cells from the gland necks (fig. 461). Mitotic figures were prominent (fig. 462).

Fig. 439. Group of unequivocal adenocarcinoma cells. × 545.

Fig. 440. Group of unequivocal carcinoma cells from gastric washings. Note large cytoplasmic vacuole containing polymorphs in upper cell. × 545.

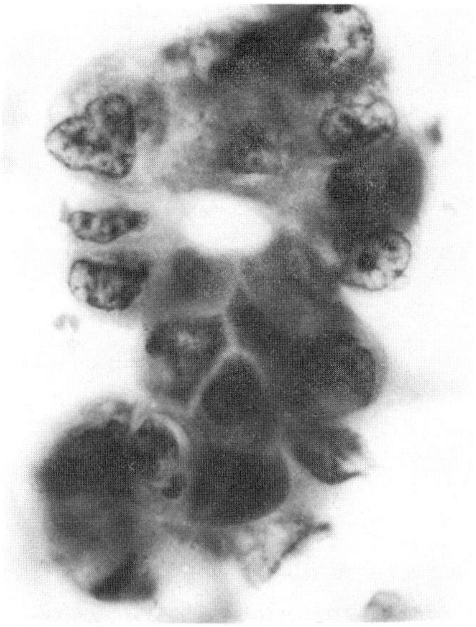

439

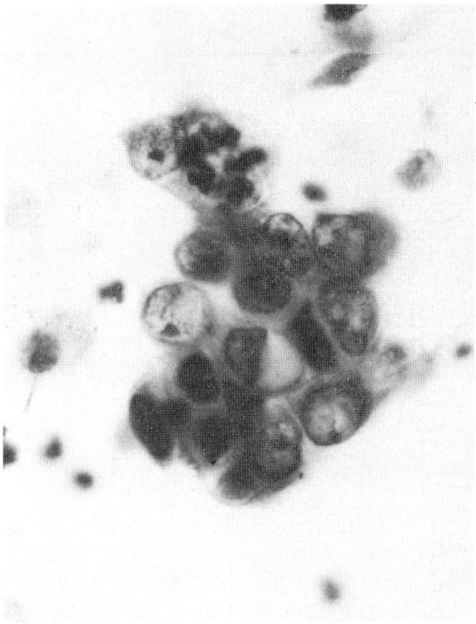

440

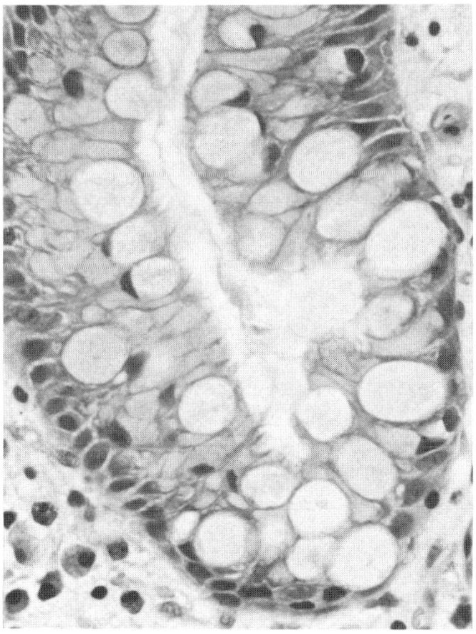

441

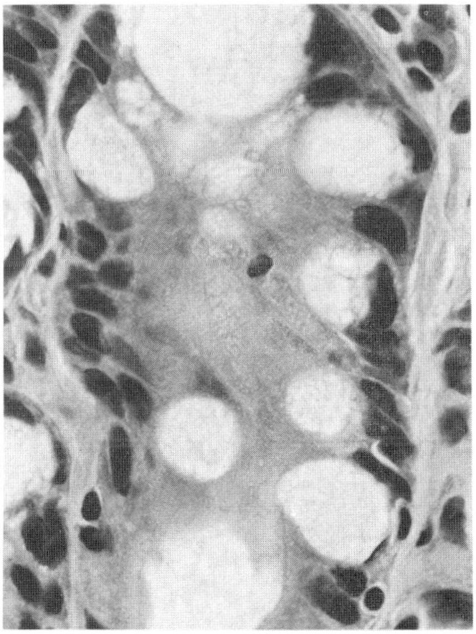

442

Fig. 441. Biopsy of stomach showing prominent intestinal metaplasia.

Fig. 442. Intestinal metaplasia with marked nuclear abnormalities – compare figure 436. ×435.

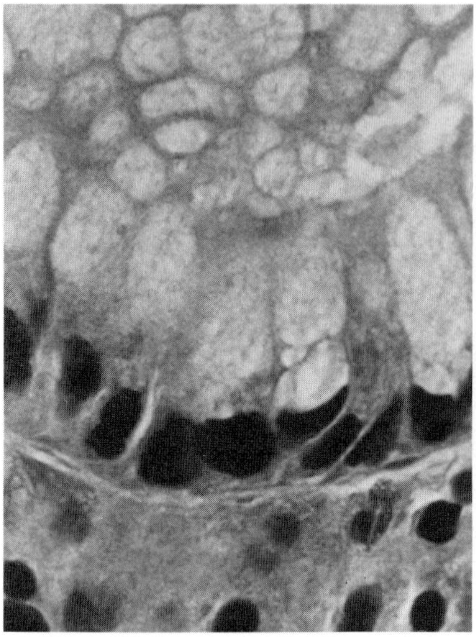

443

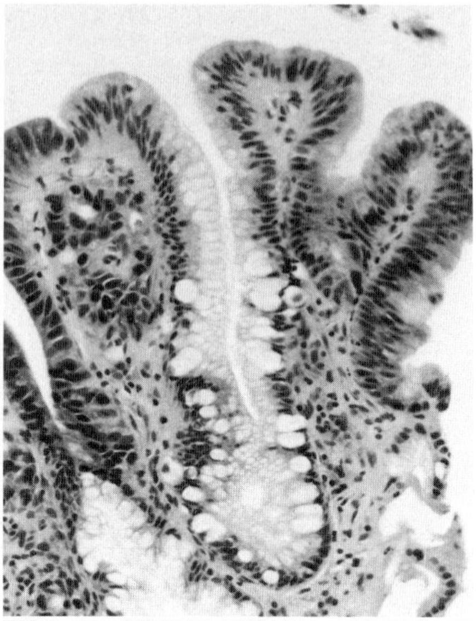

444

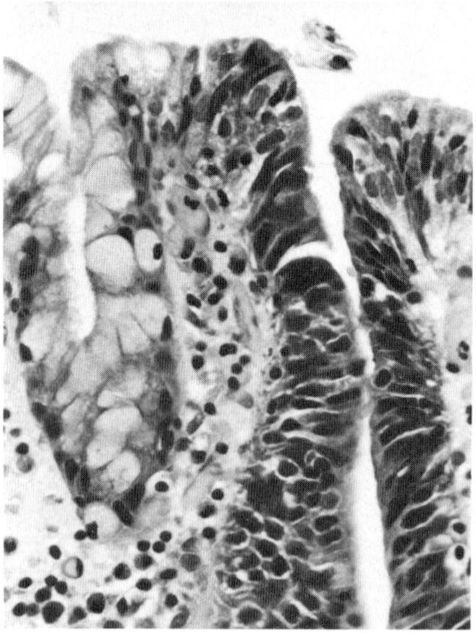

445

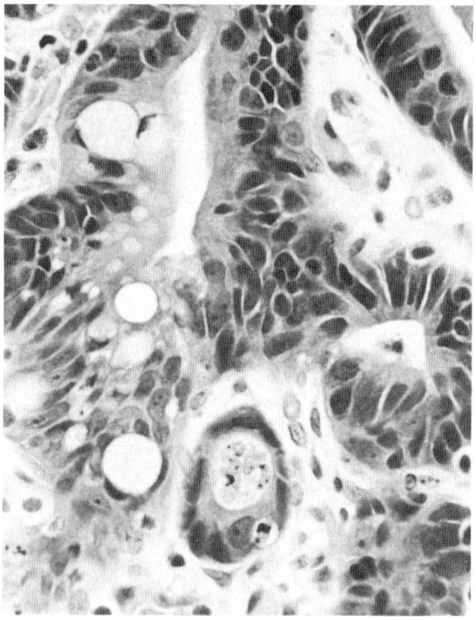

446

Fig. 443. Intestinal metaplasia with marked nuclear abnormalities – compare figure 437. × 545.

Fig. 444. Surface carcinoma of stomach arising in area of intestinal metaplasia. × 140.

Fig. 445. Abrupt transition between area of intestinal metaplasia and surface carcinoma. × 220.

Fig. 446. Surface carcinoma of stomach with evidence of pre-existing intestinal metaplasia. × 345.

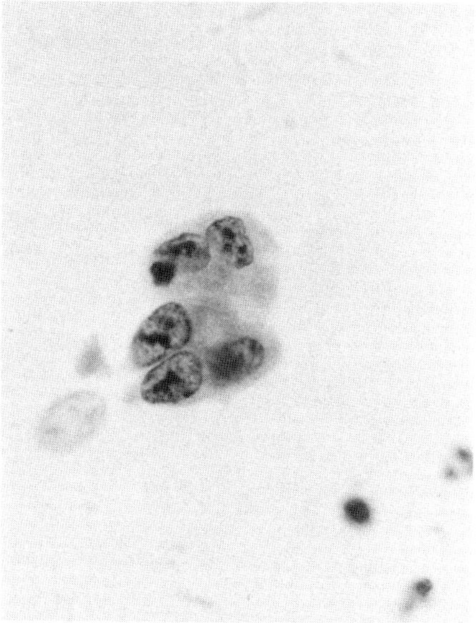

447

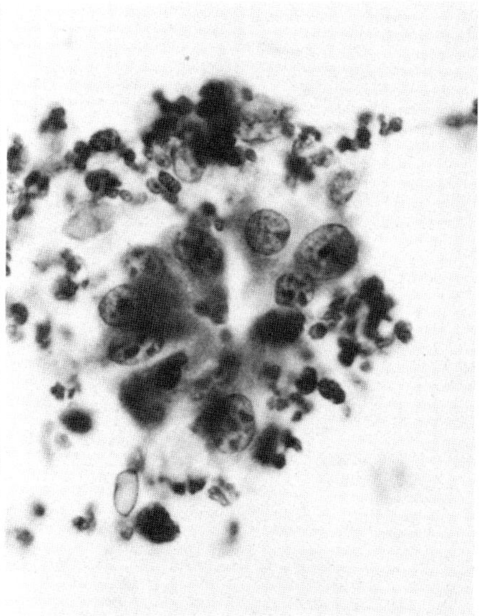

448

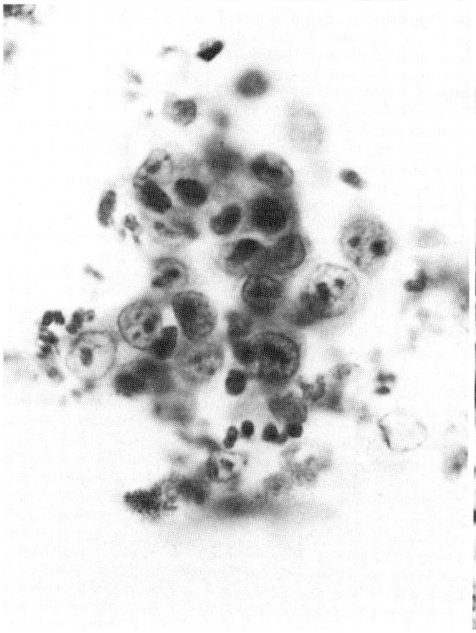

449

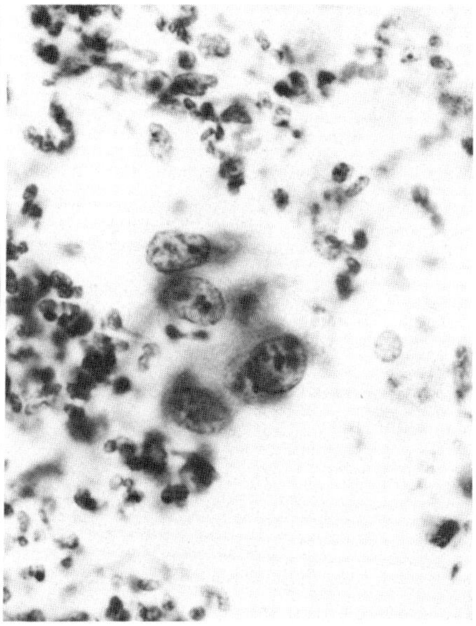

450

Fig. 447. Small group of abnormal cells from patient with surface carcinoma of stomach. Note slight pleomorphism and, in upper photograph, disturbance of cell polarity. × 545.

Fig. 448. Cells from patient with surface carcinoma of stomach. Many polymorphs in background but no cell debris. × 545.

Fig. 449. Cells from surface carcinoma of stomach. Cellular disarray and pleomorphism is evident. × 545.

Fig. 450. Cells from surface carcinoma of stomach. × 545.

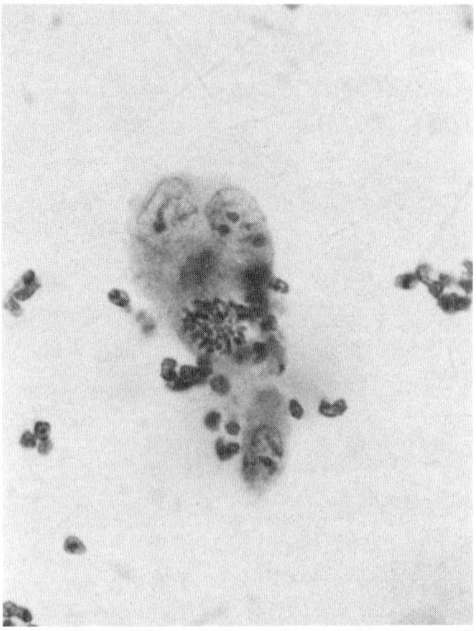

451

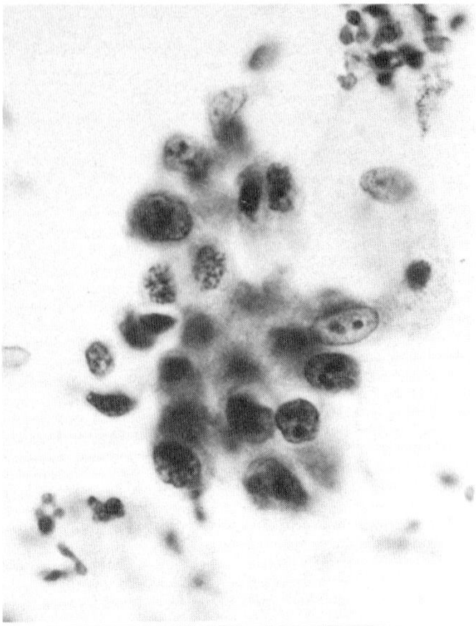

452

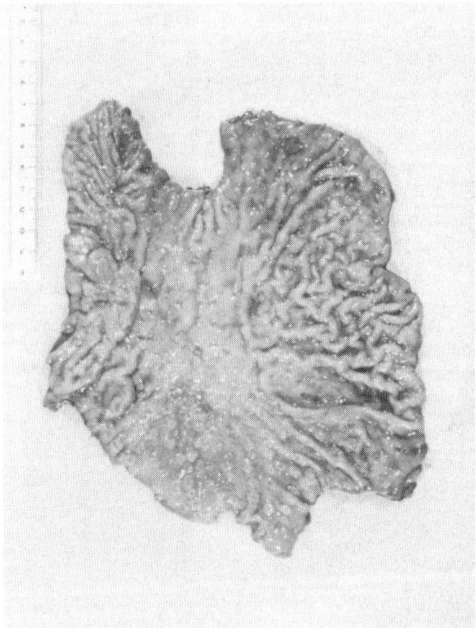

453

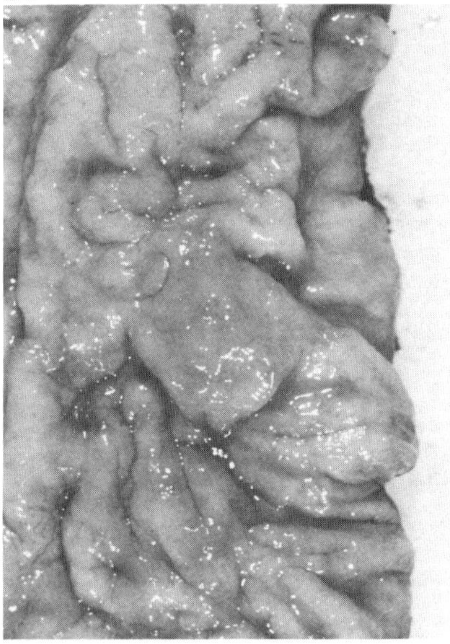

454

Fig. 451. Cells from surface carcinoma of stomach with abnormal mitotic figure. × 545.

Fig. 452. Cells from surface carcinoma of stomach with abnormal mitotic figures. × 545.

Fig. 453. Gastrectomy specimen showing exaggerated rugal pattern with diffuse mucosal nodularity.

Fig. 454. Ulcerated nodules of gastric mucosa giving umbilicated appearance.

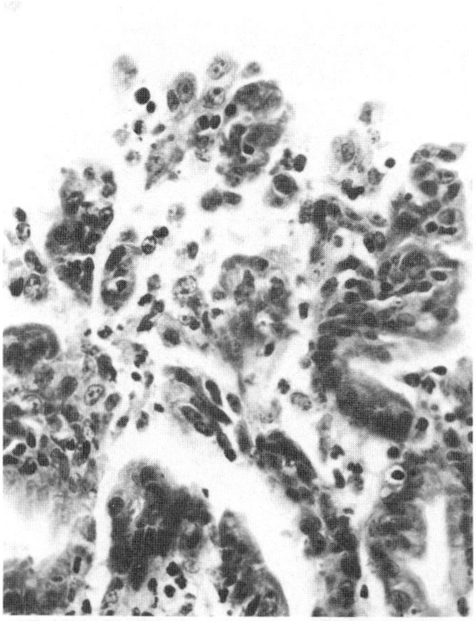

455

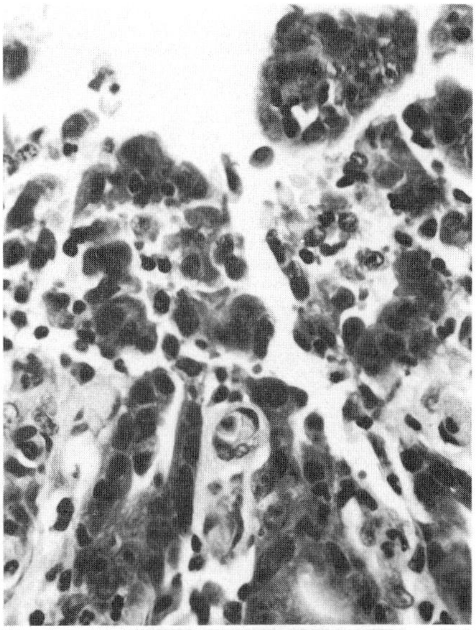

456

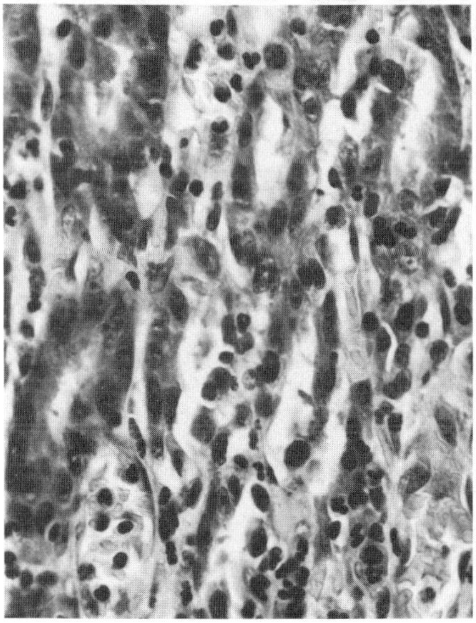

457

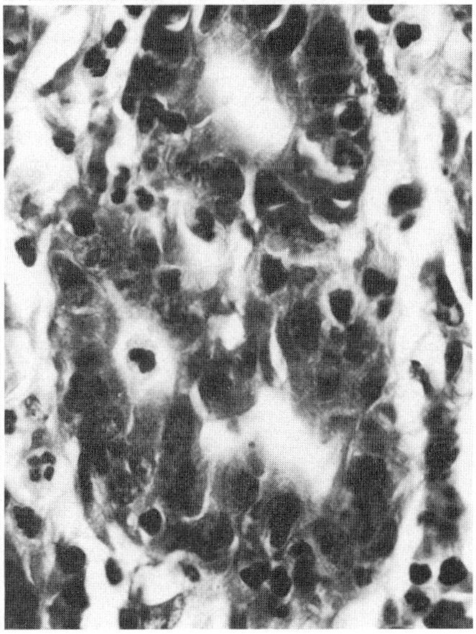

458

Fig. 455. Surface of superficial gastric mucosal carcinoma. × 345.
Fig. 456. Surface of superficial gastric mucosal carcinoma. × 345.
Fig. 457. Intramucosal carcinoma of stomach showing coexisting intense gastritis. × 345.
Fig. 458. Intramucosal carcinoma of stomach showing 'fragmentation' of glands. × 345.

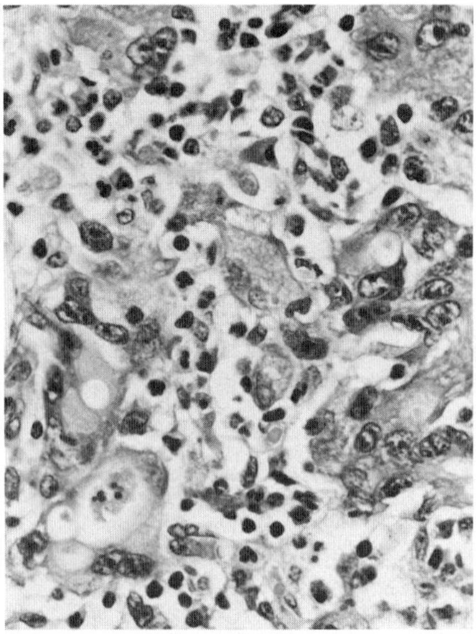

459

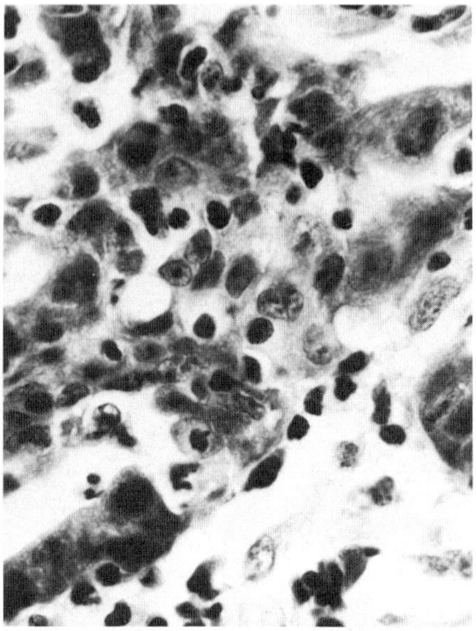

460

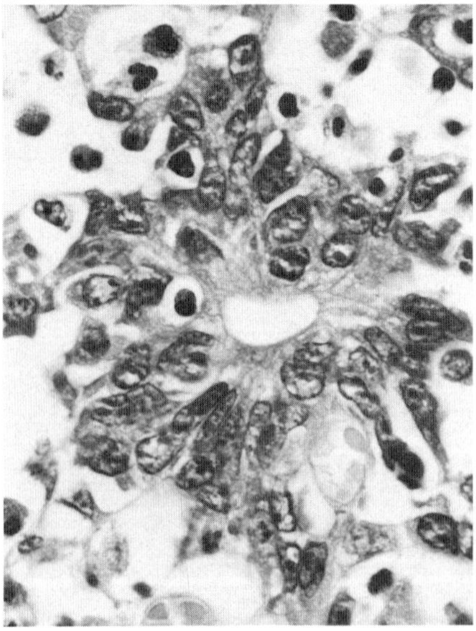

461

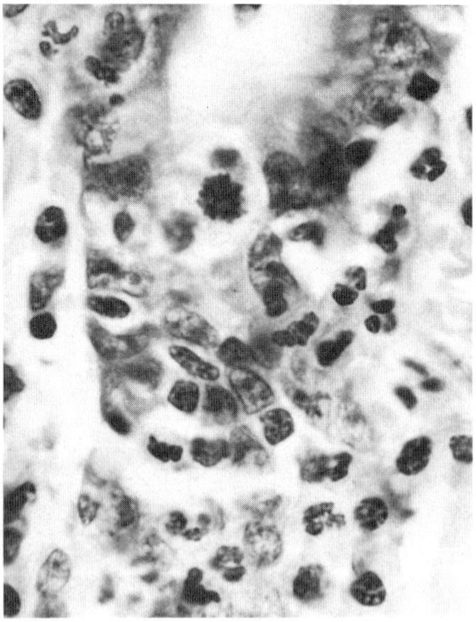

462

Fig. 459. Intramucosal carcinoma of stomach with fragmenting abnormal glands and cells spreading into lamina propria. × 345.

Fig. 460. Fragmenting abnormal glands. × 345.

Fig. 461. Carcinoma cells 'budding off' gland necks. × 345.

Fig. 462. Superficial gastritis and intramucosal carcinoma with prominent mitotic figures. × 435.

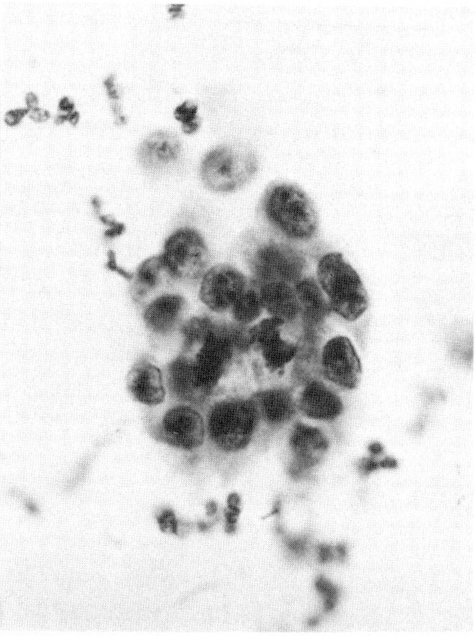

463

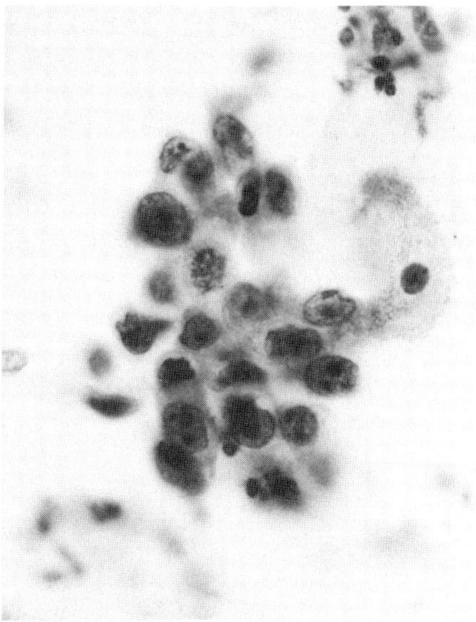

464

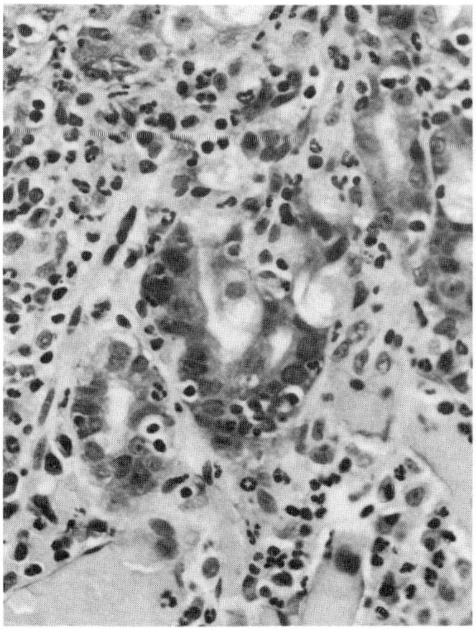

465

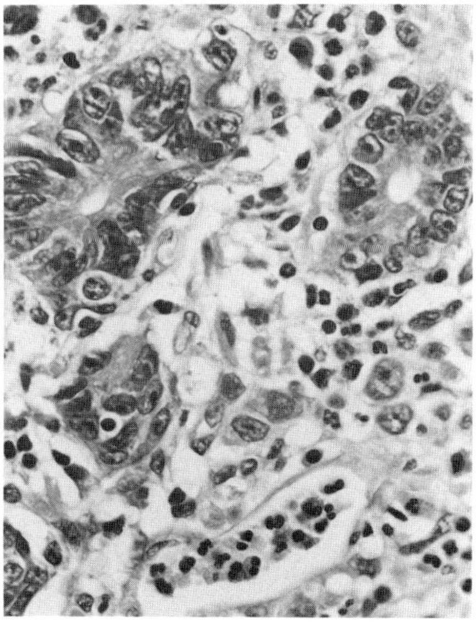

466

Fig. 463. Cells in gastric brushings from patient with surface carcinoma of stomach. × 545.

Fig. 464. Cells in gastric brushings from patient with surface carcinoma of stomach. Note abnormal mitotic figure. × 545.

Fig. 465. Gastric biopsy showing intense gastritis and fragmentation of gastric glands. × 275.

Fig. 466. Intense gastritis and intramucosal carcinoma of stomach. × 345.

These cytologic and histologic findings have now been seen in a number of other cases of superficial carcinoma of diffuse type as exemplified in figures 463–466. The material in this instance was derived from a man, aged 67, with a known hiatus hernia who was admitted to hospital following an episode of haematemesis.

It is therefore postulated that there are indeed two histogenetically separate types of gastric carcinoma. One arises in an area of intestinal metaplasia in accordance with the cancer model already discussed at length. The other is not associated with intestinal metaplasia but, contrary to what has been claimed, does not arise in 'normal' gastric mucosa but rather in an area of chronic active superficial gastritis with associated intense epithelial regenerative activity.

References

1 Berry, A.V.; Baskind, A.F.; Hamilton, D.G.: Cytologic screening for esophageal cancer. Acta cytol. *2:* 135–141 (1981).

2 Bishop, D.; Lushipan, A.; Louis, C.: The cytology of carcinoma in situ and early invasive carcinoma of the esophagus. Acta cytol. *21:* 298–300 (1977).

3 Co-Ordinating Group for the Research of Esophageal Carcinoma, Chinese Academy of Medical Sciences and Honan Province: The early detection of carcinoma of the esophagus. Scientia sinica *16:* 457–463 (1973).

4 Correa, P.; Cuello, C.; Duque, E.: Carcinoma and intestinal metaplasia of the stomach in Colombian migrants. J. natn. Cancer Inst. *44:* 297–306 (1970).

5 Correa, P.; Cuello, C.; Duque, E.; et al.: Gastric cancer in Colombia. III. Natural history of precursor lesions. J. natn. Cancer Inst. *57:* 1027–1035 (1976).

6 Correa, P.: The epidemiology and pathogenesis of chronic gastritis: three etiologic entities. Front gastroent. Res. *6:* 98–108 (1980).

7 Crespi, M.; Munoz, N.; Grassi, A.; et al.: Oesophageal lesions in Northern Iran: a premalignant condition? Lancet *ii:* 217–221 (1979).

8 Dreyer, L.: The incidence of dysplasia and associated epithelial lesions in the oesophageal mucosa of South African blacks. S. Afr. med. J. *58:* 406–408 (1980).

9 Gibbs, D.D.: Exfoliative cytology of the stomach (Butterworths, London 1968).

10 Imbriglia, J.E.; Lopusniak, M.S.: Cytologic examination of sediment from esophagus in case of intraepidermal carcinoma of esophagus. Gastroenterology *13:* 457–463 (1949).

11 Jass, J.R.: A classification of gastric dysplasia. Histopathology *7:* 181–193 (1983).

12 Koss, L.G.: Diagnostic cytology and its histopathologic bases; 3rd ed. (Lippincott, Philadelphia 1979).

13 Kurokawa, T.; Kajitani, T.; Oota, K.: Carcinoma of the stomach in early phase (Nakayama-Shoten, Tokyo 1967).

14 Maimon, H.N.; Dreskin, R.B.; Cocco, A.E.: Positive esophageal cytology without detectable neoplasm. Gastrointest. Endosc. *20:* 156–159 (1974).

15 Morson, B.C.: The Japanese classification of early gastric cancer; in The gastrointestinal tract. International Academy of Pathology Monograph (Williams & Wilkins, Baltimore 1977).
16 Morson, B.C.; Sobin, L.H.; Grundmann, E.; et al.: Precancerous conditions and epithelial dysplasia in the stomach. J. clin. Path. *33:* 711–721 (1980).
17 Munoz, N.; Crespi, M.; Grassi, A.; et al.: Precursor lesions of oesophageal cancer in high-risk populations in Iran and China. Lancet *i:* 876–879 (1982).
18 Munoz, N.; Crespi, M.: in Sherlock, Morson, Barbara, Veronesi, Precancerous lesions of the gastrointestinal tract (Raven Press, New York 1983).
19 Murakami, T.: Pathomorphological diagnosis. Definition and gross classification of early gastric cancer. Gann Monogr. Cancer Res. *11:* 53–55 (1971).
20 Murakami, T.: Early gastric cancer. Gann Monogr. Cancer Res. *11* (1971).
21 Nagayo, T.: Histological diagnosis of biopsied gastric mucosa with special reference to the borderline cases. Gann Monogr. Cancer Res. *11:* 245–256 (1971).
22 Nagayo, T.: Precancerous changes of the stomach from the aspect of dysplasia of the gastric mucosa. Histologic study; in Sherlock, Morson, Barbara, Veronesi, Precancerous lesions of the gastrointestinal tract (Raven Press, New York 1983).
23 Nakamura, K.; Sugano, H.; Maruyama, N.; et al.: Histopathological study of primary locus of linitis plastica. Stomach Intestine *10:* 79–86 (1975).
24 Nakamura, K.; Sugano, H.; Takagi, K.: Carcinoma of stomach in incipient phase, its histogenesis and histological appearances. Gann *59:* 251–258 (1968).
25 Nakamura, K.; Sugano, H.; Takagi, K.; Kumakura, K.: Conception of histogenesis of gastric carcinoma. Stomach Intestine *6:* 9–21 (1971).
26 Prolla, J.C.; Kirsner, J.B.: Handbook and atlas of gastrointestinal exfoliative cytology (University of Chicago Press, Chicago 1972).
27 Schade, R.O.K.: Gastric cytology. Principles, methods and results (Edward Arnold, London 1960)
28 Seifert, E.; Borst, H.H.; Ostertag, H.: Carcinoma in situ of the esophagus (early esophageal carcinoma). Endoscopy *5:* 147–153 (1973).
29 Serck-Hanssen, A.: Precancerous lesions of the stomach. Scand. J. Gastroent. *54:* suppl., pp. 104–105 (1979).
30 Sotus, P.C.; Majumdar, B.; Symbas, P.N.: Carcinoma in situ of the esophagus. J. Am. med. Ass. *239:* 335–336 (1978).
31 Strickland, R.C.; Mackay, I.R.: A reappraisal of the nature and significance of chronic atrophic gastritis. Am. J. dig. Dis. *18:* 426–440 (1973).
32 Suckow, E.E.; Yokoo, H.; Brock, D.R.: Intraepithelial carcinoma concomitant with esophageal carcinoma. Cancer *15:* 733–740 (1962).
33 Takeda, M.: Atlas of diagnostic gastrointestinal cytology (Igaku-Shoin, New York 1983).
34 Takeda, M.; Gomi, K.; Lewis, P.L.; et al.: Two histological types of early gastric carcinoma and their cytologic presentation. Acta cytol. *25:* 229–236 (1981).
35 Ushigome, S.; Spjut, H.J.; Noon, G.P.: Extensive dysplasia and carcinoma in situ of esophageal epithelium. Cancer *20:* 1023–1029 (1967).

7. Results of Gastro-Esophageal Cytology

Analyses of performance in gastro-esophageal cytology have now been published by many laboratories throughout the world. It is difficult to compile such analyses and even more difficult to assess their validity. To a large extent they suggest a black and white situation and do not take into account the subtle nuances of communication between the pathologist and clinician. It is important to have a clear understanding of the method of reporting used within a particular laboratory and also the diagnostic philosophy of the laboratory. In the practice of diagnostic cytology there is a constant balancing of sensitivity and accuracy. Thus the diagnostic criteria can be set at such a level that 'false-positive' diagnoses will virtually never occur. Conversely, however, in such circumstances the incidence of 'false-negative' reports may rise to an unacceptable level. In particular there will be a tendency to miss some cases of surface or intramucosal carcinoma where, as already indicated, the cell yield may be small and the abnormalities difficult to distinguish from those of dysplasia or even florid gastritis.

The basis for the 'final' or definitive diagnosis is also important in the evaluation of results. Traditionally histopathology has been regarded as the final arbiter, the accuracy or otherwise of the cytological diagnosis being determined by the histologic investigation. Whilst this may be a reasonable approach when a gastrectomy or autopsy specimen is available for examination it is not necessarily so when the only tissue available is a minute endoscopically directed biopsy – particularly if taken, for example, from the inflamed or necrotic tissue comprising the floor of an ulcer. Even when an operative specimen is available the way in which this specimen is examined may be of crucial importance. Whilst most carcinomas are readily observed and sampled a minute intramucosal carcinoma may be missed if the specimen is not subjected to meticulous examination and extensive sampling. A parallel situation exists in gynaecological cytology where an inadequate examination of a cone biopsy specimen from the uterine cervix may deny the cytological diagnosis. Similarly a minute colposcopically directed biopsy may not necessarily sample accurately the abnormality present.

Within our laboratory gastro-esophageal cytology is reported by means of a summary diagnosis and a detailed comment. The latter is by far the most valuable part of the report and is worded in a manner similar to that of a report on a tissue specimen. Thus the abnormality, if present, is described and a statement is made as to the pathological process that is present.

The summary diagnosis is a 'shorthand' method of communication but its main value is that it lends itself to the statistical manipulations to be described below. It comprises a single word – 'negative', 'inconclusive' or 'positive'. The word inconclusive is used in its literal sense to indicate that a confident diagnosis of 'positivity' or 'negativity' has not been possible on the material available for examination. However, the wording of the comment that accompanies this summary diagnosis almost invariably indicates an index of suspicion and indeed is usually heavily weighted towards a diagnosis of malignancy or the contrary. As will be indicated, therefore, it is reasonable when analyzing results to divide the inconclusive reports into 'high inconclusive', where a diagnosis of malignancy was favoured, and 'low inconclusive' where the comment indicated that the abnormality present was almost certainly benign. Indeed it would not be unreasonable to group the high inconclusive reports with the positive ones and the low inconclusive with the negatives.

It is often stated that terms such as 'suspicious', 'doubtful', and 'inconclusive' are of no value and may be misleading to the clinician. It is important to realize, however, that the cytological investigation is only one part of a diagnostic exercise and the results of this investigation must be evaluated in the light of all the other clinical and laboratory evidence that is available. A cytological suspicion of malignancy may support evidence from other sources and lead to further definitive investigation. Conversely, the decision to report a cytological specimen as negative merely because an unequivocal diagnosis of malignancy is not possible may influence the clinician to disregard other evidence and delay further investigations.

In summary, therefore, there is no doubt that a detailed descriptive report and diagnostic prediction is essential for the adequate communication of cytological results and their application to patient management. However, a summary report is equally essential if 'objective' analyses of performance are to be undertaken.

Results from Other Laboratories

With the increasing application of the techniques of diagnostic cytology to the investigation of the esophagus and stomach over the past 30–40 years many reports of individual laboratory results have been published [1–13, 15, 16, 25, 28–32]. It would be tedious and unnecessary to reproduce these results here. Indeed it could well be counterproductive since, as already indicated in the earlier part of this chapter, a knowledge of the method and philosophy of reporting for each laboratory is desirable if an analysis of their results is to be meaningful. Hence reference to the original publication is commended. The following publications are selected because they illustrate aspects of considerable interest and importance in the overall consideration of the techniques of gastro-esophageal cytology.

The pioneering work of *Schade* [23, 24] is quoted both for its historical interest and because his results illustrate some fundamental points. Although *Schade* tried a variety of mucolytic agents he found these of no value and hence advocated the use of a simple lavage technique employing either saline or Ringer's solution. He stressed the need for careful preparation of the patient and was also quite prepared to carry out two or three washings on an individual patient to ensure the collection of an adequate specimen.

During the five years 1954–1958 inclusive *Schade* and his colleagues examined 2,443 patients, a total of 3,280 lavage procedures being performed on these. In summary, the following results were achieved:

	Total patients	Cytology positive	Cytology negative	Cytology unsatisfactory
Proven carcinoma	282	252	6	24
Benign conditions	276	13	251	12

For those patients with carcinoma, therefore, the diagnostic accuracy was 97.6% whilst an accuracy of 94.8% was achieved for the benign group. Of considerable significance was the diagnosis of 29 cases of gastric cancer that were not demonstrable by radiological techniques this group including 16 cases which, on clinical and radiological grounds, were not even suspected of malignancy. Subsequent investigation showed these 16 patients to be suffering from surface carcinoma. An interesting aspect is the relatively high level of 'false positive' diagnoses – 5.2% in this series. However, undoubtedly this was due to the commendable objective of detecting 'early' carcinomas and was fully acceptable on this basis. Obviously close collaboration between pathologist and clinician would be essential in these circumstances.

Another group particularly active in the earlier days of gastro-esophageal cytology was that of *Raskin and Kirsner* [19] and *Raskin* et al. [18, 20, 21]. Their results are of interest as they used lavage techniques only and achieved a high degree of accuracy. In 1959 they reported a series of 151 patients who had been examined by esophageal lavage. Of the 151 patients examined 69 had esophageal carcinoma and of these the diagnosis of carcinoma was made cytologically in 66 – an accuracy of 95%. Of the 82 patients without cancer, 80 were negative on cytological examination (97.6% accuracy) and 2 'false-positive' diagnoses were made.

This report also included details of the examination, by gastric lavage, of 871 patients 131 of whom were eventually proved to have gastric adenocarcinoma or lymphoma. A correct cytological diagnosis was made preoperatively in 125 of the 131 patients, an accuracy of 95%, whilst the cytological examination was negative in 736 of the 740 patients without cancer – an accuracy of over 99%. Nevertheless it is important to note that 4 'false-positive' reports were issued in this group. All 4 of these patients were suffering from peptic ulceration. The importance of this series is that it demonstrates the efficacy of the lavage techniques in both esophageal and gastric cytology although it is stressed that the authors were meticulous in their approach to all phases of the procedure. There is no doubt that if a laboratory is unable to achieve comparable results it is the laboratory that is at fault and not the techniques.

Another interesting aspect of this early report is that *Raskin* and his colleagues encountered 10 cases of gastric lymphoma in their series of 131 gastric malignancies and pathognomonic cells were demonstrated in 9 of these.

Prolla et al. [14] and *Prolla and Kirsner* [17] have also published details of their experiences in the cytological diagnosis of gastric lymphomas. This group had been interested in this problem for many years the first results from their laboratory being published in 1954 [22]. In 1970 they reported on a series of 42 patients with lymphomatous involvement of the stomach, adequate gastric cytology being obtained by saline washings in 39 of these. In 27 patients, or 64% of the total series, a positive cytological diagnosis was made and the lymphomatous nature of the malignancy was identified correctly.

Their results are tabulated as follows:

Histologic type	Cytology results			Total
	positive	negative	unsatisfactory	
Lymphocytic	11	4	–	15
Histiocytic	12	6	2	20
Hodgkin's	4	2	1	7
Total	27 (64%)	12	3	42

These results are of considerable interest on a number of counts. The overall diagnostic accuracy is impressive although considerably lower than that achieved in the diagnosis of carcinoma. Some of the reasons for this have already been discussed in an earlier chapter. The authors conclude that diagnostic cytology is of little use in distinguishing between the different histological types of lymphoma.

Most recently published series refer to the evaluation of material collected by endoscopic means, either brushing or washing. Representative of the results achieved by these methods are those of *Shida* [26, 27]. Particular attention is paid to the diagnosis of early or superficial carcinoma a detection rate for this lesion of approximately 90% being achieved.

Results from Prince Henry's Hospital

Within our own laboratory gastro-esophageal cytology has been practiced for over 20 years, commencing first in 1962. For the purposes of this discussion a detailed analysis of the 5-year period, 1979–1983, inclusive, is presented:

Esophageal brushings 1979–1983

Cytological diagnosis	Number of specimens	Histology benign	Histology malignant	No follow-up available
Negative	48	46	0	2
Positive	25	0	25	0
Inconclusive	10	4	6	0
'Low'	3	3	0	0
'High'	7	1	6	0
Unsatisfactory	0	0	0	0

Total number of specimens = 83; false positive = 0; false negative = 0.

Esophageal washings 1979–1983

Cytological diagnosis	Number of specimens	Histology benign	Histology malignant	No follow-up available
Negative	11	9	0	2
Positive	10	0	10	0
Inconclusive	5	3	2	0
'Low'	3	3	0	0
'High'	2	0	2	0
Unsatisfactory	1	1	0	0

Total number of specimens = 27; false positive = 0; false negative = 0.

Gastric brushings 1979–1983

Cytological diagnosis	Number of specimens	Histology benign	Histology malignant	No follow-up available
Negative	168	163	5	0
Positive	51	0	51	0
Inconclusive	27	24	3	0
Unsatisfactory	5	5	0	0

Total number of specimens = 251; false positive = 0; false negative = 5 (2.0%).

Gastric washings 1979–1983

Cytological diagnosis	Number of specimens	Histology benign	Histology malignant	No follow-up available
Negative	104	90	8	6
Positive	10	1	9	0
Inconclusive	16	15	1	0
Unsatisfactory	4	3	1	0

Total number of specimens = 134; false positive = 1 (0.7%)[1]; false negative = 8 (6.0%).
[1] Patient had 6 endoscopically directed biopsies all of which were negative for malignancy. However, clinical condition is deteriorating but patient has refused surgery.

The immediate criticism that could be, and undoubtedly will be, made of these results is that there are too many 'inconclusive' reports. However, as indicated in the introductory comments, our laboratory is not philosophically opposed to the 'inconclusive' report, the cytological investigation being regarded as only one part of the total diagnostic exercise. In each case, as already emphasized, a detailed descriptive report is given in an attempt to convey to the clinician the level of suspicion that exists. Nevertheless, every effort should be made to eliminate unnecessary 'inconclusive' reports. It is our experience that they occur more frequently when the diagnostic work is shared between a larger number of technologists and pathologists. Such sharing does not influence significantly the accuracy of the reporting, as judged by the occurrence of 'false positive' cases although the sensitivity may suffer, the less experienced technologist, in particular, failing to detect the elusive single lymphoma or undifferentiated carcinoma cells. This problem has already been discussed.

The 'inconclusive' reports do complicate the detailed analyses of results. As far as the esophageal procedures, both washings and brushings, are concerned it is relatively easy to allocate the 'inconclusive' results into 'high' and 'low' categories. In all cases in the series analyzed above the comment that accompanied the 'high inconclusive' report indicated that the cells present were 'strongly suggestive of' or 'almost certainly from' a carcinoma. In each case an unequivocal diagnosis was not made either because of a paucity of diagnostic material or degenerative changes in the material that was present. As can be seen from the tables, with only one exception the cases reported as 'high inconclusive' were shown subsequently to have a carcinoma whilst all cases reported as 'low inconclusive' were shown to be suffering from a benign disease on further investigation. The cytology report on the one exception stated that there was necrotic debris present and an occasional markedly abnormal cell that suggested the possibility of a squamous cell carcinoma. The biopsy showed benign non-specific esophageal ulceration only.

Analysis of the 'inconclusive' reports in the gastric series is much more difficult as the range of abnormalities encountered is much greater than in the esophagus and the reports as a consequence more complex. In particular, it is difficult to find a valid criterion on which to base a separation into high and low categories. In the majority of cases reported as 'inconclusive' the comment suggested that the abnormalities present were probably due to intense regenerative activity associated with chronic superficial gastritis or peptic ulceration. However, because the abnormalities in some cells were so marked, the report was summarized as 'inconclusive' to indicate that malignancy was not excluded by the cytological investigation. In this context it is of interest to note that, of the 43 cases reported as 'inconclusive' only 4 were shown subsequently to have a gastric malignancy. In 2 of these the presence of carcinoma cells was suggested but a 'positive' diagnosis withheld because of the paucity of these cells. In the other 2 specimens 'markedly abnormal' cells were reported but an unequivocal diagnosis was not made because of degenerative changes.

The single 'false positive' case reported requires further evaluation. Although, as indicated, multiple endoscopically directed biopsies failed to reveal evidence of carcinoma, nor indeed an explanation of the cytological abnormalities, the patient's clinical condition is stated to be deteriorating in a manner consistent with malignant disease, but he has refused further investigation or treatment.

The 13 'false negative' results recorded in the analyses of the combined series of gastric brushings and washings represent only 3.4% of the 385 cases that comprise this series. Nevertheless, they are a cause for concern and require explanation. Of these 13 cases 8 were very poorly differentiated or completely undifferentiated malignancies whilst 1 case was that of an intramucosal carcinoma. The other 4 cases were adenocarcinomas of varying degrees of differentiation and in 2 of these the comment had been made that food debris tended to obscure cytological detail. On review 7 of the 13 cases were re-classified as 'positive', malignant cells being detected. In the remaining 6 cases no cytological evidence of malignancy was present and hence these must be regarded as problems of sampling. The re-interpretation of abnormal cells in the knowledge of histological evidence of malignancy is completely lacking in validity but it is important to note that, in 4 of the 7 cases re-classified to 'positive', the diagnostic cells were not detected in the initial evaluation. In each case the cells were few in number and occurred as single cells. These cases emphasize the problems of detection or 'screening' already discussed.

In summary, it would appear that a 'positive' cytological report on both esophageal and gastric specimens is very reliable evidence indeed that a malignancy is present. As far as esophageal specimens are concerned a 'negative' report can be regarded with an equal degree of confidence. With gastric specimens, however, there is a small but significant incidence of 'false negative' reports and hence the clinician should not be deterred from further investigations if these are indicated by the results of other procedures or for clinical reasons. However, this is, of course, true of all laboratory techniques. The status of the 'inconclusive' report has already been discussed. Whilst every effort must be made to minimize the frequency with which this report is issued, it is inevitable that cases will be encountered where an unequivocal diagnosis is not possible on the available morphologic evidence. The report may indicate the need for close observation, follow-up cytology, or immediate further investigation the action taken being based frequently on the detailed descriptive report. It is again emphasized that this detailed comment is a far more valuable means of communication than the single word 'negative', 'inconclusive' or 'positive'.

References

1 Behmard, S.; Sadeghi, A.; Bagheri, S.A.: Diagnostic accuracy of endoscopy with brushing cytology and biopsy in upper gastrointestinal lesions. Acta cytol. *22:* 153–154 (1978).
2 Brandborg, L.L.; Taniguchi, I.; Rubin, C.E.: Is exfoliative cytology practiced for more general use in the diagnosis of gastric cancer? Cancer *14:* 1074–1080 (1961).
3 Carney, J.A.: Gastric exfoliative cytology. Surg. Clins N. Am. *51:* 979 (1971).
4 Goldenberg, I.S.; Vidone, R.A.: Gastric cytology. Ann. Surg. *176:* 721 (1972).
5 Graham, R.M.; Ulfelder, H.; Green, T.H.: The cytologic method as an aid in the diagnosis of gastric cancer. Surgery Gynec. Obstet. *86:* 257 (1948).
6 Halter, F.; Witzel, L.; Gretillat, P.A.; et al.: Diagnostic value of biopsy, guided lavage and brush cytology in esophagogastroscopy. Am. J. dig. Dis. *22:* 129 (1977).
7 Hishon, S.; Smithies, A.; Lovell, D.; et al.: Cytology in the diagnosis of esophageal cancer. Lancet *i:* 296 (1976).

8 Hughes, H.E.; Lee, F.D.; Mackenzie, J.F.: Endoscopic cytology and biopsy in the upper gastrointestinal tract. Clin. Gastroent. *7:* 375–395 (1978).

9 Kasugai, T.: Evaluation of gastric lavage cytology under direct vision by the fibergastroscope employing Hanks' solution as a washing solution. Acta cytol. *12:* 345–351 (1968).

10 Kasugai, T.; Kobayashi, S.: Evaluation of biopsy and cytology in the diagnosis of gastric cancer. Am. J. Gastroent. *60:* 199–203 (1974).

11 Kobayashi, S.; Kasugai, T.: Brushing cytology for the diagnosis of gastric cancer involving the cardia or the lower esophagus. Acta cytol. *22:* 155–157 (1978).

12 Mackenzie, J.F.; Rodgers, I.M.; Moule, B.: Comparison of double contrast radiology, standard radiology, endoscopy, also of histology and cytology in the diagnosis of gastric cancer. Gut *18:* 416A (1977).

13 Papanicolaou, G.N.; Cooper, W.A.: The cytology of the gastric fluid in the diagnosis of carcinoma of the stomach. J. natn. Cancer Inst. *7:* 357–360 (1946/1947).

14 Prolla, J.C.; Kobayashi, S.; Kirsner, J.B.: Cytology of malignant lymphoma of the stomach. Acta cytol. *14:* 291–296 (1970).

15 Prolla, J.C.; Reilly, R.W.; Kirsner, J.B.; et al.: Direct vision endoscopic cytology and biopsy in the diagnosis of esophageal and gastric tumours: current experience. Acta cytol. *21:* 399 (1977).

16 Prolla, J.C.; Yoshii, Y.; Xavier, R.G.; Kirsner, J.B.: Further experience with direct vision brushing cytology of malignant tumors of upper gastro-intestinal tract histopathologic correlation with biopsy. Acta cytol. *15:* 375–378 (1971).

17 Prolla, J.C.; Kirsner, J.B.: Handbook and atlas of gastrointestinal exfoliative cytology (University of Chicago Press, Chicago 1972).

18 Raskin, H.F.; Kirsner, J.B.; Palmer, W.L.; et al.: Gastrointestinal cancer: definitive diagnosis by exfoliative cytology. Archs intern. Med. *101:* 731–740 (1958).

19 Raskin, H.F.; Kirsner, J.B.: Application of exfoliative cytology to benign and malignant disease of the esophagus and stomach. Am. J. dig. Dis. *6:* 978–982 (1961).

20 Raskin, H.F.; Kirsner, J.B.; Palmer, W.L.: Role of exfoliative cytology in the diagnosis of cancer of the digestive tract. J. Am. med. Ass. *169:* 789–791 (1959).

21 Raskin, H.F.; Palmer, W.L.; Kirsner, J.B.: Benign and malignant exfoliated gastrointestinal mucosal cells. Archs intern. Med. *107:* 872–884 (1961).

22 Rubin, C.E.; Massey, B.W.: Preoperative diagnosis of gastric and duodenal malignant lymphoma by exfoliative cytology. Cancer *7:* 271–288 (1954).

23 Schade, R.O.K.: Gastric cytology – Principles, methods and results (Edward Arnold, London 1960).

24 Schade, R.O.K.: A critical review of gastric cytology. Acta cytol. *3:* 7–14 (1959).

25 Seybolt, J.F.; Papanicolaou, G.N.; Cooper, W.A.: Cytology in the diagnosis of gastric cancer. Cancer *4:* 286 (1951).

26 Shida, S.: Biopsy smear cytology with the fibergastroscope for direct observation. Gann Monogr. Cancer Res. *11:* 223–231 (1971).

27 Shida, S.: Cytological diagnosis of gastric cancer by gastroendoscopical method with fibergastroscope. Gastroenterol. jap. *2:* 101–107 (1967).

28 Shida, S.; Takamura, S.: Cytological diagnosis of early gastric cancer. Gann Monogr. *3:* 189–198 (1968).

29 Von Haam, E.: A comparative study of the accuracy of cancer cell detection by cytological methods. Acta cytol. *6:* 508 (1962).

30 Winawer, S.J.; Melamed, M.; Sherlock, P.: Potential of endoscopy, biopsy and cytology in the diagnosis and management of patients with cancer. Clin. Gastroent. *3:* 575 (1976).

31 Yamada, T.; Matsumoto, S.; Sankawa, H.I.; et al.: Clinical evaluation of proteolytic enzyme lavage method in the gastric cytodiagnosis, especially in the detection of early cancer of the stomach. Acta cytol. *8:* 27–33 (1964).

32 Yamada, T.; Murohisa, B.; Muto, Y.; et al.: Point, minute and small cancers of the stomach at the early developmental stage detected by improved chymotrypsin lavage method for diagnostic cytology. Acta cytol. *22:* 460–469 (1978).

8. Epilogue: Gastro-Esophageal Cytology in Perspective

With the complete exploitation of the newer fiberoptic endoscopic techniques gastro-esophageal cytology has probably now realized its maximum potential. It is widely accepted as an extremely sensitive, accurate, and valuable procedure for the diagnosis of malignancy of both organs, and can be used for the investigation of patients with symptoms referable to these organs or, alternatively, for the screening of selected population groups. Whilst a variety of non-neoplastic conditions have specific cytological manifestations, and these have been described and illustrated in considerable detail, the recognition of these manifestations is largely of value in the exclusion of cancer rather than for the positive diagnosis of the various non-cancerous lesions. Perhaps an exception to this statement is the use of cytology for the diagnosis of the specific forms of esophagitis which, as indicated, are becoming of increasing importance in clinical practice.

Whilst conceding that the advent of fiberoptic endoscopy has improved significantly the collection of cytological specimens from both esophagus and stomach a strong plea has been made for the retention of traditional lavage techniques. Every cytology laboratory with an interest in gastro-esophageal cytology should be capable of carrying out these procedures with the skill and meticulous attention to detail that, in experienced hands, can give excellent diagnostic results. Although the lavage techniques have, quite justifiably, taken a secondary place, they remain an effective means of resolving difficult diagnostic problems and continue to provide the only practical method of cytological screening for the detection of early gastric cancer.

The theoretical basis for the practice of gastro-esophageal cytology is relatively simple. In both organs the major diagnostic problem is the distinction between inflammatory processes and malignant neoplastic disease. As far as the esophagus is concerned malignancies are almost exclusively squamous cell carcinomas and adenocarcinomas. In the stomach adenocarcinomas predominate, with malignant lymphoma being a less common but nonetheless important neoplasm. It is in the stomach particularly that diagnostic difficulties are encountered. Here the epithelial cell changes associated with intense gastritis and peptic ulceration, and due to coexistent cellular degeneration and regeneration, may be virtually indistinguishable from those of malignancy. As emphasized, this cytologic diagnostic problem is by no means unique to the stomach but may be encountered in a variety of other body sites. However, the conditions can be distinguished in most cases by the recognition of a number of cytological features, all of which have been described in detail and illustrated profusely. It is difficult to single out any one of these features as the most important but, in the author's experience, assessment is most reliable when the diagnostic criteria are applied to small groups of cells or even individual cells. Conversely the assessment of sheets of cells is fraught with diagnostic dangers.

In the final evaluation of the accuracy and sensitivity of the techniques of gastro-esophageal cytology it is important not to lose perspective. There is a danger in the practice of diagnostic cytology, as indeed in all morphologic techniques, to seek infallibility – to desire always to achieve a black or white situation. The truth is, of course, that there will always be a grey area. Skill and experience will narrow this area but will never lead to its complete elimination. However, the cytologic result should never be evaluated in isolation. It is one part, admittedly a most important one, of the diagnostic profile upon which is based the final decision regarding patient management. Even if an unequivocal cytological result is not possible in an individual case the level of suspicion of malignancy or otherwise may be of considerable value as an adjunct to other investigations and the clinical evaluation. Basing an unequivocal cytological diagnosis on inadequate morphologic evidence is like riding a bicycle without hands – a good trick if it comes off but disastrous if it does not.

Subject Index

Numbers in *italics* refer to illustrations

Malignant Neoplasms

Carcinoma

Carcinoma is the most common malignant tumour occurring in the stomach. Although the incidence of the disease has fallen recently in a number of countries, notably the United States of America and Great Britain, it remains a common neoplastic disease and hence its cytologic diagnosis is of considerable importance.

Site and Macroscopic Features

Carcinomas of the stomach occur most commonly in the pre-pyloric region, pyloric antrum, and on the lesser curvature. Detailed analyses of several large series of cases [12, 68] show some variation in distribution but, in general, approximately 55% of gastric cancers are situated in the pyloric region, 20% along the lesser curvature, and 10% in the region of the cardia. The remaining tumours are distributed randomly on the greater curvature, the fundus, or the anterior or posterior walls, whilst in some cases the tumour may be so large, and involve so much of the gastric surface, that it is impossible to define a point of origin. As already indicated, those arising in the region of the cardia frequently extend into the esophagus rendering impossible a distinction between an esophageal and gastric origin of the tumour.

Many different classifications of the gross or macroscopic features of gastric carcinomas have been proposed but perhaps the most useful is that devised by *Ming* [41], based on gross morphologic features of the tumour and a consideration of its biologic and behavioural characteristics. Thus *Ming* proposes the following types:

Superficial carcinoma
Polypoid carcinoma
Ulcerated infiltrative carcinoma
    Prominent deep ulceration
    Shallow ulcer on a broad base
Fungating carcinoma
Diffusely infiltrative carcinoma

Fig. 320. Large leiomyoma of stomach with central mucosal ulceration.

Fig. 321. Radiological demonstration of ulcerated gastric leiomyoma. The larger straight arrow indicates the edge of the tumour mass whilst the smaller curved arrow shows the ulcer crater filled with contrast medium.

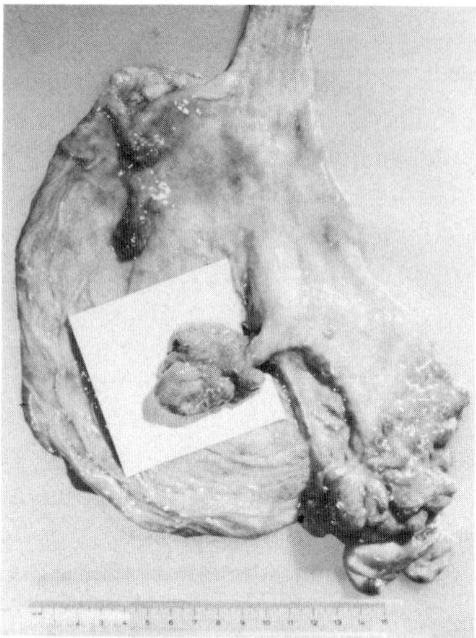

316

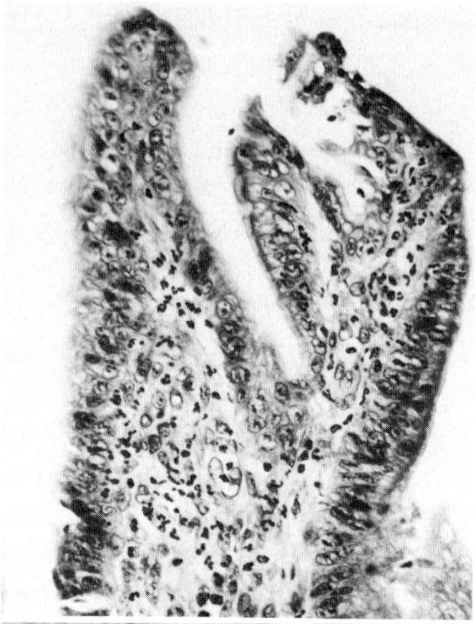

317

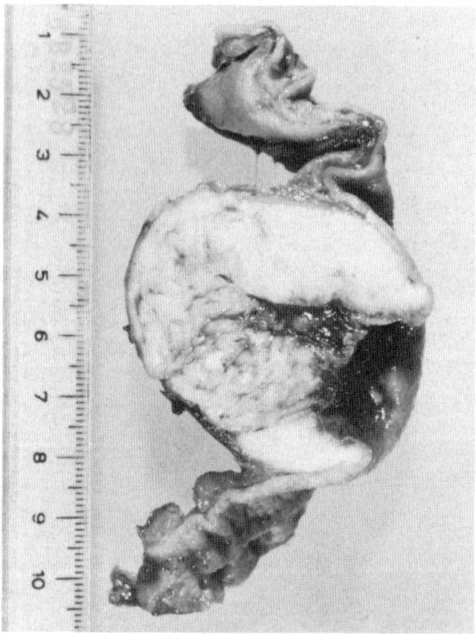

318

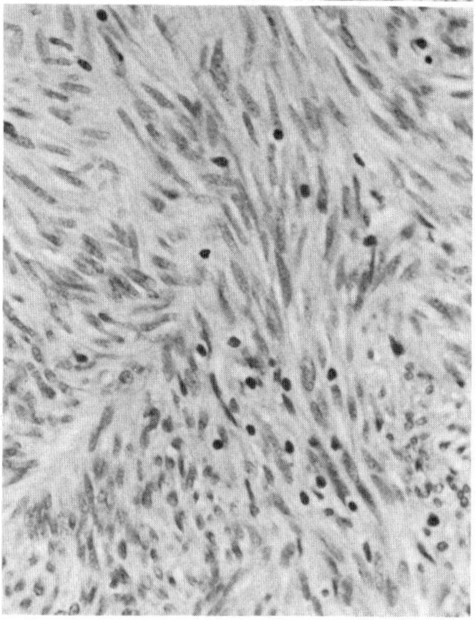

319

Fig. 318. Cut surface of large leiomyoma of stomach showing characteristic whorled appearance.

Fig. 319. Section of gastric leiomyoma showing irregular bundles of smooth muscle cells. ×205.